Biomedical Visualization

Volume 11

Series Editor

Paul M. Rea, Anatomy Facility, School of Medicine, Dentistry and Nursing, College of Medical, Veterinary and Life Sciences, University of Glasgow, Glasgow, UK

Aims and Scope

The Biomedical Visualization book series invites contributions related to visualization and imaging in the biomedical sciences and related fields such as medicine, dentistry, veterinary surgery, informatics, and the allied health professions.

This can encompass work in image analysis, workflow methodologies, photogrammetry in science, animations, digital reconstructions and applications, big data and visualizations, educational methodologies, usability and evaluations, augmented and virtual reality, and 3D and 4D technologies, including 3D printing, as well as informatics, e-tutorials, MOOCs, HCI and public engagement. It can be from macroscopic to microscopic but show how we can view data and information related to the biomedical field in a much more accessible, innovative and engaging way using technology.

This series is intended for researchers, clinicians, students, and teaching staff in biomedical and medical schools who use and develop visualization and imaging techniques in medical education, patient care, and biomedical and medical research.

Peer review description:

Proposal review

The book proposal developed by the invited volume editors is reviewed by the series editors and in-house editors at Springer to meet the scope of the series. Editors can make suggestions for content changes and contributing authors at this stage, which are then included in the proposal.

Manuscript review

1st round of revision: Submitted manuscripts are first reviewed by the volume editor(s), who are expert(s) in the field of the volume. In cases of particular concern, a system of single-blind review is applied, in which independent external experts who know the identity of the authors can be asked for advice.

2nd round of revision: The manuscripts approved by the volume editor (s) then undergo a technical review by Springer and are then forwarded to the series editor for additional content review.

Type of review: The review of manuscripts by the volume editor(s) and series editor includes a check for the scientific soundness and coherence of the manuscript, and adherence to the guidelines for the Biomedical Visualization series regarding style, content, and level of writing. The authors are asked to address any recommended revisions and resubmit to the editors. The resubmitted manuscripts are again reviewed by the volume editor and series editor and, once approved, sent to Springer. The technical review by Springer comprises a plagiarism check, a check for completeness of the chapter, and an assessment of its compliance with Springer's and the series' guidelines for chapter authors.

Volume and series editors jointly decide on the suitability of the manuscript for the *Biomedical Visualization* series based on their reviews.

The book series *Biomedical Visualization* publishes only solicited contributions, which are invited by the volume or series editors.

Kumar Satish Ravi
N. B. Pushpa • Apurba Patra
Editors

Visualization in Neuroanatomical Sciences

Springer

Editors
Kumar Satish Ravi
Department of Anatomy
National Institute of Medical Sciences
and Research
Jaipur, Rajasthan, India

Department of Anatomy
KMC Mangalore, MAHE
Manipal, Karnataka, India

N. B. Pushpa
Department of Anatomy
JSS Medical College, JSSAHER
Mysore, Karnataka, India

Apurba Patra
Department of Anatomy
All India Institute of Medical Sciences
Bathinda, Punjab, India

ISSN 2731-6130 ISSN 2731-6149 (electronic)
Biomedical Visualization
ISBN 978-3-032-07379-2 ISBN 978-3-032-07380-8 (eBook)
https://doi.org/10.1007/978-3-032-07380-8

Cover Illustration: Catherine MacRobbie

This Springer imprint is published by the registered company Springer Nature Switzerland AG
The registered company address is: Gewerbestrasse 11, 6330 Cham, Switzerland

Preface

The human brain is truly one of nature's most stunning feats of engineering. It's a complex web of structures and functions that has long captivated anatomists, neurologists, and the broader neuroscience community. Despite our progress, its unparalleled intricacy continues to challenge our understanding and fuel scientific exploration. Interestingly, the ways we study this enigmatic organ have changed drastically in recent decades. While dissection and microscopy remain fundamental, they have been enhanced—and, in many senses, redefined—by advanced imaging methods, computational models, and detailed digital maps.

Visualization in Neuroanatomical Sciences originated from the desire to connect the enduring principles of traditional neuroanatomy with the groundbreaking visualization technologies that are transforming modern neuroscience. This volume is organized as a well-structured journey, starting with fundamental concepts in neuroanatomy and neurohistology and advancing to cutting-edge techniques such as quantitative neuroimaging, digital reconstructions, and comprehensive diagnostic platforms. Each chapter aims for a balance of clarity and depth, serving as both a foundation for learning and an introduction to current advancements in the field.

However, our goal extends beyond just technical knowledge; we aim to highlight visualization's practical role within brain science. From enhancing diagnostic accuracy and updating educational methods to inspiring new research, visualization has become essential, not just supplementary. This book should serve a wide audience: students gaining fundamental knowledge, educators revising course content, researchers examining structure-function links, and clinicians applying neuroimaging in patient care.

By bringing together the wisdom of the past and the promise of new technologies, we hope this book not only shares knowledge but sparks curiosity and new ways of thinking. We're truly thankful to all the contributors whose dedication, insights, and hard work made this project possible. It's our genuine hope that this volume becomes a helpful companion—one that supports learning, inspires ideas, and encourages continued discovery in the fascinating and ever-changing world of brain science.

Jaipur, Rajasthan, India — Kumar Satish Ravi
Mysore, Karnataka, India — N. B. Pushpa
Bathinda, Punjab, India — Apurba Patra

Acknowledgments

The creation of *Visualization in Neuroanatomical Sciences* has been more than the completion of a book. It has been a journey of curiosity, collaboration, and shared purpose. At its heart, this work reflects our desire to make neuroanatomy not only academically rigorous but also engaging and accessible through the transformative power of visuals. What lies in these pages is not the product of a single effort but the collective vision and dedication of many extraordinary people to whom we owe our deepest gratitude.

First and foremost, we are deeply grateful to Professor Paul M. Rea, our esteemed Series Editor. His leadership, mentorship, and constant encouragement have been the guiding force behind this work. His pioneering vision in digital anatomy laid the foundation for this project, and his support at every stage reminded us of the power of innovation, collaboration, and passion in advancing anatomical sciences.

Our heartfelt thanks also go to our Co-Editors, Dr. Pushpa N. B. and Dr. Apurba Patra. Their dedication and insight shaped this book in profound ways. Dr. Pushpa's meticulous attention to detail and clarity enriched every chapter, while Dr. Patra's analytical approach and thoughtful feedback strengthened the book's coherence and overall scholarship. More than colleagues, they have been remarkable collaborators, and their contributions have elevated this work immeasurably.

We extend our warmest appreciation to all our contributing authors, without whom this book could not exist. Beyond sharing their original research and specialized expertise, they brought unique perspectives, intellectual depth, and cultural richness to every chapter. Their willingness to work across disciplines and geographies has truly embodied the spirit of global collaboration. This volume is not just a collection of writings; it is a tapestry of ideas, stitched together with passion, knowledge, and a shared commitment to advancing neuroanatomical science. For that, we remain deeply indebted.

We are grateful to Dr. Ina Karen Stoeck and the Springer Nature team for their insightful guidance and support, and to Bibhuti Bhusan Sharma for his patient and professional handling of the production process. Their efforts helped transform this work into the book we are proud to share.

On a more personal note, we owe our deepest thanks to our families, students, and supportive colleagues. Their patience, encouragement, and quiet strength carried us through late nights, long hours, and the challenges that accompany any academic endeavor. They reminded us daily of why we do this work, not just to publish, but to teach, to inspire, and to contribute meaningfully to future generations of learners and thinkers.

This book is dedicated to all who share a passion for discovery, for those who teach, who learn, and who continually push the boundaries of understanding in neuroanatomy. We hope it serves as both a resource and an inspiration, sparking new ideas and fueling the curiosity that drives science forward.

Jaipur, Rajasthan, India Kumar Satish Ravi

Contents

Editors and Contributors

About the Series Editors

Paul M. Rea Professor Rea is an internationally recognized leading expert, in the applications of digital technologies to biomedical/medical visualization in education, clinical settings, and industry linked research. Initially he trained in Medicine, undertaking roles in general medicine, pediatric surgery, and emergency medicine, followed by a period as a pathologist. He then changed direction, pursuing a clinical anatomy educator role as Full Professor of Digital and Anatomical Education. It was in academia he realized the potential digital technologies had for education, simulation, training, and public/civic engagement. He led the strategic partnership with the School of Simulation and Technology (SIT), The Glasgow School of Art (GSA), and founded the MSc Medical Visualisation and Human Anatomy, following significant grant funded research. This degree improved graduate employability and had significant high profile media coverage. He was the founding and now Series Editor (Editor-in-Chief equivalent) for the internationally acclaimed Biomedical Visualization book series with Springer Nature. He holds strategic leadership roles for innovation, engagement, and enterprise; body donation; education; and previously project management for large-scale infrastructural developments. He is a Licensed Teacher of Anatomy, licensed by the Scottish Government, and has held this role for 19 years.

He is also the Director of Innovation, Engagement and Enterprise within the School of Medicine, Dentistry and Nursing, a role he has held since August 2022. He is responsible for leading, delivering, and managing the three pillars of the strategy: driving innovation and enterprise; employability for careers and growing entrepreneurs; and regional economic and societal impact.

About the Volume Editors

Kumar Satish Ravi is Additional Principal and Controller and Professor and Head of Anatomy at National Institute of Medical Sciences and Research, Jaipur, Adjunct Professor, Department of Anatomy, KMC, Mangalore, MAHE, Manipal, Karnataka and Editor-in-Chief of the *National Journal of Clinical Anatomy*. With over 19 years of teaching experience, he has held academic leadership roles at AIIMS Gorakhpur, AIIMS Rishikesh, and internationally at Spartan Health Sciences University. He holds an MBBS, MD (Anatomy), PhD (Cytogenetics), and FAIMER fellowship. He has authored 5 book chapters, edited renowned anatomy textbooks, and published over 110 papers with 1700+ citations. A recipient of multiple national awards, he is a respected member of editorial boards and scientific societies, nationally and internationally.

N. B. Pushpa is Associate Professor of Anatomy at JSS Medical College, Mysore, with over 13 years of teaching experience. A graduate of Mysore Medical College, she completed her post-graduation at KIMS, Bangalore. Renowned for innovative teaching, she received the Best Teacher Award in 2024. She has over 70 publications, 3 book chapters, and has co-authored and edited leading histology textbooks. She serves as Assistant Editor of the *National Journal of Clinical Anatomy* and Treasurer of SOCA. A recipient of the Research Excellence Award (2022), she actively contributes to several anatomical societies, shaping education and research in the field.

Apurba Patra is Associate Professor of Anatomy at AIIMS, Bathinda, holding qualifications including MBBS, MS, DNB, MNAMS, and FIMSA. A Fellow of the International Medical Sciences Academy and Member of the National Academy of Medical Sciences, he is actively involved in several scientific societies. With over 100 publications, he has made notable contributions to anatomical and biomedical research. He serves as Associate Editor for *Cureus Journal of Medical Science* and Assistant Editor for the *National Journal of Clinical Anatomy*. His research focuses on translational and clinical anatomy, emphasizing innovations in anatomical visualization and medical education.

Contributors

Juman Almasaad Department of Basic Medical Sciences, College of Medicine, King Saud bin Abdulaziz University for Health Sciences, Jeddah, Saudi Arabia

King Abdullah International Medical Research Centre (KAIMRC), King Abdulaziz Medical City, National Guard Health Affairs, Jeddah, Saudi Arabia

Mohammed Alsharif Department of Professional and Medical Education, Meharry Medical College, Nashville, TN, USA

Ken W. S. Ashwell Department of Anatomy, School of Medical Sciences, The University of New South Wales, Sydney, NSW, Australia

Andre Bongers Biological Resources and Imaging Laboratory, The University of New South Wales, Sydney, NSW, Australia

Dibakar Borthakur Department of Anatomy, All India Institute of Medical Sciences, New Delhi, India

Triinetrra Suresh Chanuvegowda Sir Seewoosagur Ramgoolam Medical College, University of Mauritius, Reduit MU, Belle Rive, Mauritius

Rima Dada Department of Anatomy, All India Institute of Medical Sciences, New Delhi, India

Srijit Das Department of Human & Clinical Anatomy, College of Medicine & Health Sciences, Sultan Qaboos University, Muscat, Oman

Hosam Eldeen Elsadig Gasmalla Brunel Medical School, Brunel University London, London, UK

Vykuntaraju K. Gowda Department of Pediatric Neurology, Indira Gandhi Institute of Child Health, Bangalore, Karnataka, India

Jyothi Thittamranahalli Kariyappa Department of Anatomy, University of Sydney, School of Rural Health, Dubbo, NSW, Australia

Mohamed Al Mushaiqri Department of Human & Clinical Anatomy, College of Medicine & Health Sciences, Sultan Qaboos University, Muscat, Oman

Amena Nayyer Pediatric Radiology Fellow Department of Pediatric Radiology, Indira Gandhi Institute of Child Health, Bangalore, India

Apurba Patra Department of Anatomy, All India Institute of Medical Sciences, Bathinda, Punjab, India

N. B. Prajwala Associate Software Engineer, The Boeing Company, Bengaluru, Karnataka, India

N. B. Pushpa Department of Anatomy, JSS Medical College, JSSAHER, Mysore, Karnataka, India

Kumar Satish Ravi Department of Anatomy, National Institute of Medical Sciences and Research, Jaipur, Rajasthan, India

Department of Anatomy, KMC Mangalore, MAHE, Manipal, Karnataka, India

Gulgun Sengul Department of Anatomy, Ege University, School of Medicine, Bornova, Izmir, Turkey

Varunvenkat M. Srinivasan Department of Pediatric Neurology, Indira Gandhi Institute of Child Health, Bangalore, Karnataka, India

Sankara Reddy Thamma Technical Delivery Master II, Deloitte Consulting LLP, Charlotte, NC, USA

Lydia Tong Taronga Wildlife Hospital, Taronga Zoo, Taronga Conservation Society Australia, Sydney, NSW, Australia

C. Viveka Santhosh Reddy Department of Pediatric Neurology, Indira Gandhi Institute of Child Health, Bangalore, Karnataka, India

Simone Zanoni Biological Resources and Imaging Laboratory, The University of New South Wales, Sydney, NSW, Australia

1 Introduction to Neuroanatomy and Neurohistology

N. B. Pushpa, Apurba Patra, and Kumar Satish Ravi

Abstract

The human nervous system, with its intricate network of neurons and supporting cells, governs every sensation, thought, and action. Understanding its structure is fundamental to medical science and in clinical practice. Neuroanatomy and neurohistology deal with the anatomical divisions and microscopic architecture of the complex nervous system and serve as the gateway to this vast domain of complexity. It introduces the types of neurons and neuroglia, the structural differences between gray and white matter, and the fundamental histological features observed under the microscope. By establishing a clear connection between neuronal structure and function, this chapter provides a solid foundation to explore higher neural functions and associated clinical conditions in subsequent sections of neuroscience education. The primary focus of this chapter is to familiarize medical graduates, researchers, and clinicians with the basic organization of the central and peripheral nervous systems, identify key neural structures, and comprehend the cellular components that make up nervous tissue.

Keywords

Axon · Brain · Neurons · Neuroglia · Tracts

1.1 Introduction

Neuroanatomy is a branch of anatomy which deals with the structure and organization of the nervous system (Thau et al. 2024). The nervous system stands as the most intricate outcome of biological evolution. Billions of interconnected units generate constantly shifting activity patterns, forming the fundamental physical foundation of all human behavior and experience. Multidisciplinary research on the nervous system is at the forefront of modern biology and medicine, with rapid advancements offering promising prospects for improved prevention and treatment of numerous neurological disorders. Therefore, a critical knowledge of the anatomy and physiology of the nervous system is crucial for unraveling the pathology of neurological diseases, facilitating accurate diagnosis, and enhancing treatment strategies (Ludwig et al. 2022).

The nervous system controls all other systems of the body. Living organisms perceive the

N. B. Pushpa
Department of Anatomy, JSS Medical College, JSSAHER, Mysore, Karnataka, India

A. Patra
Department of Anatomy, All India Institute of Medical Sciences, Bathinda, Punjab, India

K. S. Ravi (✉)
Department of Anatomy, National Institute of Medical Sciences and Research, Jaipur, Rajasthan, India

Department of Anatomy, KMC Mangalore, MAHE, Manipal, Karnataka, India

K. S. Ravi et al. (eds.), *Visualization in Neuroanatomical Sciences*, Biomedical Visualization 11,
https://doi.org/10.1007/978-3-032-07380-8_1

changes (stimuli) in the internal and external environment and respond to the stimulus via the nervous system. This functional potential of the nervous system is a result of its innumerable intercommunicating neurons (Galli et al. 2003; Kennea and Mehmet 2002).

This chapter is designed to provide a comprehensive overview of the nervous system—its divisions, development, components, and functions, thus emphasizing the importance of visualization in neuroanatomy.

1.2 Division

The nervous system is broadly categorized into the central nervous system (CNS) and peripheral nervous system (PNS). The brain and the spinal cord are the principal organs of the CNS, while the PNS includes cranial and spinal nerves as well as their related ganglia. Functionally, the nervous system is divided into the somatic nervous system and autonomic nervous system (ANS). The former innervates skeletal muscle, and it mainly helps to respond to the external environment. The autonomic nervous system includes two divisions—sympathetic and parasympathetic systems. The ANS innervates glands, cardiac muscle, and smooth muscles of the viscera. The ANS regulates the internal environment of the body, and the wall of the gastrointestinal tract is innervated by enteric nervous system (Stranding 2020) (Fig. 1.1).

There are 12 pairs of cranial nerves (CN) which originate directly from the brain. The first cranial nerve is the olfactory nerve, followed by the optic, oculomotor, trochlear, trigeminal, abducent, facial, vestibulocochlear, glossopharyngeal, vagus, spinal accessory, and hypoglossal nerves in chronological order. There are 31 pairs of spinal nerves arising from the spinal cord. Each spinal nerve consists of sensory (afferent) and motor (efferent) fibers, contributing to both somatic and autonomic functions (Kandel 2013; Vilensky et al. 2015; Sonne and Lopez-Ojeda 2019).

1.3 Development of CNS

The nervous system develops from the neural plate, which is an ectodermal thickening in front of the primitive node. The lateral borders of the neural plate become elevated and form neural folds. These neural folds approach each other to form the neural tube. This process of formation of the neural tube from is called neurulation. The folds fuse in the cranio-caudal direction. The open end in the cranial end forms the cranial or anterior neuropore. At the caudal end, the opening is the caudal or posterior neuropore. At the cranial end of the neural tube, there are three

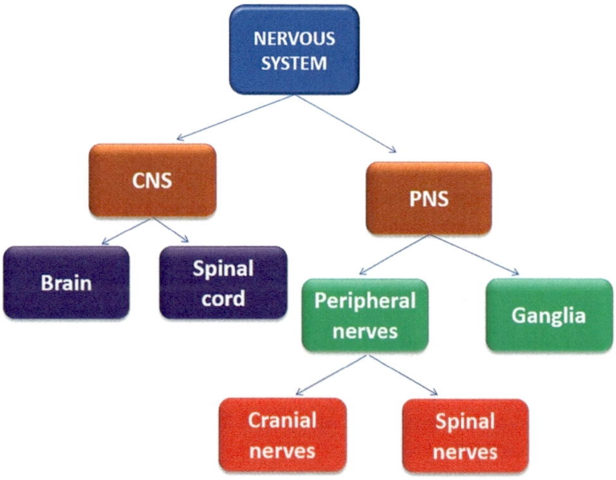

Fig. 1.1 Divisions of the nervous system

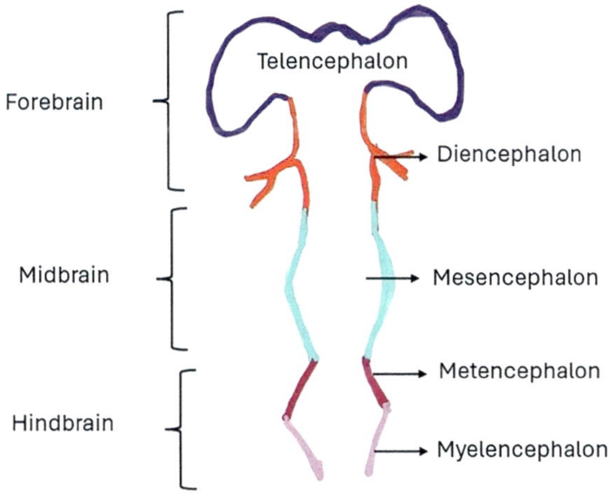

Fig. 1.2 Development of nervous system from neural tube

successive dilatations called brain vesicles. The cranial most dilatation is the forebrain (prosencephalon), followed by the midbrain (mesencephalon), and the caudal most vesicle is the hindbrain (rhombencephalon) (Fig. 1.2). The vesicles are demarcated by flexures, which are two in number. The proximal flexure is called the cephalic flexure in the midbrain territory. The distal flexure is also the cervical flexure and is found at the junction of the hindbrain and spinal cord. The prosencephalon has two parts—the *telencephalon* and *diencephalon* with the telencephalon forming the two cerebral hemispheres (Gilbert 2014). The diencephalon forms the thalamus and related structures like the hypothalamus, epithalamus, and subthalamus. The m*esencephalon* forms the midbrain. The *rhombencephalon* consists of two parts—the *metencephalon* and *myelencephalon*. The metencephalon forms the pons and cerebellum, while the myelencephalon forms the spinal cord (de Vellis and Carpenter 1999; Sadler 2019).

During development, the neural tube wall thickens significantly with an intact central lumen. In the spinal cord, this lumen remains as the narrow canal, while in the brain it expands into a network of interconnected cavities known as ventricles forming the ventricular system. At certain sites in the forebrain and hindbrain, the roof of the neural tube forms thin, folded sheets which are highly vascular called choroid plexuses. These plexuses are responsible for producing cerebrospinal fluid (CSF), and it is the ventricles which are filled with this CSF. The cavity within the rhombencephalon expands and forms the fourth ventricle, which is located behind the pons and the rostral part of the medulla. Distally, the fourth ventricle connects to a canal within the caudal medulla that continues into the spinal central canal. The fourth ventricle, at its upper end, narrows into the cerebral aqueduct which is a narrow passage that passes through the midbrain. The cerebral aqueduct then opens into a narrow, slit-like cavity into the third ventricle, bounded on either side by diencephalon. Toward the rostral end of the third ventricle, a small opening on each side leads into the lateral ventricles, which initially form within the prosencephalon and later become enclosed within the cerebral hemispheres (Sadler 2019).

1.3.1 Central Nervous System

The central nervous system includes the majority of the neuronal cell bodies of the nervous system. The clustered cell bodies within the CNS are referred to as gray matter. The gray matter is separate from the white matter, and it is in this white matter that axons of the neurons are found. Smaller, functionally related clusters of neuronal cell bodies are referred to as nuclei. Neuronal dendrites and synaptic interactions are primarily located within gray matter. In contrast, axons are typically organized in the white matter, named for its paler appearance due to the presence of myelin sheaths (Arnould-Taylor 1998). Axons traveling between similar sources or destinations within the CNS are arranged into distinct pathways or tracts, which frequently cross the midline. As a result, one side of the brain often receives information from the opposite side of the body and thus controls it (Thau et al. 2022; Stranding 2020).

The CNS comprises the cerebral hemispheres, diencephalon, midbrain, pons, medulla oblongata, cerebellum, and spinal cord. The cerebral hemispheres along with the diencephalon are collectively known as the forebrain. The midbrain, pons, and medulla oblongata collectively are referred to as the brainstem (Purves et al. 2001).

1.3.2 Brain

The brain (encephalon) is the chief organ of the CNS and is located within the cranial cavity enclosed by protective coverings called the meninges. The meninges are arranged in three strata: the outermost, called the dura mater; the middle layer, referred to as the arachnoid mater; and the innermost being the pia mater. The cerebrum constitutes the largest component of the brain and occupies the anterior and middle cranial fossae. It is composed of two cerebral hemispheres that are partially separated from each other by the median longitudinal fissure, which houses fold of dura mater called the falx cerebri. The two halves of the cerebrum are connected by a structure called the corpus callosum. The cerebral hemispheres are highly convoluted gray matter, beneath which lies the white matter. The crest of each convolution is called a gyrus (plural: gyri), separated by a depression called a sulcus (plural: sulci). Few of the sulci and gyri are prominent and constant, though the cerebral hemispheres are divided into mainly four lobes (Fig. 1.3a, b). They are the frontal, temporal, parietal, and occipital lobes (the insular lobe and limbic lobes are defined based on a physiological concept, with the insula located at the insular fossa). The lobes are named after the overlying bones (Shipp 2007; Rosdahl and Kowalski 2008; Stranding 2020).

1.3.2.1 Frontal Lobe

The frontal lobe is the most anterior part of the cerebrum. It is bordered posteriorly by the central sulcus, which demarcates it from the parietal lobe, and inferoposteriorly by the lateral sulcus, which demarcates the frontal lobe from the temporal lobe. The frontal lobe is mainly involved in higher intellectual functions, mood, social behavior, personality, and language processing (specifically on the dominant hemisphere side) (Bui and Das 2023; Stinnett et al. 2023).

1.3.2.2 Parietal Lobe

The parietal lobe is situated between the frontal and occipital lobes. It is separated from these lobes by the central sulcus and the parieto-occipital sulcus. Superior to the temporal lobe, it is divided from it by the lateral sulcus. The parietal lobe is concerned with language and mathematical calculations on the dominant hemisphere, while on the nondominant side it is involved in visuospatial functions, such as two-point discrimination (Altabakhi and Liang 2023).

1.3.2.3 Temporal Lobe

The temporal lobe is located beneath the frontal and parietal lobes. It is demarcated from them by the lateral sulcus. It is responsible for memory and language, as well as auditory processing, since it houses the primary auditory cortex (Javed et al. 2023).

1.3.2.4 Occipital Lobe

The occipital lobe represents the most posterior region of the cerebrum, with its underside supported by the tentorium cerebelli, which separates it from the cerebellum below. The parieto-occipital sulcus delineates the occipital lobe from the anteriorly positioned parietal and temporal lobes. This lobe contains the primary

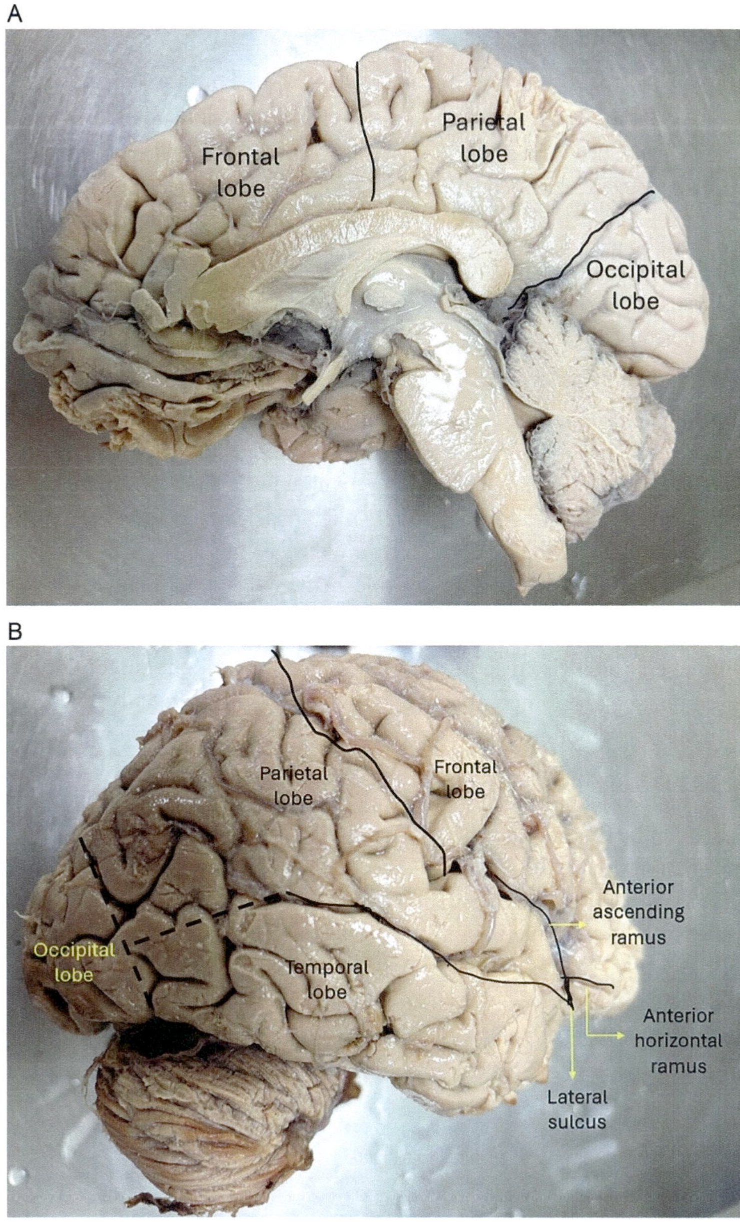

Fig. 1.3 (**a**) Major sulci on the medial surface of brain. (**b**) Major sulci on the superolateral surface of brain

visual cortex (V1) and is primarily responsible for processing visual information (Rehman and Al Khalili 2023).

1.4 White Matter of Cerebrum

In the brain, the white matter lies deep to the gray matter which in itself has various gray matter regions. It contains three main types of fibers:

- Association fibers: These link functional areas within the same cerebral hemisphere. They are either short fibers, linking nearby gyri, or long fibers, connecting distant gyri, forming bundles like the uncinate fasciculus, superior/inferior longitudinal fasciculi, and cingulum.
- Commissural fibers: These link corresponding functional areas between the two cerebral hemispheres. The corpus callosum is the largest commissural fiber. Other examples of these

fibers include the anterior and posterior commissures.

- Projection fibers: These link the cortex with deeper brain structures, the brainstem, and spinal cord. Corticofugal fibers descend from the cortex, and corticopetal fibers ascend to the cortex. Projection fibers form the corona radiata, which becomes the internal capsule.

1.5 Internal Capsule

The internal capsule is located on the inferomedial surface of the cerebral hemisphere. It includes ascending and descending myelinated nerve fibers that traverse through the basal ganglia to connect the cerebral cortex with subcortical structures. It plays a key role in neural signal transmission to and from the cerebral cortex, and damage to it can cause severe disability or death (Emos et al. 2023).

On a transverse section, the internal capsule appears V-shaped (Fig. 1.4), with its concavity facing medially. It has five parts: the anterior part, called the anterior limb; the middle part, called the genu; the posterior part, called the posterior limb; and also the retrolenticular and sublenticular parts. The caudate nucleus is positioned medial to the anterior limb, while the thalamus lies medial to the posterior limb. The lentiform nucleus is located lateral to both the anterior and posterior limbs (Fig. 1.4).

The internal capsule contains motor and sensory fibers. The motor fibers descend from the cerebral cortex to various parts of the brain and spinal cord. The sensory fibers, after being relayed in the thalamus, ascend and are projected to specific sites of the cerebral cortex. The location of different fibers within the internal capsule is as follows:

- The anterior limb accommodates pontine fibers and anterior thalamic radiations.
- The genu contains corticobulbar fibers from the motor cortex to cranial nerve motor nuclei (mainly contralateral) and anterior fibers of the superior thalamic radiation.
- The posterior limb accommodates the corticospinal tract and fibers of the superior

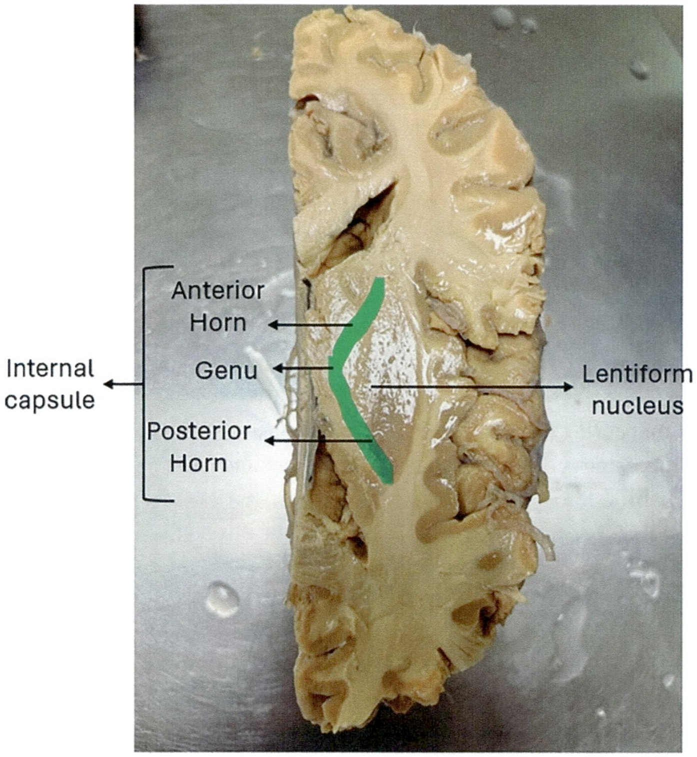

Fig. 1.4 Parts of internal capsule

thalamic radiation (somatosensory fibers to the postcentral gyrus). Other descending fibers include the frontopontine and corticorubral fibers.
- The retrolenticular part accommodates the parietopontine, occipitopontine, and occipitotectal fibers as well as the posterior thalamic radiation and optic radiation.
- The sublenticular part accommodates the temporopontine and some parietopontine fibers, as well as the auditory radiation (Chowdhary et al. 2010; Costa et al. 2018).

1.6 Basal Ganglia

The basal ganglia, or basal nuclei, are groups of subcortical structures located within the cerebrum. They are primarily composed of the dorsal and ventral striatum. The dorsal striatum includes the putamen and caudate nucleus, while the ventral striatum consists of the nucleus accumbens and olfactory tubercle and is associated with structures like the ventral pallidum, globus pallidus, substantia nigra, and subthalamic nucleus. Positioned at the base of the forebrain and extending into the upper midbrain, the basal ganglia form complex networks with the cerebral cortex, thalamus, brainstem, and other brain areas. They are essential for voluntary motor control, procedural and associative learning, habit formation, and ocular movement and contribute to cognitive and emotional processes (Snell 2006).

1.7 Diencephalon

The diencephalon primarily includes structures that develop on either side of the third ventricle. Its lateral walls give rise to the epithalamus at the superior most part, the thalamus in the center, and the subthalamus and hypothalamus inferiorly. In the adult brain, the epithalamus includes the anterior and posterior paraventricular nuclei, the stria medullaris thalami, the medial and lateral habenular nuclei, and the pineal gland. The thalamus contains multiple groups of nuclei that form reciprocal connections with the cerebral cortex. The subthalamic region comprises the subthalamic nucleus, zona incerta, and the fields of Forel. The subthalamic nucleus is closely associated with the basal ganglia and is regarded as the part of basal ganglia circuitry. The hypothalamic rudiment develops into most of the subdivisions of the adult hypothalamus (Stranding 2020).

1.8 Brainstem

The brainstem comprises the medulla oblongata, pons, and midbrain (Fig. 1.5). It connects the cerebrum to the spinal cord and cerebellum and

Fig. 1.5 Sagittal section brain

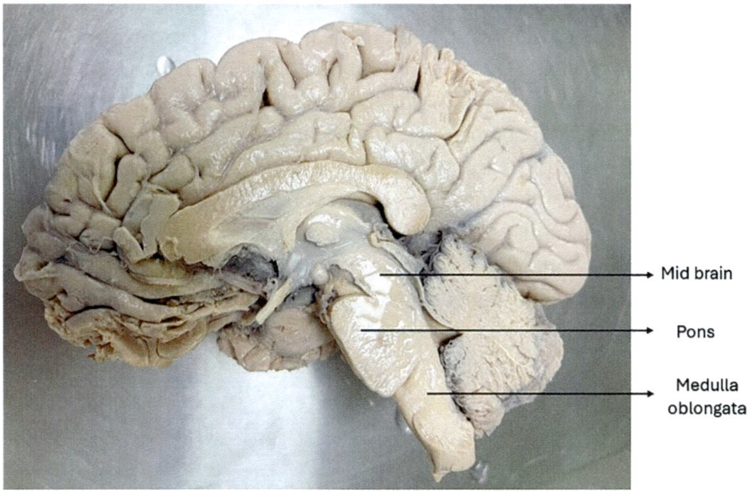

is located in the posterior cranial fossa. The ventral surface of the brainstem rests on the clivus. Within the brainstem, numerous intrinsic neuronal cell bodies and their processes exist, some of which serve as counterparts to spinal neuronal groups. The brainstem has special groups of nerve cells called cranial nerve nuclei. These include sites where axons entering or exiting the brainstem via cranial nerves terminate or originate, providing sensory, motor, and autonomic innervation, primarily to the head and neck. However, autonomic fibers originating from the brainstem extend more widely throughout the body. Additionally, specialized neuronal groups process inputs related to hearing, taste, and vestibular function. The reticular formation, an extensive yet often ill-defined neuronal network, spans the length of the brainstem and continues caudally into the spinal cord. Some of the nuclei of brainstem nuclei regulate cardiac, respiratory, and digestive functions, while others participate in various neural processes or form extensive afferent and efferent connections with the cerebellum (Ackerman 1992).

The brainstem serves as a key relay center for ascending and descending fibers. The spinothalamic (spinal lemniscal), medial lemniscal, and trigeminal pathways ascend through the brainstem to their destined site in the thalamus, while major corticospinal tracts descend through it, and corticobulbar fibers terminate within it (Snell 2006).

Clinically, brainstem damage is often severe and life-threatening due to its compact structure, where even minor lesions can disrupt vital cardiac and respiratory centers, impair forebrain motor control, and sever sensory input pathways essential for consciousness, perception, and cognition. Complete damage of neural respiratory and cardiac centers in the medulla leads to irreversible respiratory and cardiac arrest. This condition, known as brainstem death, requires precise diagnosis, particularly in patients on life-support machines, where artificial maintenance of respiratory and cardiac functions can be prolonged indefinitely (Snell 2006).

1.8.1 Midbrain

The midbrain links the pons to the diencephalon and communicates posteriorly with the cerebellum through the superior cerebellar peduncles. Its anterior section contains the crus cerebri (Fig. 1.4), which houses the corticospinal, corticonuclear, and pontine fiber tracts, with the interpeduncular fossa situated between them. Centrally, the cerebral aqueduct runs through the midbrain, linking the third and fourth ventricles, surrounded by the periaqueductal gray. Structurally, the midbrain is divided by the aqueduct into the posterior tectum and anterior tegmentum (Ackerman 1992).

The posterior midbrain features the corpora quadrigemina, consisting of two sets of paired colliculi—superior and inferior (Fig. 1.6). The superior colliculi are involved in visual reflexes and interact with the lateral geniculate body through the superior brachium. The inferior colliculi handle auditory processing and connect to the medial geniculate nuclei via the inferior brachium. The trochlear nerve exits just below the inferior colliculi, while the oculomotor nerve (CN III) arises via the oculomotor sulcus in the interpeduncular cistern (Snell 2006).

Key midbrain nuclei include the substantia nigra, which regulates movement via the basal ganglia, and the red nucleus, which is involved in motor control and cerebellar connections. The ventral tegmental area contains dopaminergic neurons linked to reward pathways, while the raphe nuclei are rich in serotonergic neurons. These neurons extend their projections broadly across the brain. The gray matter around the aqueduct is associated with pain modulation, and the medial longitudinal fasciculus, located in front of it, coordinates eye movements (Haines and Mihailoff 2018).

1.8.2 Pons

The pons is located superior to the medulla oblongata and below the midbrain. It serves as a crucial neural bridge between the cerebrum and

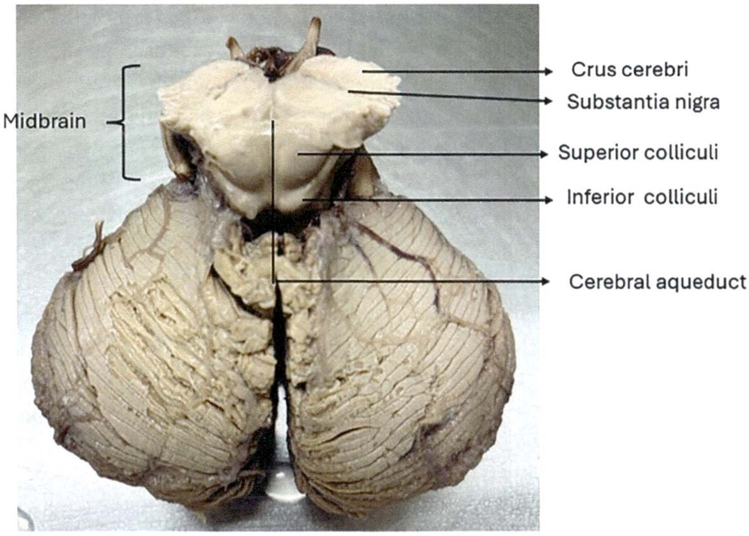

Fig. 1.6 Parts of midbrain

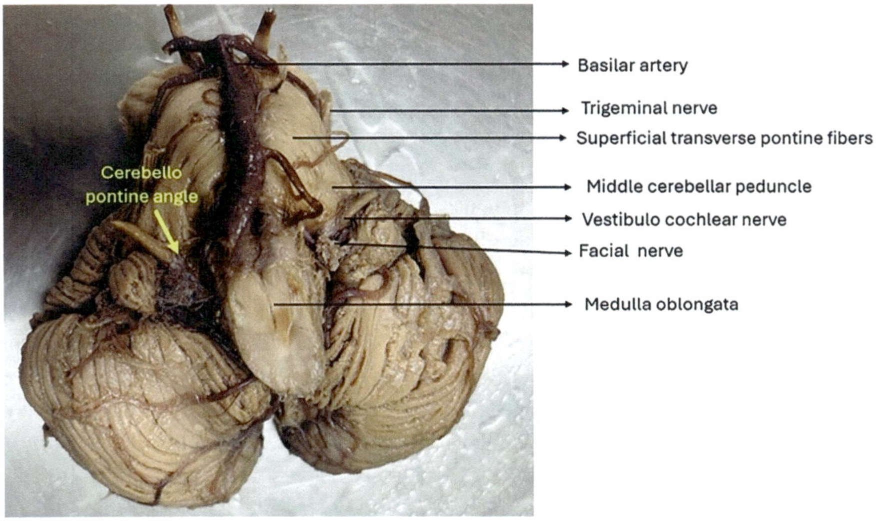

Fig. 1.7 Ventral aspect of pons

cerebellum—its name comes from the Latin word for "bridge." On its ventral (anterior) surface, the pons features a noticeable bulge created by transverse pontocerebellar fibers. Running along the midline of this surface is the basilar groove, which houses the basilar artery (Carpenter 1985). The pontomedullary junction, marking the point where the lower edge of the pons meets the upper edge of the medulla, serves as a significant anatomical landmark. Several cranial nerves emerge from the ventral surface of the pons: CN V (trigeminal), VI (abducens), VII (facial), and VIII (vestibulocochlear) (Sinnatamby 2011) (Fig. 1.7).

The pons connects to the cerebellum via the middle cerebellar peduncles. Its dorsal surface, together with the medulla, forms the floor of the fourth ventricle. The important features of this floor include the medial eminence, which marks the midline, and the facial colliculus, a bulge formed by the looping of facial nerve fibers around the nucleus of cranial nerve VI, the abducens nucleus. The stria medullaris—a band of transverse fibers—crosses from the lateral side into the midline, marking the posterior boundary between the pons and medulla (Carpenter 1985).

The cerebellopontine angle is a significant site at the junction of the pons, medulla, and cerebellum. This region contains the cerebellar flocculus, the choroid plexus of the fourth ventricle, and the emerging cranial nerves VII and VIII, all surrounding the foramen of Luschka (lateral apertures of the fourth ventricle) (Stranding 2020).

1.8.3 Medulla Oblongata

The medulla oblongata is the lowest part of both the brain and the brainstem. It connects to the midbrain via the pons and continues downward into the spinal cord. It plays a vital role in relaying signals between the spinal cord and higher brain centers, and it regulates essential autonomic functions such as heart rate and breathing (Snell 2006).

The medulla has two primary regions: the ventral medulla (anterior) and the dorsal medulla, also called the tegmentum (posterior). The ventral medulla features two pyramid-shaped structures called pyramids, which house the pyramidal tracts. These include the corticospinal tract, carrying motor signals from the cerebral cortex to the spinal cord, and the corticobulbar tract, which transmits impulses from the motor cortex to the cranial nerves in the brainstem (Ackerman, 1992).

Near the junction with the spinal cord, about 80–90% of the corticospinal fibers cross over at a point known as the pyramidal decussation. Lateral to the pyramids are the olivary bodies, which contain the inferior olivary nuclei involved in motor coordination. The upper dorsal medulla contributes to the lower part of the IV ventricle, formed by the widening of the spinal cord's central canal. Like the spinal cord, the surrounding structure has white matter externally and gray matter internally. The dorsal medulla also gives rise to the final seven cranial nerves (VI–XII), though most exit from its ventral surface.

The medulla contains both myelinated (white matter) and unmyelinated (gray matter) fibers, which are interspersed—unlike the typical layering seen in other brain regions. This typical arrangement is part of the reticular formation, a network of interconnected neurons vital for transmitting motor and sensory signals. Key nuclei in the gray matter include those responsible for cranial nerves IX–XII, the cardiovascular and respiratory centers. Neurons in the medulla are involved in autonomic regulation (e.g., breathing, heart function, digestion), motor control, sensory relay, and the modulation of arousal and sleep (Webb 2017).

Damage to the medial medulla can lead to medial medullary syndrome, which may cause weakness or paralysis of the body on the contralateral side, impaired touch and position sense, and partial tongue paralysis. Injury to the lateral medulla can result in lateral medullary syndrome, presenting symptoms such as loss of pain and temperature sensation, difficulty swallowing, absent gag reflex, vertigo, vomiting, and coordination issues (Snell 2006).

1.9 Spinal Cord

The spinal cord is located within the vertebral canal of the vertebrae. It extends from the medulla oblongata to the L1–L2 vertebral level in adults. It is responsible for conducting sensory and motor signals between the brain and different parts of the body, particularly the trunk and limbs. The spinal cord is surrounded by three protective meningeal layers: the outer dura mater, the middle arachnoid mater, and the inner pia mater. Denticulate ligaments secure the cord laterally. It is divided into four regions—cervical, thoracic, lumbar, and sacral—with two enlargements, the cervical enlargement C3 to T1 (upper limbs) and lumbar enlargement L1 to S2 (lower limbs) (Carpenter 1985).

The spinal cord has 31 segments, each segment giving rise to a pair of spinal nerves. There are eight cervical, 12 thoracic, five lumbar, five sacral, and one coccygeal spinal nerves from these spinal segments. Each spinal nerve forms from dorsal (sensory) and ventral (motor) roots. These roots emerge through the intervertebral foramina. Due to differential growth rates during development, the lower end of spinal cord lies at L1–L2, but spinal nerves continue to exit at their original vertebral levels. This results in the formation of a bundle of descending lumbar and sacral nerve roots called the cauda equina. The first cervical root exits above the vertebrae, the second cervical root leaves between the first and second cervical vertebrae, and the rest of the nerve roots leave below the level of the vertebrae.

A dermatome refers to a specific area of skin innervated by sensory fibers from a single dorsal root ganglion. There is usually some degree of overlap between adjacent dermatomes. Understanding the segmental innervation of skin and muscles is essential for accurately locating neurological injuries (Canbay 2014; Snell 2006) (Fig. 1.7).

In a transverse (horizontal) section of the adult spinal cord, the white matter forms the outer layer, while the gray matter lies centrally (Fig. 1.8). At the core is the central canal, a narrow channel filled with cerebrospinal fluid (CSF) and lined by ependymal cells, which are either simple cuboidal or columnar in form. Surrounding this canal is the gray matter, which contains neuronal cell bodies and typically exhibits an H-shaped or butterfly-like appearance. The two lateral extensions of gray matter are connected by the dorsal gray commissure, while the white commissure lies just below it. The morphology and size of the gray matter differ across various spinal cord levels. Lower spinal segments exhibit a higher proportion of gray matter relative to white matter due to a decline in the number of ascending and descending fiber tracts. The gray matter is subdivided into three major regions:

Dorsal horn: Present at all levels of the spinal cord, it contains sensory neurons that process somatosensory input and give rise to ascending pathways targeting the midbrain and diencephalon.

Lateral horn: Contains autonomic neurons that innervate visceral organs, particularly prominent in thoracic and upper lumbar regions.

Ventral horn: Houses motor neurons responsible for innervating skeletal muscles.

Within the gray matter, the neurons are multipolar and show variability in shape. Two primary types are:

Golgi type I neurons, which have long axons that exit the spinal cord via ventral roots or pass into the white matter tracts.

Golgi type II neurons, which have short axons and dendrites, remaining confined within the gray matter and forming local synaptic circuits (Carpenter 1985; Moore and Agur 2007).

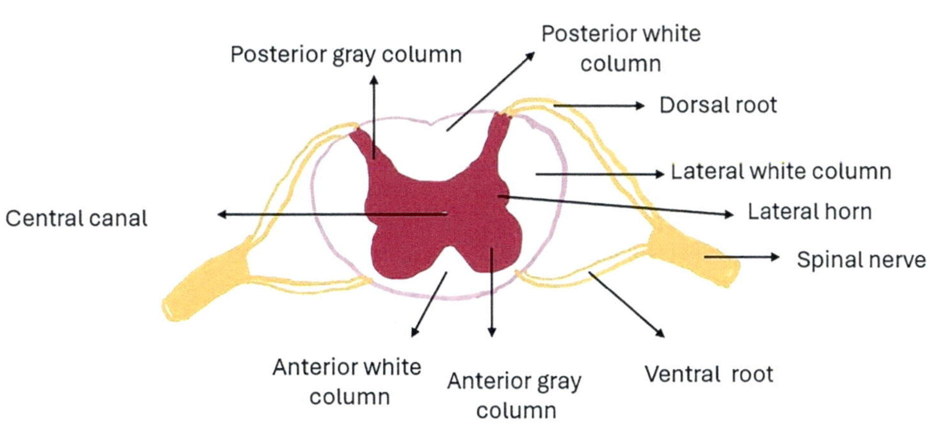

Fig. 1.8 Transverse section of spinal cord

1.9.1 Nuclei

The spinal cord contains several key nuclear groups arranged dorsally to ventrally, each with distinct functions:

- Marginal zone (posterior marginal nucleus): Thin layer at the dorsal horn tip, present at all levels; contributes to the lateral spinothalamic tract relaying pain and temperature.
- Substantia gelatinosa: Found at all spinal levels in the dorsal horn cap; relays pain, temperature, and light touch. Its neurons project to Rexed layers IV–VII and contribute to spinothalamic tracts.
- Nucleus proprius: Located beneath the substantia gelatinosa; involved in mechanical and temperature sensation. Its axons contribute to the spinothalamic and spinocerebellar tracts.
- Dorsal nucleus of Clarke: Found from C8 to L3, in the dorsal horn base; forms the dorsal spinocerebellar tract, relaying unconscious proprioception. The medullary counterpart is the accessory cuneate nucleus.
- Intermediolateral nucleus: Located in the lateral horn (C8–L3) and S2–S4; contains preganglionic sympathetic (thoracolumbar) and parasympathetic (sacral) neurons, sending fibers to the autonomic ganglia.
- Lower motor neuron nuclei: Found in the ventral horn at all levels; consist of α, β, and γ motor neurons that innervate skeletal and visceral muscles, forming the final motor output of the CNS (Carpenter 1985; Stranding 2020).

1.9.2 White Matter

The white matter surrounds the spinal cord's gray matter and contains myelinated and unmyelinated fibers that conduct ascending (sensory) and descending (motor) signals. It is organized into three regions: the dorsal (posterior), lateral, and ventral (anterior) columns. The anterior white commissure, located centrally, contains crossing fibers from the spinothalamic, spinocerebellar, and anterior corticospinal tracts (Snell 2006).

1.9.3 Tracts of Spinal Cord

The white matter of the spinal cord has ascending and descending tracts responsible for transmitting sensory and motor information (Moore and Agur 2007; Standing 2020).

1.9.3.1 Ascending Tracts (Sensory Pathways)

These tracts transmit sensory input from receptors in the periphery to the higher centers in the CNS. The first-order neurons arise from the dorsal root ganglion (DRG).

Dorsal Column (Dorsal Funiculus) The fasciculus gracilis and fasciculus cuneatus carry sensations of fine touch, pressure, vibration, two-point discrimination, conscious proprioception, and sense of position.

Lateral Column The lateral spinothalamic tract conveys pain, crude touch, and temperature from both somatic and visceral receptors. Next to it are dorsal and ventral spinocerebellar tracts which transmit unconscious proprioception from joints and lower limbs to the cerebellum.

Ventral Column

- Anterior spinothalamic (paleospinothalamic) tract: Conveys pain, temperature, and touch to the brainstem and diencephalon.
- Spino-olivary tract: Conveys signals from Golgi tendon to the cerebellum.
- Spinoreticular and spinotectal tracts: Involved in arousal and reflexive responses to sensory stimuli.
- Fasciculus proprius: A thin ring of intersegmental fibers around the gray matter; carries pain signals across spinal levels to higher centers.

1.9.3.2 Descending Tracts (Motor Pathways)

These tracts originate from the cerebral cortex and brainstem nuclei, transmitting commands to regulate motor control, posture, balance, muscle tone, and reflexes.

- Lateral column: Consists of lateral corticospinal tract (major pathway for voluntary motor control) and rubrospinal tract (assists in voluntary movement of the limbs).
- Ventral column: Includes anterior corticospinal, reticulospinal, and vestibulospinal tracts (coordinate posture and balance).
- Lissauer's tract is located in the posterolateral region of the spinal cord. It carries descending fibers from the dorsolateral funiculus that modulate pain signals at the spinal level and include intersegmental connections (Standing 2020).

1.10 Cerebellum

The cerebellum is situated in the posterior cranial fossa. It has the fourth ventricle anteriorly. The tentorium cerebelli, a fold of dura mater, separates it from the cerebrum. Structurally, the cerebellum consists of two hemispheres which are connected by the central vermis. Each cerebral hemisphere has anterior, posterior, and flocculonodular lobes (Fig. 1.9a, b). These lobes are demarcated by transverse fissures. The V-shaped primary fissure separates the anterior and posterior lobes, while the posterolateral fissure distinguishes the posterior lobe from the flocculonodular lobe (Fig. 1.9a). A deep horizontal fissure in the posterior lobe distinguishes the superior from the inferior cerebellar surfaces. Rich in neurons, the cerebellum contains about 80% of the brain's neurons, densely packed in its thin, folded cortical layer (Snell 2006).

The cerebellar cortex is a single, sheet-like structure less than 1 mm thick, folded into accordion-like ridges that merge at the midline. The cerebellum features an outer layer of highly folded gray matter (the cerebellar cortex) encasing a tree-like arrangement of white matter called as the arbor vitae. Within this central white matter are three pairs of deep cerebellar nuclei. From medial to lateral, the nuclei are the fastigial, the interpositus nucleus (comprising the globose and emboliform nuclei), and, the largest, the dentate nucleus. Axons from the dentate, emboliform, and globose nuclei travel through the superior cerebellar peduncle, whereas those from the fastigial nucleus pass through the inferior cerebellar peduncle (Moore and Agur 2007).

Functionally, the vermis controls movements of the axial body—neck, shoulders, thorax, abdomen, and hips. The intermediate zone of the hemispheres regulates movement of the distal limbs, while the lateral zones contribute to planning sequential body movements and help in consciously assessing motor errors (Roostaei et al. 2014; Witter and De Zeeuw 2015; Standing 2020).

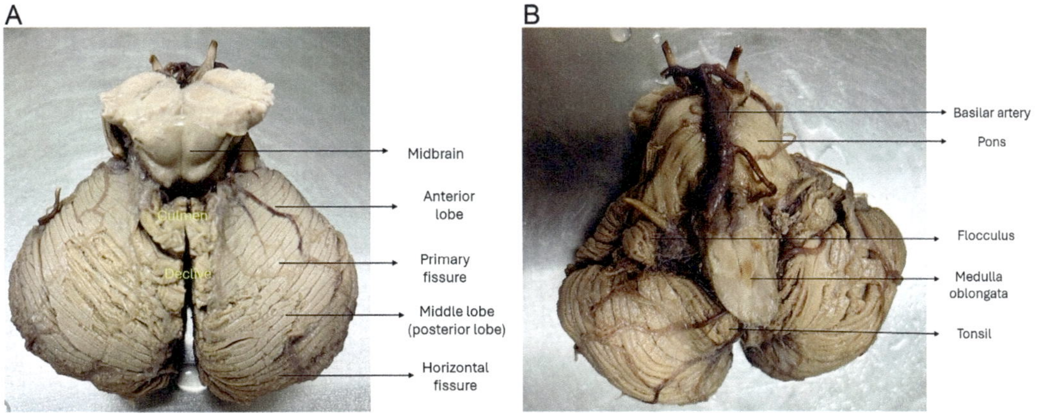

Fig. 1.9 (**a**) Superior view of cerebellum (**b**) Inferior view of cerebellum

1.11 Neurohistology

The nervous system has neurons, supported by neuroglial cells. The neuron is the structural and functional unit of the nervous system (Ravi and Pushpa 2024). Neurons/nerve cells are specialized cells which conduct impulses. A collection of nerve cell bodies within the CNS is called as nucleus, and such collections outside the CNS are called ganglia (Carpenter 1985).

1.11.1 Structure of a Neuron

The cell body (soma) is the enlarged part of a neuron containing the nucleus, which is euchromatic with prominent nucleoli (Fig. 1.9). The surrounding cytoplasm has typical organelles, but a key feature is the presence of dark-staining ribosomes called Nissl bodies. The axon hillock, where the axon originates, lacks major organelles. The axon is the longest process, carrying impulses away from the cell body. It lacks Nissl bodies, Golgi apparatus, and free ribosomes. Axon branches, called axon terminals, form synapses with other neurons. The axon is enclosed by the axolemma, and many are wrapped in a myelin sheath, a lipid-rich layer that insulates the axon. Nodes of Ranvier, gaps in the myelin, facilitate saltatory conduction. The axon hillock and branching ends (axonal arborization) are unmyelinated. Dendrites are shorter, branched processes that receive signals. They are usually unmyelinated and form a complex dendritic tree. Dendritic spines, small protrusions on dendrites, are common in excitatory neurons (Ross and Pawlina 2020; Ravi and Pushpa 2024).

1.11.2 Classification of Neurons

Based on the functions, three types of neurons have been identified. They are sensory (*afferent*), *motor* (*efferent*), *and interneurons*. Sensory neurons detect internal and external changes, supporting general senses like touch, pressure, pain, vibration, temperature, and proprioception and special senses like vision, hearing, olfaction, taste, and balance. Motor neurons transmit signals from the central nervous system (CNS) to the effector organs (muscles and glands). Interneurons, found only in the CNS, outnumber the other types and enable analysis, integration, and information storage (Ravi 2021).

Based on the number of processes, neurons are classified as multipolar neurons, which have single axon and two or more dendrites (e.g., motor neurons and interneurons); bipolar neurons, which have single axon and single dendrite (e.g., bipolar neuron of retina, ganglia of VIII cranial nerve) (Fig. 1.10); and pseudo-unipolar/unipolar neurons which have developed from bipolar neurons but their two processes (axon and dendrite) have fused proximally. Among the two peripheral processes, one extends to the periphery as dendritic branch and the other extends to the CNS as axonal branch. Pseudo-unipolar neurons are usually sensory in function (e.g., neurons of dorsal root ganglia).

Based on the length of the axons, neurons are classified as Golgi type I neurons (they have long axons arising from the motor nerve nuclei of the CNS, e.g., Purkinje cells, pyramidal cells, motor neurons) and Golgi type II neurons (they are nothing but the interneurons of CNS with short axons, e.g., granule cells of cerebellum).

1.11.3 Myelination

Myelination refers to the formation of a myelin sheath around an axon. The process begins when an axon comes into contact with a Schwann cell, resulting in two membrane regions: the abaxonal (outer) and adaxonal (inner or peri-axonal) membranes. As the Schwann cell wraps itself around the axon, a mesoaxon forms—linking the inner and outer membranes and enclosing extracellular space. Progressive wrapping leads to the formation of multiple spiraling layers (lamellae) around the axon, thus creating the myelin sheath. Each Schwann cell contributes to a single segment of myelin, and numerous such cells are required to cover the length of a single axon. The number of spirals correlates with axonal diameter. Nodes of Ranvier, the gaps between myelinated segments, enable saltatory conduction. Even unmyelinated axons are supported by

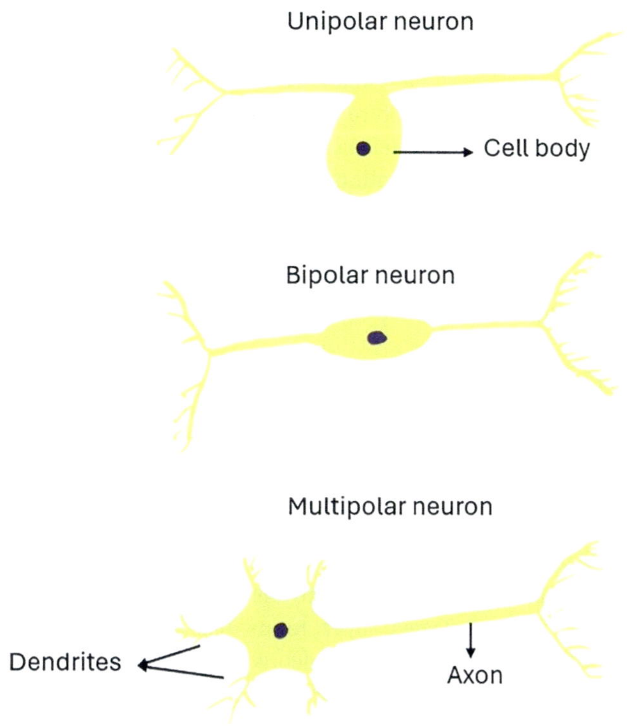

Fig. 1.10 Types of neurons based on polarity

Schwann cells, though these are not wrapped in lamellae; instead, multiple axons occupy separate invaginations within a single Schwann cell (Stranding 2020).

1.11.4 Synapse

These are the specialized sites where the axons meet and communicate. Synapses are also formed between the axon and the receptors of target cells. Three types of synapses have been defined:

1. Axoaxonic: Synapse between a presynaptic and postsynaptic axon.
2. Axodendritic: Synapse between an axon and a dendrite.
3. Axosomatic: Synapse between an axon and the soma (Snell 2006).

1.11.5 Glial Cells

Neurons are supported by *glial cells/gliocytes* (glia: glue). These cells were discovered by Rudolf Virchow. They are nonconducting; small cells present intimately related to the neurons. Based on their location, they are broadly classified as shown below (Ravi and Pushpa 2024).

1.11.5.1 Central Neuroglia

Central neuroglial cells are astrocytes, ependymal cells, oligodendrocytes, and microglia. Among these cells, the first three are broadly called macroglia. Microglia are the only neuroglial cells derived from the mesoderm, while all other types originate from the ectoderm:

1. *Astrocytes*: Large, specialized cells with numerous projections. Most abundant type of glial cell and are of two types: fibrous astrocytes (found in white matter, have few slender processes) and protoplasmic astrocytes (found in gray matter, have granular cytoplasm with many thick processes). Astrocytes mainly provide physical support to neurons.
2. *Ependymal cells* are simple cuboidal or columnar cells that line the brain's ventricles and the central canal within the spinal cord. They are

mainly responsible for the formation and circulation of cerebrospinal fluid (CSF).
3. *Oligodendrocytes* are small cells with few processes and are involved in the formation of myelin and its maintenance in the CNS.
4. *Microglia* are the smallest neuroglial cells and have dark, oval nuclei. They are immune cells of the CNS that act as phagocytes, clearing debris and pathogens in the CNS (Ross and Pawlina 2020).

1.11.5.2 Peripheral Neuroglia

Peripheral neuroglia are present in the peripheral nervous system. They are:

1. Schwann cells (neurolemmocytes): Responsible for myelination of axons in the PNS. Mature Schwann cells wrap around axons; their inner layers form the myelin sheath, and the outer nucleated layer forms the neurilemma. They also remove debris in the PNS and support nerve regeneration.
2. Satellite cells are cuboidal cells like Schwann cells but surround neuron cell bodies in peripheral ganglia. They form a complete sheath around the soma with neuronal processes that form synapses. They provide electrical insulation for the cell body and regulate the microenvironment through gap junctions, allowing metabolic exchange.
3. Enteric neuroglia: Glial cells found in the enteric nervous system, closely associated with neurons and their processes. They resemble astrocytes in both structure and function.
4. Tissue-specific neuroglia: Glial cells confined to specific organs or tissues. Examples include teloglia, Müller cells, olfactory ensheathing cells, and pituicytes (Ross and Pawlina 2020).

1.12 Structure of Peripheral Nerve

The peripheral nervous system includes peripheral nerves and ganglia. A peripheral nerve consists of many nerve fibers bundled together by connective tissue called the perineurium. A nerve fiber refers to any nerve cell process (axon, axon with myelin sheath and Schwann cell, or dendrite). The cell bodies of peripheral nerves may be located in the CNS or ganglia outside the CNS. In a cross section, the Schwann cell nuclei are visible. Schwann cells support peripheral nerves by forming myelin sheaths around axons. Each nerve fiber and its Schwann cells are surrounded by endoneurium, a loose connective tissue with minimal blood supply. A group of nerve fibers forms a nerve fascicle, which is enclosed by perineurium, a specialized connective tissue layer with perineurial cells that create a blood-nerve barrier. Multiple nerve fascicles are held together by the epineurium, a dense irregular connective tissue layer containing adipocytes and blood vessels, which penetrate the perineurium (Ravi and Pushpa 2024).

1.13 Ganglia

Ganglia refer to collection of cell bodies of the neuron outside the CNS. Based on the type of neuron and function, two types of ganglia are defined (Ross and Pawlina 2020).

1.13.1 Sensory Ganglia/Spinal Ganglia/Dorsal Root Ganglia

These include ganglia of spinal nerves and cranial nerves (V, VII, VIII, IX, X). They contain pseudo-unipolar sensory neurons with centrally located nuclei. Myelinated fibers are arranged in bundles, forming clusters of cell bodies. These cell bodies are surrounded by an inner capsule made by satellite cells and an outer capsule of connective tissue with fibroblasts. No synapses between adjacent cell bodies exist. Examples include trigeminal ganglion, geniculate ganglion, dorsal root ganglia, and vestibular ganglion.

1.13.2 Sympathetic Ganglia/Autonomic Ganglia

These include ganglia of sympathetic and parasympathetic (postsynaptic motor) neurons. They contain multipolar neurons with eccentric nuclei.

The cell bodies are scattered among unmyelinated fibers. Satellite cells sparsely surround cell bodies, and neural processes penetrate these boundaries to form synapses. Examples for sympathetic ganglia are sympathetic trunk ganglia, prevertebral ganglia (celiac, superior, and inferior mesenteric), and the adrenal medulla. Examples for parasympathetic ganglia are the ciliary, otic, submandibular, and pterygopalatine ganglia.

1.14 Conclusion

Understanding neuroanatomy and neurohistology forms the cornerstone for comprehending the structure and function of the nervous system. Neuroanatomy examines the organization of both the central and peripheral nervous systems, with a focus on vital structures such as the brain, spinal cord, and associated components. In contrast, neurohistology investigates the microscopic architecture of the nervous system, particularly neurons and their supporting glial cells. These elements are essential for processes such as signal generation, transmission, support, and protection. Specialized features like synaptic junctions and the myelin sheath facilitate efficient communication throughout the body. Mastery of these foundational structures is critical for understanding both normal neural function and the pathological changes that underlie neurological disorders.

References

Ackerman S (1992) Major structures and functions of the brain. In: Discovering the brain. National Academies Press, Washington DC

Altabakhi IW, Liang JW (2023) Gerstmann syndrome. In: StatPearls [Internet]. StatPearls, Treasure Island

Arnould-Taylor W (1998) A textbook of anatomy and physiology. Nelson Thornes

Bui T, Das JM (2023) Neuroanatomy, cerebral hemisphere [Updated 2023 Jul 24]. In: StatPearls [Internet]. StatPearls, Treasure Island, 2025 Jan. https://www.ncbi.nlm.nih.gov/books/NBK549789

Canbay S, Gürer B, Bozkurt M, Comert A, Izci Y, Başkaya MK (2014) Anatomical relationship and positions of the lumbar and sacral segments of the spinal cord according to the vertebral bodies and the spinal roots. Clin Anat 27(2):227–233

Carpenter M (1985) Core text of neuroanatomy, 3rd edn. Williams & Wilkins

Chowdhury F, Haque M, Sarkar M, Ara S, Islam M (2010) White fiber dissection of brain; the internal capsule: a cadaveric study. Turk Neurosurg 20(3):314–322

Costa M, Braga VL, Yağmurlu K, Centeno RS, Cavalheiro S, Chaddad-Neto F (2018) A technical guide for fiber tract dissection of the internal capsule. Turk Neurosurg 28(6):934–939

de Vellis J, Carpenter E (1999) General development of the nervous system. In: Siegel GJ, Agranoff BW, Albers RW et al (eds) Basic neurochemistry: molecular, cellular and medical aspects, 6th edn. Lippincott-Raven, Philadelphia. Available from: https://www.ncbi.nlm.nih.gov/books/NBK28253/

Emos MC, Khan Suheb MZ, Agarwal S (2023) Neuroanatomy, internal capsule [Updated 2023 Jun 5]. In: Stat Pearls [Internet]. Stat Pearls, Treasure Island, 2025 Jan. https://www.ncbi.nlm.nih.gov/books/NBK542181/

Galli R, Gritti A, Bonfanti L, Vescovi AL (2003) Neural stem cells: an overview. Circ Res 92(6):598–608

Gilbert SF (2014) Developmental biology, 10th edn. Sinauer, Sunderland

Haines D, Mihailoff G (2018) Fundamental neuroscience for basic and clinical applications, 5th edn. Elsevier

Javed K, Reddy V, Das JM, Wroten M (2023) Neuroanatomy, Wernicke area. In: StatPearls [Internet]. StatPearls, Treasure Island

Kandel ER (2013) Principles of neural science, 5th edn. Appleton and Lange/McGraw Hill, pp 1019–1036. ISBN 978-0-07-139011-8

Kennea NL, Mehmet H (2002) Neural stem cells. J Pathol 197(4):536–550

Ludwig PE, Reddy V, Varacallo MA (2022) Neuroanatomy, Central Nervous System (CNS) [Updated 2022 Oct 10]. In: Stat Pearls [Internet]. Stat Pearls, Treasure Island, 2025 Jan. https://www.ncbi.nlm.nih.gov/books/NBK442010/

Moore K, Agur A (2007) Essential clinical anatomy, 3rd edn. Lippincott Williams & Wilkins

Purves D, Augustine GJ, Fitzpatrick D et al (eds) (2001) The subdivisions of the central nervous system. In: Neuroscience, 2nd edn. Sinauer Associates, Sunderland. Available from: https://www.ncbi.nlm.nih.gov/books/NBK10926/

Ravi SK (ed) (2021) Snell's clinical neuroanatomy, South Asian edn. Wolters Kluwer

Ravi KS, Pushpa NB (2024) Textbook of histology, 1st edn. Elsevier India

Rehman A, Al Khalili Y (2023) Neuroanatomy, occipital lobe. In: StatPearls [Internet]. StatPearls, Treasure Island, Jul 24, 2023

Roostaei T, Nazeri A, Sahraian MA, Minagar A (2014) The human cerebellum: a review of physiologic neuroanatomy. Neurol Clin 32(4):859–869

Rosdahl C, Kowalski M (2008) Textbook of basic nursing, 9th edn. Lippincott Williams & Wilkins

Ross MH, Pawlina W (2020) Histology: a text and atlas: with correlated cell and molecular biology, 8th edn. Wolters Kluwer Health
Sadler TW (2019) Langman's medical embryology, 14th edn. Lippincott Williams & Wilkins
Shipp S (2007) Structure and function of the cerebral cortex. Curr Biol 17(12):R443–R449
Sinnatamby CS (2011) Last's anatomy, 12th edn. Elsevier Australia
Snell RS (2006) Clinical neuroanatomy, 6th edn. Lippincott Williams & Wilkins
Sonne J, Lopez-Ojeda W (2019) Neuroanatomy, cranial nerve 0 (terminal nerve). StatPearls
Standring S (2020) Gray's anatomy: the anatomical basis of clinical practice, 42nd edn. Elsevier
Stinnett TJ, Reddy V, Zabel MK (2023) Neuroanatomy, Broca area. In: StatPearls [Internet]. StatPearls, Treasure Island
Thau L, Reddy V, Singh P (2022) Anatomy, central nervous system [Updated 2022 Oct 10]. In: Stat Pearls [Internet]. Stat Pearls, Treasure Island, 2024 Jan. Available from: https://www.ncbi.nlm.nih.gov/books/NBK542179/
Vilensky J, Robertson W, Suarez-Quian C (2015) The clinical anatomy of the cranial nerves: the nerves of "on Olympus towering top". Wiley-Blackwell, Ames
Webb WG (2017) 2 – organization of the nervous system I. In: Webb WG (ed) Neurology for the speech-language pathologist, 6th edn. Mosby
Witter L, De Zeeuw CI (2015) Regional functionality of the cerebellum. Curr Opin Neurobiol 33:150–155

Traditional Methods of Neurovisualization

2

Apurba Patra, N. B. Pushpa, and Kumar Satish Ravi

Abstract

Neurovisualization techniques have progressed remarkably, but classic methods proved relevant and pioneered the way to modern techniques. The chapter starts with the history and the basic methods of neurovisualization: histological stain methods, imaging by X-ray, and the early electrophysiological methods. First, we introduce a historical perspective of histological methods that were used to explore the nervous system starting from the first silver staining methods introduced by Camillo Golgi and later developed by Santiago Ramón y Cajal to visualize previously unattainable structural details of neuronal architecture. We subsequently review the development of X-ray imaging of the brain with an overview of the initial in vivo visualization of brain anatomy by way of pneumoencephalography and cerebral angiography. Also discussed are preceding electrophysiological methods (e.g., electroencephalogram [EEG] and magnetoencephalogram [MEG]) used to measure neural activity. Throughout this chapter, we highlight the contributions and limitations of these traditional techniques, acknowledging their enduring impact on the field of neuroscience. By understanding these foundational methods, we gain insights into the evolution of neurovisualization and appreciate the technological advancements that have since transformed our ability to explore the brain.

Keywords

Anatomy education · Curriculum · Cadavers · Remote learning · Virtualization

2.1 Introduction

Neuroscience has long addressed the brain's structure and function, and neurovisualization technologies have enabled this. The traditional techniques of neurovisualization which have evolved over the last century set the basis for our present interpretation of the nervous system. These technological landmarks have provided not only foundational knowledge but also led to the development of further advanced imaging technologies in use today. Histological staining methods, where it all started from, permitted scientists to visualize single neurons and

A. Patra
Department of Anatomy, All India Institute of Medical Sciences, Bathinda, Punjab, India

N. B. Pushpa
Department of Anatomy, JSS Medical College, JSSAHER, Mysore, Karnataka, India

K. S. Ravi (✉)
Department of Anatomy, National Institute of Medical Sciences and Research, Jaipur, Rajasthan, India

Department of Anatomy, KMC Mangalore, MAHE, Manipal, Karnataka, India

K. S. Ravi et al. (eds.), *Visualization in Neuroanatomical Sciences*, Biomedical Visualization 11,
https://doi.org/10.1007/978-3-032-07380-8_2

networks between them. The development of the silver chromate staining method (commonly known as Golgi staining) by Camillo Golgi provided the foundation for assessing neuronal structures and their processes and related lesions in exquisite detail (Golgi 1873).

This same technique was further developed by Santiago Ramón y Cajal, bringing to light the neuron doctrine, the neuron being the basic unit of the nervous system (Ramón 1909). X-ray imaging techniques like pneumoencephalography and cerebral angiography were introduced in the early twentieth century. These techniques allowed for the in vivo visualization of brain structures and blood vessels. The use of air for increased contrast in X-ray images became known as pneumoencephalography, a term introduced by one of the pioneers, Walter Dandy, who had first used the direct introduction of air into the ventricular system (Dandy 1918). Egas Moniz developed cerebral angiography by injecting contrast agents to visualize cerebral vasculature, a fundamental method to study vascular anatomy and pathology of the brain (Moniz 1927).

Electrophysiological techniques such as EEG and MEG were developed to measure electrical and magnetic activity in the brain. The discovery of the EEG by Hans Berger provided a noninvasive method to record electrical brain activity from the scalp, revolutionizing the study of brain function and disorders (Berger 1929). Later, MEG was developed, which measured the magnetic fields generated by neural activity and allowed for high-temporal-resolution imaging of brain function (Cohen 1968).

Computed tomography (CT) and nuclear medicine, including positron emission tomography (PET) and single-photon emission computerized tomography (SPECT), constitute the critical shift from old to new imaging technologies. Detailed cross-sectional images of brain anatomy were made possible with the invention of CT scanning by Godfrey Hounsfield and Allan Cormack (Hounsfield 1973). Functional imaging capabilities are provided by PET and SPECT, respectively, which give the capability to study brain metabolism and blood flow.

Such conventional methods, despite their limitations, have contributed so much to neurovisualization. Understanding these conventional techniques not only highlights the extent of the historical progression of neurovisualization but also emphasizes the importance of technological innovation toward the progress of neuroscience.

2.1.1 Brief Overview of Neurovisualization

Neurovisualization consists of several techniques that enable the visualization of nervous system structure and function. These methods are central to identifying brain anatomy, neural pathways, and brain activity for research and clinical diagnostics.

Histological Staining

Histological staining was among the first developed techniques to visualize neurons and their labyrinthine networks. Golgi and Nissl stains highlighted different cellular components of the nervous system. After Cajal modified the Golgi method, it facilitated a more detailed study of neuronal morphology and connectivity between them (Golgi 1873; Ramón 1909).

X-Ray Imaging

X-ray imaging techniques like pneumoencephalography and cerebral angiography yielded the first in vivo images of the brain. Pneumoencephalography, first introduced by Walter Dandy, injected air into the ventricular system to improve X-ray contrast (Dandy 1918). Cerebral angiography, developed by Egas Moniz, utilized contrast agents to visualize the brain's vascular structures (Moniz 1927).

Electrophysiological Approaches

Electrophysiological techniques, including electroencephalography (EEG) and magnetoencephalography (MEG), assess the electrical and magnetic activity produced via neuronal activity. EEG, outlined by Hans Berger, put forward a noninvasive means to record brain wave patterns, thereby drastically changing how researchers studied brain function and disorders (Berger 1929). MEG measures magnetic fields produced by neuronal activity and affords excellent temporal resolution imaging of brain function (Cohen 1968).

Computed Tomography (CT)
Computed tomography (CT), developed by Godfrey Hounsfield and Allan Cormack, takes advantage of X-rays to render detailed cross-sectional images of the brain. CT scans provide a good representation of brain anatomy and are most helpful in detecting structural abnormalities (Hounsfield 1973).

Positron Emission Tomography (PET)
Positron emission tomography involves the injection into the venous circulation of an agent for visualization of metabolic processes in the brain by detecting emitted positrons. This will support investigations on the brain metabolism and function (Hoffman et al. 1973).

Single-Photon Emission Computed Tomography (SPECT)
Single-photon emission computed tomography (SPECT) uses gamma-emitting radioisotopes to obtain three-dimensional images of cerebral blood flow. Nonetheless, SPECT is advantageous over other imaging methods when used to determine brain perfusion and function (Kuhl and Edwards 1963).

2.1.2 Importance of Traditional Methods in the Development of Present Methods

The old neurovisualization methods have greatly contributed to the modern map of neuroscience. Their importance does not lie in their foundational knowledge only, but equally importantly they directly gave way to the emergence of modern imaging techniques. Here are some ways that old methods gave rise to present methods:

To build basic knowledge, scientists created techniques such as Golgi staining first introduced by Camillo Golgi (1873) but perfected by Santiago Ramón y Cajal (1909). These techniques allowed scientists to see single neurons for the first time, and these methods exposed the complexity and diversity in neuronal structure that were required to formulate neuron doctrine, which states the neuron is the unit of the brain.

Early Diagnostic Equipment In vivo imaging, specifically pneumoencephalography and cerebral angiography, offered the first chance to observe the structures inside the brain in the patient's body. These were, in fact, precursors to later more enhanced modalities for imaging, and they proved the technological feasibility and clinical utility of in vivo imaging of the brain (Srisuchinnawong et al., 2021).

Progressive Functional Insighting Electrophysiological methods, EEG and MEG, which measure electrical and magnetic activity from the brain, were among the first tools providing real-time evidence about the functioning of the brain. These techniques are still very much in vogue today, impacting subsequent functional imaging methods, including functional MRI and PET.

Driving Technological Innovation Computed Tomography (CT): Based on the physics of X-ray imaging, the concept of cross-sectional imaging was introduced with CT scans and became a cornerstone for several modern imaging techniques, including MRI and advanced CT technologies. Metabolic imaging, such as PET and SPECT, based on principles of radiotracer imaging, has influenced newer imaging techniques that also depend on metabolic and functional markers.

Development of Imaging Techniques The lack of resolution and specificity of the earlier techniques paved the way for techniques with superior resolutions to capture clear images, which includes high-resolution MRI and diffusion tensor imaging (DTI).

Development of Multimodal Approach Structural and functional information from traditional methods showed that a multimodal approach is essential for enhancing the clarity in understanding brain function and structure (Xue et al. 2015).

2.2 Historical Background

2.2.1 Evolution of Neurovisualization Methods

Neurovisualization has made incredible leaps since the beginning of this century. Major findings started early with techniques like X-ray imaging and electroencephalography in the early twentieth century, which were later used to determine brain structure and function. These were later advanced in the 1970s with the development of computed tomography and magnetic resonance imaging, which ushered in a new renaissance in the imaging of the brain at detailed anatomical levels. Positron emission tomography (PET) and single-photon emission computed tomography (SPECT) further developed functional imaging by the development of metabolic activity visualization. The above conventional techniques have played a role in the development of modern neuroimaging techniques, especially fMRI and DTI (Chiu and Yen 2023).

2.2.2 Special Stains for Neurons and Neuroglia

Special stains are necessary for the differentiation of neurons and neuroglia in histology. Neurons are detectable with Nissl stains like cresyl violet and toluidine blue, which stain Nissl substance in the cytoplasm, and thus are helpful in evaluating neuronal integrity. The Bodian stain, a silver method, stains axons and neurofibrillary structures, whereas the Golgi stain selectively impregnates a small group of neurons, demonstrating their entire morphology. In addition, NeuN immunostaining is highly specific for mature neurons and has broad application in neurodegenerative studies (Lin et al. 2016). Glial fibrillary acidic protein (GFAP) immunostaining is astrocyte-specific and has an important application in the diagnosis of gliosis and astrocytomas (Duan et al. 2016). Olig2 immunostaining labels oligodendrocytes and oligodendrocyte precursors. Luxol fast blue (LFB) is a myelin-specific stain, and it finds application in the identification of demyelinating diseases like multiple sclerosis (Ghasemi-Kasman et al. 2018). Moreover, Cajal's gold sublimate method differentially stains astrocytes and glial networks, which helps in neuropathological studies. Such stains are a must for the diagnosis of neurological diseases, neurodevelopmental studies, and the study of neurodegenerative conditions (Klüver and Barrera 1953).

2.2.3 Key Milestones and Discoveries

Important milestones in neurovisualization date from 1895, with the discovery of X-rays, the first noninvasive glimpse of the brain with useful information about its structure (Hounsfield 1973). The 1920s saw the development of electroencephalography (EEG) by Hans Berger, allowing the registration of the electrical activity that the brain produces (Berger 1929). The outset of the 1970s marked an era of detailed CT scans imaging the brain with slices through the head (Schulz et al. 2021). MRI was the next step forward, and it could produce detailed anatomic images with no radiation whatsoever (Mansfield 1977). In the late twentieth century, functional imaging with positron emission tomography (PET) and single-photon emission computed tomography (SPECT) expanded imaging capacities to visualize metabolic and functional processes in the brain (Tan et al. 2022; Cerqueira and Jacobson 1989). The development of functional MRI (fMRI) in the 1990s represented a further giant leap: with fMRI, one could see brain activity as it happened in real time (Ogawa et al. 1990).

They provided the avenues into our understanding of brain structure and function.

2.3 Overview of Traditional Modalities

2.3.1 Neurohistology

Neurohistology focuses on the microscopic structure of the nervous system, examining the organization and function of neural tissues, including the brain and spinal cord. Key techniques include Nissl staining, which visualizes rough endoplasmic reticulum in neurons, causing cell bodies to appear dark blue or purple (Kadar et al. 2009). Golgi staining highlights entire neurons, displaying them in black against a light background (Golgi 1873). Immunohistochemistry (IHC) detects specific proteins using antibodies, resulting in brown or fluorescent signals (Magaki et al. 2019). Myelin staining highlights myelin sheaths, appearing blue or black, while silver staining visualizes axons (Cohen 1972). Hematoxylin and eosin (H&E) staining provides an overview of tissue structure, with nuclei appearing blue and cytoplasm pink (Fix et al. 1996). Other techniques, like fluorescent staining and Timm staining, offer insights into specific cellular components. These methods are crucial for understanding both normal and pathological states of the nervous system (Schröder et al. 2020).

2.3.2 X-Ray Imaging

2.3.2.1 Pneumoencephalography

Pneumoencephalography was a neuroimaging technique wherein air or gas was infused into the ventricular system of the brain, filling the space designed by the cerebrospinal fluid (CSF). The replacement of CSF by the air or gas allowed better X-ray images to be produced regarding the brain's structure.

The procedure was first introduced by Walter Dandy, an American neurosurgeon in 1918. It was the first technique to give complete-scale images of the human brain and its ventricular system that contributed much to the diagnosis of many brain disorders (Dandy 1918).

In this procedure, a patient would usually have local or general anesthesia. A lumbar puncture or other method was used to remove a portion of cerebrospinal fluid. Then, air or another gas (e.g., oxygen) was introduced into the ventricles via the same puncture, and X-rays were taken in various positions to ascertain the visualization of the brain structures as the air or gas spread through the ventricles (Ladley-O'Brien 2011).

This method was also most frequently used in the diagnosis of tumors, cysts, and abscesses, determining ventricular system blockages and surveying hydrocephalus, among other conditions.

The procedure is very painful. Many patients claim to have severe headaches and dizziness accompanied by nausea after treatment. A few complications include infection, hemorrhage, and intracranial hypertension.

The noninvasive imaging techniques that began to penetrate the market using CT and MRI during the 1970s and 1980s made pneumoencephalography redundant. Although it poses a lot of risks, pneumoencephalography played a significant role in the neurology and neurosurgery field and is considered one of the pioneering techniques discovered (Leeds and Kieffer 2000).

2.3.2.2 Cerebral Angiography

Cerebral angiography, also referred to as cerebral arteriography, is an imaging technique for the visualization of blood vessels in the brain. The technique involves intraluminal injection of a contrast agent into the cerebral arteries followed by X-ray imaging of the patient. Such detailed pictures of the blood vessels aid in the diagnosis and management of various cerebrovascular conditions (Dowd 2021) (Fig. 2.1).

It was first performed in the year 1927 by Portuguese neurologist Egas Moniz to help provide detailed images of cerebral vasculature and consequently aid diagnosis for vascular diseases, aneurysms, arteriovenous malformations (AVMs), and other disorders of the

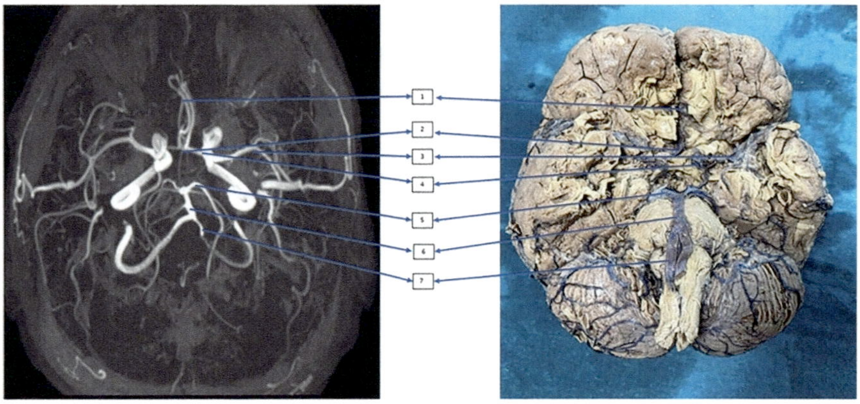

Fig. 2.1 Anatomy of the circle of Willis (dissected specimen). Gross findings compared with MRI findings. (1) Anterior cerebral artery, (2) anterior communicating artery, (3) middle cerebral artery, (4) internal carotid artery, (5) posterior cerebral artery, (6) basilar artery, and (7) vertebral artery. (Picture courtesy Department of Anatomy and Radiology, All India Institute of Medical Sciences, Bathinda, India. Choudhary et al. 2022)

cerebrovascular systems. Normally, while performing the procedure, a local anesthesia is administered to the patient in the area where catheters are inserted. This is most commonly performed in the groin region. A catheter is then inserted into the femoral artery that is passed along the vascular system up to the carotid or vertebral artery. The contrast dye is infused from the catheter within the cerebral arteries. X-rays or fluoroscopy images are taken along the blood movement through the cerebral vessels (Kaufmann and Kallmes 2008).

Such a procedure is used to identify aneurysms and AVMs; it detects stenosis or blockages in arteries, evaluates blood vessel malformations or tumors, and assists in planning for surgeries or other interventions such as endovascular procedures. A few major drawbacks with such procedures include allergic reaction to the contrast dye, bleeding or bruising at the catheter insertion site, possibility of a stroke or transient ischemic attack (TIA) due to dislodgement of atherosclerotic plaques, and patient discomfort during the contrast injection (Heiserman et al. 1994).

Recently, advances and alternatives for cerebral angiography, such as digital subtraction angiography (DSA), noninvasive alternatives, magnetic resonance angiography (MRA), and computed tomography angiography (CTA), were developed (Little et al. 1982; Nichols et al. 1994). Cerebral angiography remains a vital tool in neurovascular imaging, offering detailed and precise images essential for accurate diagnosis and treatment planning. Although noninvasive alternatives are available, cerebral angiography is often preferred for its superior image resolution and ability to guide therapeutic interventions.

2.3.3 Computed Tomography (CT) Scans

Computed tomography (CT) scans are broadly used in neurovisualization, showing detailed images of the inside of the brains. Rotating X-rays and detectors make this possible. Since its discovery in the 1970s, CT scans had been critical in the review of acute neurological conditions, such as traumatic brain injury and stroke, mainly because of its speed and accuracy. It assists in anatomical disease identification, disease process, and surgical intervention support (Ter-Pogossian 1977). There exist different types of CT, including spiral CT, electron beam tomography (EBT), and dual-energy CT, with each of them applied in a different clinical application for unique benefits (Stiller 2018). The

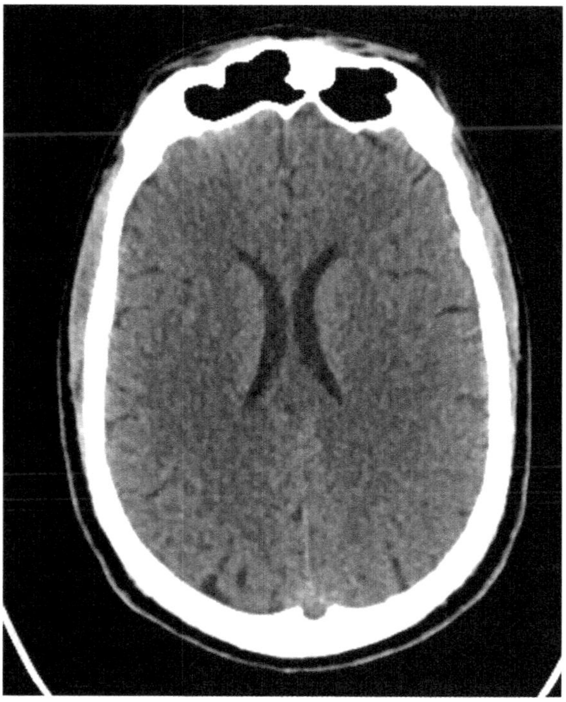

Fig. 2.2 CT scan brain showing bilateral lateral ventricles within normal limit

application of CT perfusion imaging will enhance the condition of blood flow measurement in stroke management (Cviková et al. 2024). Overall, the CT scans are a very important tool both in the diagnosis and in the management of neurological disorders. It does provide a clue to the anatomy and pathology within the brain while being noninvasive (Fig. 2.2).

2.3.4 Direct Visualization Techniques

Direct visualization of the human brain can be achieved through several methods, each with different applications in research, clinical diagnosis, and education.

2.3.4.1 Gross Anatomical Dissection

Traditional cadaveric dissection is still the gold standard for direct observation of the human brain, providing unmatched accuracy in exploring macroscopic features like sulci, gyri, ventricles, and white matter tracts. This technique enables precise comprehension of neuroanatomical associations, which is imperative in medical training and neurosurgical education (Standring 2020). Research has demonstrated that cadaveric dissection offers a three-dimensional view of brain structures that cannot be fully achieved by imaging or virtual models (Felle et al. 1995) (Fig. 2.3).

In addition, extensive studies employing dissection methods such as Klingler's fiber dissection allow improved visualization of white matter tracts by selectively disconnecting myelinated tracts (Martino et al. 2011). Klingler's fiber dissection is a neuroanatomical method specifically created by Ludwig Klingler in 1935 to investigate the three-dimensional arrangement of white matter tracts. The technique consists of fixation of the brain in formalin and then freezing (−10 °C to −20 °C) and thawing, which loosens the extracellular matrix so that myelinated fiber tracts can be manually separated. This method allows accurate visualization of large tracts like the arcuate fasciculus, superior longitudinal fasciculus, and corticospinal tract and is thus useful in neurosurgical planning and teaching neuroanatomy (Martino et al. 2011). In contrast to diffusion tensor imaging (DTI), Klingler's dissection offers

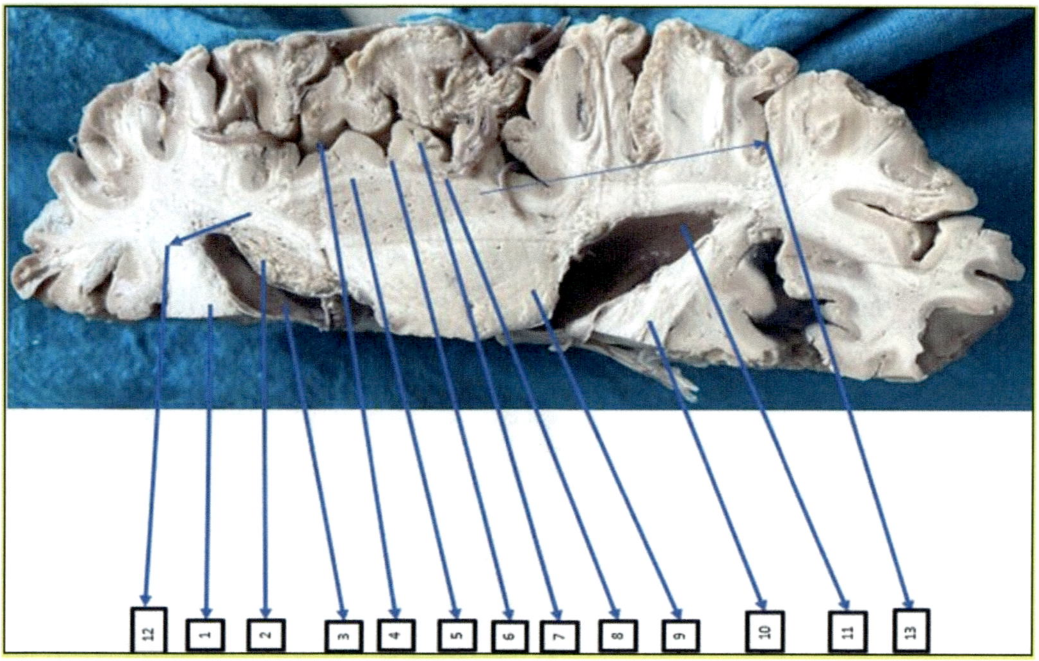

Fig. 2.3 Showing axial section at basal ganglia level, (1) genu of corpus callosum, (2) head of caudate nucleus, (3) frontal horn of lateral ventricle, (4) Sylvian fissure, (5) external capsule, (6) insula, (7) extreme capsule, (8) claustrum, (9) thalamus, (10) splenium of corpus callosum, (11) atrium of lateral ventricle, (12) anterior limb of internal capsule, and (13) lentiform nucleus

direct anatomical confirmation of fiber tracts, though it is labor-intensive and expertise-dependent. Even with limitations, such as artificial separations resulting from freezing artifacts, it is still a critical instrument to grasp brain connectivity and surgical anatomy (Klingler 1935). These methods have greatly influenced neurosurgical procedures, especially in operations concerning deep-seated brain structures (Latini et al. 2017). Moreover, studies comparing dissection with imaging methods have established that hands-on cadaveric dissection enhances spatial comprehension and memorization of anatomy among medical students (Evans and Watt 2005).

2.3.4.2 Plastination Techniques for Neurovisualization

Plastination is a preservation method pioneered by Gunther von Hagens in 1977 that promotes neurovisualization through the substitution of water and lipids in biological tissue with curable polymers like silicone, epoxy, or polyester resin. This results in odorless, hard, and dry specimens with retained microscopic and macroscopic details that are very useful for the teaching of neuroanatomy and for research purposes (von Hagens et al. 1987). The conventional silicone plastination method produces lifelike and flexible brain specimens capable of handling and dissection, whereas epoxy plastination yields transparent brain sections for high-resolution imaging of deep structures like white matter tracts and ventricles (Arnts et al. 2014; Janelle et al. 2022). Plastinated brain specimens have been found to improve the learning of neuroanatomy in medical students, filling the gap between cadaveric dissections and histological slides (Latorre et al. 2007). Plastination also conserves pathological specimens, which facilitate research into neurodegenerative conditions such as Alzheimer's and Parkinson's. The method is time-consuming and calls for specialized machinery, but the fact that it

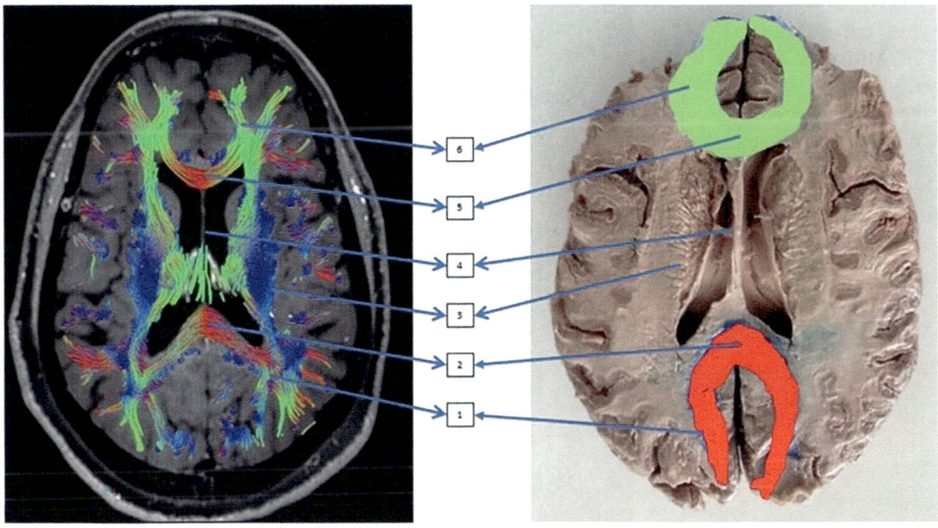

Fig. 2.4 Axial section of the brain at supraganglionic level. Cadaveric section compared with MRI image at the same plane. (1) Forceps major, (2) splenium of corpus callosum, (3) corona radiata, (4) septum pellucidum, (5) genu of corpus callosum, and (6) forceps minor

yields long-lasting, high-quality anatomical models makes it an invaluable resource in contemporary neuroanatomy education and research (Neha and Dhingara 2013).

2.3.5 Magnetic Resonance Imaging

An MRI is one of the critically significant imaging tools that are used to make detailed anatomical and physiological images without any ionizing radiation, which tends to be safer than using CT or PET (Brown and Semelka 2010). MRI detects the movements of water and the fat content in the body by applying strong magnetic fields along with radio waves to compose images of tissues based on their high contrast (McRobbie et al. 2008). Other methods are applied, such as structural MRI, in which tumors and demyelination are detected; fMRI shows activity by changes in blood flow inside the brain. Diffusion tensor imaging detects tracts in the white matter mainly for traumatic brain injury cases, while MR spectroscopy assesses brain metabolism for the presence of tumors or epilepsy (Yeh et al. 2021). It is due to the superior resolution and versatility of MRI that it has become critical in diagnosis in neurological conditions and in surgical planning. In scientific research, MRI is indispensable in studying brain functions, connectivity, and structural changes associated with cognitive and psychological disorders (Brown and Semelka 2010). MRI imaging not only allows visualization of the brain parenchyma and its associated pathologies (Fig. 2.4) but also enables detailed assessment of the brain ventricles, including measurement and identification of conditions such as hydrocephalus or ventricular tumors (Fig. 2.5).

2.3.6 Positron Emission Tomography (PET)

Positron emission tomography was invented in the 1970s and is considered one of the most powerful neuroimaging techniques to image brain function and metabolism by detecting radioactive tracers, thus providing many critical insights into biochemical processes in vivo (Cherry et al. 2013). Through its ability to image functional aspects, PET can study

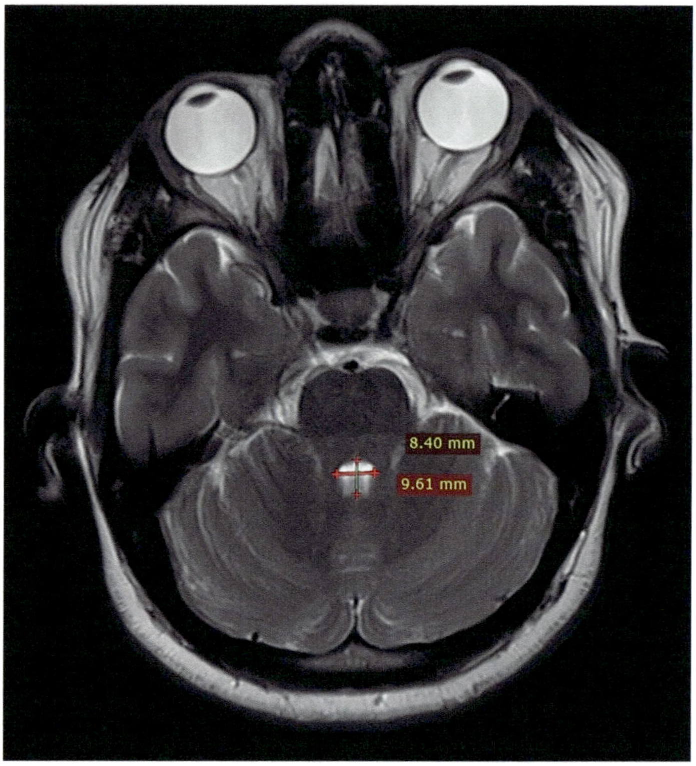

Fig. 2.5 Axial section of T2-weighted turbo spin echo (TSE) MRI at the level of posterior fossa of brain displaying measurement of the fourth ventricle of a healthy subject. Green marking points toward the anteroposterior width of the fourth ventricle, and red marking shows the transverse width of the fourth ventricle

metabolic processes, neurotransmitter activity, and blood flow, making it a crucial tool in both research and clinical practice (Placzek et al. 2016).

The measure of glucose metabolism, through tracers such as fluorodeoxyglucose (FDG), is critical in diagnosing Alzheimer's disease, epilepsy, and brain tumors (Hoffman et al. 1973). Furthermore, it can visualize neurotransmitter activity and receptor distribution, which helps explain and advance the treatment of psychiatric disorders (Cervenka et al., 2022). It is also a critical technique in the early diagnosis of neurodegenerative diseases, such as Alzheimer's, through the detection of amyloid plaques and tau tangles before it symptomatically becomes apparent (Jack et al. 2018).

PET measures cerebral blood flow and perfusion, which mark the hallmarks of assessment for stroke and for management strategies (Dastur 1985). Distinguishing between types and following response to therapy defines the role of PET in neuro-oncology. With its high sensitivity and giving quantitative information, PET provides invaluable functional information, advancing our knowledge concerning neurological diseases (Verger et al. 2022).

2.3.7 Single-Photon Emission Computed Tomography (SPECT)

SPECT, or single-photon emission computed tomography, is indeed a well-developed neuroimaging process aimed at studying cerebral blood flow and brain functions. Developed in the late twentieth century, SPECT facilitates a three-dimensional image of blood flow with radiopharmaceuticals emitting gamma rays; hence, it forms a substantial part in diagnosing and monitoring almost every type of neurological condition (Crisan et al. 2022).

SPECT is utilized to study regional cerebral blood flow (rCBF) because it is vital for stroke, TIAs, and vascular dementia; it assesses the brain function for neurological and psychiatric conditions that include epilepsy and schizophrenia; nowadays, in neurodegenerative diseases such as Alzheimer's and Parkinson's, once characteristically pathologic changes in the blood flow are seen, in order to facilitate earlier diagnoses and management.

It is of great value in the clinical setting to be used in the evaluation of epilepsy, especially for planning presurgical interventions, comparing interictal and ictal blood flow to localize seizure foci (Newton et al. 1995). It helps neuro-oncology specialists distinguish between tumor types and assess treatment response (Alexiou et al. 2021).

The advantages of SPECT are availability, economy, and functional imaging, which are important in assessing brain perfusion and metabolism (Pavel et al. 2006). It is used in cerebrovascular disorder, epilepsy, neurodegenerative diseases, and psychiatric research to offer the functional data in addition to anatomical imaging.

2.3.8 Electroencephalography (EEG)

Electroencephalography is a neuroimaging technique that has been used since the beginning of the twentieth century in recording the electrical activity of the brain. The technique has been successfully employed in both clinical and research fields. In both cases, it provided immediate information regarding the state of brain function (Light et al. 2010). The EEG is recording of brain activity while evaluating abnormalities such as epilepsy and sleep disorders; it is the gold standard in the management of epilepsy because it defines the nature and origin of seizure, both of which are useful for treatment planning (Niedermeyer 2011). Besides the sleep study, sleep stage determination is another paramount role of an analysis of EEG, for diagnosing conditions like insomnia and sleep apnea (Fisch 1999).

This has rendered EEG highly invaluable in research to trace brain activity while performing certain tasks, hence pursuing the cognitive processes such as attention, memory, and emotion. It also forms a basis for developing brain-computer interface (BCI). BCI technology enables direct communication between the brain and external devices by interpreting brain activity (typically recorded via EEG). Based on this technique, BCI is used in neuroprosthetics and assistive technology, as shown by Buzsaki (2006) (Britton et al.,(2016).

Second, intraoperative EEG during neurosurgery ensures the maintenance of the critical functions of the brain and keeps the surgeon informed in real time. The benefits of EEG are that it has millisecond-level temporal resolution, is noninvasive, and is safe, thus allowing its repeated use in different population groups. Also, EEG is considerably inexpensive and more accessible than MRI or PET. It was made portable, thereby increasing utility in remote and long-term monitoring (Bhavnani et al. 2022).

2.3.9 Magnetoencephalography

MEG, or magnetoencephalography, is a noninvasive neuroimaging technique that records changes in magnetic fields generated by neuronal electrical activity. Usage of superconducting quantum interference devices provide high spatial and temporal resolution. MEG is effectively suited to map out brain functions in real time (Cohen et al. 2002). It helps clinicians localize brain regions before surgery, especially for patients suffering from epilepsy, and improves the applications of neurofeedback (Forss 1997). The cognitive processes, brain oscillations, and connectivity arise from dynamic brain oscillations and are mapped via connectivity patterns using techniques like beamforming and source modeling (beamforming and source models). These methods localize neural activity by filtering EEG/MEG signals to identify active brain regions, enabling precise mapping of functional networks involved in attention, memory, and decision-making. It mainly detects magnetic fields from the cortical surface, requiring good shielding against ambient noise (magnetic shielding techniques) (Boto et al. 2018). One of the significant challenges is the inverse problem: since it deals with determining the source of brain activity from measurements taken outside, advanced algorithms such as dipole fitting must be used (inverse problem solutions). MEG has also permitted fetal brain recordings as early as 25 weeks' gestation, thus proving this to be highly flexible (Sheridan et al. 2010).

2.3.10 Cranial Ultrasound

Cranial ultrasound is a noninvasive imaging tool which has been widely applied mainly to neonates because of the specific anatomical configurations of an infant's skull, particularly the fact that fontanelles are open, allowing the sound waves to penetrate the infant's brain. It is used for providing real-time images and generally applied in studying the architecture of the brain, making diagnoses of congenital anomalies as well as other disorders like hydrocephalus, intraventricular hemorrhage, and periventricular leukomalacia (Caro-Domínguez et al. 2021). While cranial ultrasound is noninvasive and does not apply any harmful ionizing radiation, unlike other neuroimaging techniques, it is safe, portable, and inexpensive. Thus, it has found specific aptness for bedside monitoring in neonatal intensive care units (van Wezel-Meijler 2010).

The technique relies on the use of high-frequency sound waves that must be transmitted through the fontanelles and reflected back to produce images. The anterior fontanelle is the most common acoustic window, but the posterior, mastoid, and temporal fontanelles may be used to achieve alternative views (Nongena et al. 2010). In childhood development, the fontanelles close; therefore, ultrasound has a limited effectiveness on older children and adults.

On the other hand, cranial ultrasound is somewhat less effective due to its relatively lower resolution than MRI or CT but is excellent for neonatal screening and follow-up in neurodevelopmental disorders of premature and term neonates. Further advancement, such as 3D cranial ultrasound, will better illuminate the structures and increase the scope of assessment, thus improving the precise diagnostic ability of this modality. Therefore, cranial ultrasound is one of the main early neuroimaging modalities that largely contribute to neonatal care (De Vries and Groenendaal 2002; Diwakar and Khurana 2018; Nazparveen and Phirke 2017) (Fig. 2.5).

2.4 Comparison with Modern Techniques

2.4.1 Contrast Between Traditional and Modern Neurovisualization Methods

Traditional neurovisualization techniques and the more recent advances use two opposing methods in the study of the brain, each with merits and faults. Among the classic techniques, computed tomography and magnetic resonance imaging represent most of the general knowledge on the anatomy and pathology of the brain (Florkow et al. 2022). Positron emission tomography and single-photon emission computed tomography allow metabolic and biochemical research but are limited by resolution and exposure to radiation. Electroencephalography offers the best temporal resolution possible for brain dynamics but has poor spatial resolution. New technologies such as fMRI and DTI have become available for noninvasive, high-resolution imaging of brain functionality and connectivity (Le Bihan et al. 2001).

MEG provides very high resolutions in space and time, but it is also expensive and somewhat cumbersome; computational analysis that further breaks down data was more advanced and included several neuroimaging techniques (Koch and Tsuchiya 2007). To express it differently, ancient methods constituted a theoretical framework upon which brain anatomy and physiology came to be based, whereas modern innovations are quite obviously fine-tuning and building on neurovisualization capabilities.

2.4.2 Integration of Traditional Methods with Advanced Technologies

Introduction of modern technologies in conventional neurovisualization techniques revolutionized neuroscience that assists in improvement toward understanding brain structure and its function. Modern techniques of information acquisition of the brain correlate with MRI, CT, PET, SPECT, EEG, alongside DTI and MEG.

This integration helps to correlate structural and functional imaging; for example, MRI and fMRI together help decide which regions are involved in specific tasks, whereas the combination of CT with PET or SPECT correlates structural damage with corresponding functional deficits in particular oncology. The integration of EEG with MEG improves spatial resolution but without reducing temporal accuracy and thus provides better identification of brain activity during seizures. This makes it possible to trace dynamic brain activities onto the detailed anatomical structures and is very helpful in the planning phase for surgery, especially before a patient suffering from epilepsy undergoes surgery; this is because it makes it possible to localize exactly epileptogenic zones.

2.5 Future Directions

2.5.1 Potential Advancements in Traditional Methods

Advancements in traditional neurovisualization techniques will be highly important as they would bring improved accuracy and applicability of brain imaging. Ultrahigh-field scanners (7 Tesla and above) with enhanced functional MRI (fMRI) techniques promise superior spatial and temporal resolution, which will make possible enhanced diagnosis and mapping of brain anatomy (Altmeyer et al. 2016). CT with dual-energy technology and iterative reconstruction algorithms increases tissue differentiation and reduces patient dose exposure for evaluation (D'Angelo et al. 2019). Hybrid imaging systems, integrating PET or SPECT with MRI, increase the accuracy of diagnosis by providing both functional and anatomical information in a single session (Beyer et al. 2011). High-density EEG systems enhance the spatial resolution of electroencephalography in the brain activity mapping. Latest advancements in wearable technology are helpful in supporting long-term monitoring (Casson 2019). Finally, artificial intelligence (AI) and machine learning enable one to analyze neuroimaging data to increase performance in personalized diagnostics (Pinto-Coelho et al. 2023). In that regard, it forms a new understanding of, and ability in, managing neurological conditions.

2.5.2 Emerging Trends and Technologies

Emerging trends in neuroimaging are significantly enhancing our understanding of brain structure and function. High-resolution imaging techniques, such as ultrahigh-field MRI (7 Tesla and above), enable detailed visualization for early disease detection (Feinberg et al. 2023). Advanced diffusion imaging methods like diffusion spectrum imaging (DSI) and neurite orientation dispersion and density imaging (NODDI) improve insights into brain connectivity and white matter integrity (Pasternak et al. 2018). Additionally, hybrid imaging systems (PET/MRI, SPECT/MRI) integrate functional and structural assessments (Hicks et al. 2007), while wearable EEG devices facilitate long-term, real-world brain monitoring (Casson 2019).

2.6 Conclusion

2.6.1 The Ongoing Relevance of Traditional Methods in Neurovisualization

These classical neurovisualization techniques include MRI, PET, SPECT, EEG, and MEG and are vital for neuroscience as well as clinical applications. It has been seen that the techniques work remarkably well with widespread diagnostic scope and excellent resolution in terms of time and thus present a fine structural as well as functional description of the human brain. This will also become economical along with wider usability and forming the root for further education. Thus, this technique is well adaptable and can complement any of the advanced technology as they, too, would develop as more advanced science evolves in these aspects (Choudhary et al. 2022).

2.6.2 Summary of Key Points

Traditional neurovisualization methods have significantly advanced our understanding of the human brain by providing essential structural and functional insights for clinical and research applications. Despite newer technologies, these methods continue to evolve through integration with advanced computational techniques and hybrid imaging. They form the foundation of contemporary neuroimaging, ensuring their ongoing relevance. In summary, traditional methods have not only shaped present knowledge but also pave the way for future innovations in neuroscience, remaining crucial for exploring the brain's complexities.

References

Alexiou GA, Tsiouris S, Fotopoulos AD (2021) Single-photon emission computed tomography [NeuroSPECT] imaging of brain tumors. In: Dierckx RAJO, Otte A, de Vries EFJ, van Waarde A, Leenders KL (eds) PET and SPECT in neurology. Springer, pp 1071–1085. https://doi.org/10.1007/978-3-030-53168-3_40

Altmeyer W, Steven A, Gutierrez J (2016) Use of magnetic resonance in the evaluation of cranial trauma. Magn Reson Imaging Clin 24(2):305–323

Arnts H, Kleinnijenhuis M, Kooloos JG, Schepens-Franke AN, van Cappellen van Walsum AM (2014) Combining fiber dissection, plastination, and tractography for neuroanatomical education: revealing the cerebellar nuclei and their white matter connections. Anat Sci Educ 7:47–55

Berger H (1929) Über das Elektrenkephalogramm des Menschen. Eur Arch Psychiatry Clin Neurosci 87(1): 1–22

Beyer T, Freudenberg LS, Townsend DW, Czernin J (2011) The future of hybrid imaging-part 1: hybrid imaging technologies and SPECT/CT. Insights Imaging 2(2):161–169. https://doi.org/10.1007/s13244-010-0063-2

Bhavnani S, Parameshwaran D, Sharma KK, Mukherjee D, Divan G, Patel V, Thiagarajan TC (2022) The acceptability, feasibility, and utility of portable electroencephalography to study resting-state neurophysiology in rural communities. Front Hum Neurosci 16:802764. https://doi.org/10.3389/fnhum.2022.802764

Boto E, Holmes N, Leggett J, Roberts G, Shah V, Meyer SS et al (2018) Moving magnetoencephalography towards real-world applications with a wearable system. Nature 555(7698):657–661. https://doi.org/10.1038/nature26147

Britton JW, Frey LC, Hopp JL, St. Louis EK, Frey LC (2016) Electroencephalography (EEG): an introductory text and atlas of normal and abnormal findings in adults, children, and infants. American Epilepsy Society. https://www.ncbi.nlm.nih.gov/books/NBK390346/

Brown MA, Semelka RC (2010) MRI: basic principles and applications. Wiley-Blackwell

Buzsáki G (2006) Rhythms of the brain. Oxford University Press

Caro-Domínguez P, Lecacheux C, Hernandez-Herrera C, Llorens-Salvador R (2021) Cranial ultrasound for beginners. Transl Pediatr 10(4):1117–1137. https://doi.org/10.21037/tp-20-399

Casson AJ (2019) Wearable EEG and beyond. Biomed Eng Lett 9(1):53–71. https://doi.org/10.1007/s13534-018-00093-6

Cerqueira MD, Jacobson AF (1989) Assessment of myocardial viability with SPECT and PET imaging. Am J Roentgenol 153(3):477–483. https://doi.org/10.2214/ajr.153.3.477

Cervenka S, Frick A, Bodén R, Lubberink M (2022) Application of positron emission tomography in psychiatry-methodological developments and future directions. Transl Psychiatry 12(1):248. https://doi.org/10.1038/s41398-022-01990-2

Cherry SR, Sorenson JA, Phelps ME (2013) Physics in nuclear medicine, Saunders, pp 209–211

Chiu FY, Yen Y (2023) Imaging biomarkers for clinical applications in neuro-oncology: current status and future perspectives. Biomark Res 11:35. https://doi.org/10.1186/s40364-023-00476-7

Choudhary P, Peer S, Manaktala I, Singh P (2022) RadioCad: Correlative Atlas of Radiological and Cadaveric Anatomy of the Human Brain and Skull. 1st edition. Notion Press. Chennai. ISBN-13: 979-8888839188

Cohen D (1968) Magnetoencephalography: Evidence of magnetic fields produced by human brain activity. Science 161(3843):784–786

Cohen D (1972) Magnetoencephalography: Detection of the brain's electrical activity with a superconducting magnetometer. Science 175(4022):664–666

Cohen D, Schläpfer U, Ahlfors S, Hämäläinen M, Halgren E (2002) New six-layer magnetically-shielded room for MEG. In: Proceedings of the 13th international conference on biomagnetism. VDE, Jena

Crişan G, Moldovean-Cioroianu NS, Timaru DG, Andrieş G, Căinap C, Chiş V (2022) Radiopharmaceuticals for PET and SPECT imaging: a literature review over the last decade. Int J Mol Sci 23(9):5023. https://doi.org/10.3390/ijms23095023

Cviková M, Haršány M, Vinklárek J, Štefela J, Fojtová I, Mikulík R (2024) Effectiveness of computed tomography perfusion imaging in stroke management. Front Neurol 15:1390501. https://doi.org/10.3389/fneur.2024.1390501

D'Angelo T, Cicero G, Mazziotti S, Ascenti G, Albrecht MH, Martin SS, Othman AE, Vogl TJ, Wichmann JL (2019) Dual energy computed tomography virtual monoenergetic imaging: technique and clinical applications. Br J Radiol 92(1098):20180546. https://doi.org/10.1259/bjr.20180546

Dandy WE (1918) Ventriculography following the injection of air into the cerebral ventricles. Am J Roentgenol 6:1–10

Dastur DK (1985) Cerebral blood flow and metabolism in normal human aging, pathological aging, and senile dementia. J Cereb Blood Flow Metab 5(1):1–9. https://doi.org/10.1038/jcbfm.1985.1

De Vries LS, Groenendaal F (2002) Neuroimaging in the preterm infant. Ment Retard Dev Disabil Res Rev 8:273–280

Diwakar RK, Khurana O (2018) Cranial sonography in preterm infants with short review of literature. J Pediatr Neurosci 13(2):141–149. https://doi.org/10.4103/jpn.JPN_60_17

Dowd CF (2021) Cerebral angiography: techniques and practice. In: Handbook of clinical neurology, vol 176, pp 107–119. https://doi.org/10.1016/B978-0-444-64034-5.00006-7

Duan W, Zhang YP, Hou Z et al (2016) Novel insights into NeuN: from neuronal marker to splicing regulator. Mol Neurobiol 53(3):1637–1647

Egas Moniz-Biographical (1927) The official web site of the Nobel prize. http://nobelprize.org/nobel_prizes/medicine/laureates/1949/moniz-bio.html. Accessed 11 Jan 2025

Evans DJ, Watt DJ (2005) Provision of anatomical teaching in a new British medical school: getting the right mix. Anat Rec B New Anat 284(1):22–27

Feinberg DA, Beckett AJS, Vu AT et al (2023) Next-generation MRI scanner designed for ultra-high-resolution human brain imaging at 7 Tesla. Nat Methods 20:2048–2057. https://doi.org/10.1038/s41592-023-02068-7

Felle P, Lockman AK, Kellaghan P (1995) Removal of the brain for teaching and examination. Clin Anat 8:363–365

Fisch BJ (1999) Fisch and Spehlmann's EEG primer: basic principles of digital and analog EEG, 3rd edn. Elsevier Science B.V

Fix AS, Ross JF, Stitzel SR, Switzer RC (1996) Integrated evaluation of central nervous system lesions: stains for neurons, astrocytes, and microglia reveal the spatial and temporal features of MK-801-induced neuronal necrosis in the rat cerebral cortex. Toxicol Pathol 24(3):291–304. https://doi.org/10.1177/019262339602400305

Florkow MC, Willemsen K, Mascarenhas VV, Oei EHG, van Stralen M, Seevinck PR (2022) Magnetic resonance imaging versus computed tomography for three-dimensional bone imaging of musculoskeletal pathologies: a review. J Magn Reson Imaging 56(1):11–34. https://doi.org/10.1002/jmri.28067

Forss N (1997) Magnetoencephalography (MEG) in epilepsy surgery. Acta Neurochir Suppl 68:81–84. https://doi.org/10.1007/978-3-7091-6513-3_15

Ghasemi-Kasman M, Zare L, Baharvand H, Javan M (2018) In vivo conversion of astrocytes to myelinating cells by miR-302/367 and valproate to enhance myelin repair. J Tissue Eng Regen Med 12(1):e462–e472

Golgi C (1873) On the structure of the gray matter of the brain. Gazzetta Medica Italiana Lombardia 33:244–246

Heiserman JE, Dean BL, Hodak JA, Flom RA, Bird CR, Drayer BP, Fram EK (1994) Complications of diagnostic cerebral angiography and neurointerventional procedures: a review of the literature. AJNR Am J Neuroradiol 15(8):1401–1411

Hicks R, Lau E, Binns D (2007) Hybrid imaging is the future of molecular imaging. Biomed Imaging and Intervention J 3(3):e49. https://doi.org/10.2349/biij.3.3.e49

Hoffman EJ, Huang SC, Phelps ME (1973) Quantitation in positron emission computed tomography: 1. Effect of object size. J Comput Assist Tomogr 1(3):299–308

Hounsfield GN (1973) Computerized transverse axial scanning (tomography): Part 1. Description of system. Br J Radiol 46(552):1016–1022

Jack CR Jr, Bennett DA, Blennow K et al (2018) NIA-AA research framework: toward a biological definition of Alzheimer's disease. Alzheimers Dement 14(4):535–562. https://doi.org/10.1016/j.jalz.2018.02.018

Janelle F, Iorio-Morin C, D'amour S, Fortin D (2022) Superior longitudinal fasciculus: a review of the anatomical descriptions with functional correlates. Front Neurol 13:794618

Kádár A, Wittmann G, Liposits Z, Fekete C (2009) Improved method for combination of immunocytochemistry and Nissl staining. J Neurosci Methods 184(1):115–118. https://doi.org/10.1016/j.jneumeth.2009.07.010

Kaufmann TJ, Kallmes DF (2008) Diagnostic cerebral angiography: archaic and complication-prone or here to stay for another 80 years? AJR Am J Roentgenol 190(6):1435–1437. https://doi.org/10.2214/AJR.07.3522

Klingler J (1935) Erleichterung der makroskopischen Präparation des Gehirns durch den Gefrierprozess. Schweiz Arch Neurol Psychiatr 36:247–256

Klüver H, Barrera E (1953) A method for the combined staining of cells and fibers in the nervous system. J Neuropathol Exp Neurol 12(4):400–403

Koch C, Tsuchiya N (2007) Attention and consciousness: two distinct brain processes. Trends Cogn Sci 11(1): 16–22. https://doi.org/10.1016/j.tics.2006.10.012

Kuhl DE, Edwards RQ (1963) Image separation radioisotope scanning. Radiology 80(4):653–662

Ladley-O'Brien S (2011) Pneumoencephalography. In: Kreutzer JS, DeLuca J, Caplan B (eds) Encyclopedia of clinical neuropsychology. Springer. https://doi.org/10.1007/978-0-387-79948-3_60

Latorre RM, García-Sanz MP, Moreno M, Hernández F, Gil F, López O, Ayala MD, Ramírez G, Vázquez JM, Arencibia A, Henry RW (2007) How useful is plastination in learning anatomy? J Vet Med Educ 34(2):172–176. https://doi.org/10.3138/jvme.34.2.172

Le Bihan D, Mangin JF, Poupon C, Clark CA, Pappata S, Molko N, Chabriat H (2001) Diffusion tensor imaging: concepts and applications. J Magn Reson Imaging 13(4):534–546. https://doi.org/10.1002/jmri.1076

Leeds NE, Kieffer SA (2000) Evolution of diagnostic neuroradiology from 1904 to 1999. Radiology 217(2):309–318

Light GA, Williams LE, Minow F, Sprock J, Rissling A, Sharp R, Swerdlow NR, Braff DL (2010) Electroencephalography (EEG) and event-related potentials (ERPs) with human participants. Curr Protoc Neurosci Chapter 6: Unit 6.25.1–24. https://doi.org/10.1002/0471142301.ns0625s52

Lin YS, Wang HY, Huang DF, Hsieh PF, Lin MY, Chou CH, Wu IJ, Huang GJ, Gau SS, Huang HS (2016) Neuronal splicing regulator RBFOX3 (NeuN) regulates adult hippocampal neurogenesis and synaptogenesis. PLoS One 11(10):e0164164. https://doi.org/10.1371/journal.pone.0164164

Little JR, Furlan AJ, Modic MT, Weinstein MA (1982) Digital subtraction angiography in cerebrovascular disease. Stroke 13(5):557–566. https://doi.org/10.1161/01.str.13.5.557

Magaki S, Hojat SA, Wei B, So A, Yong WH (2019) An introduction to the performance of immunohistochemistry. Methods Mol Biol 1897:289–298. https://doi.org/10.1007/978-1-4939-8935-5_25

Mansfield P (1977) Multi-planar image formation using NMR spin echoes. J Phys C Solid State Phys 10(1): L55–L58

Martino J, De Witt Hamer PC, Vergani F, Brogna C, de Lucas EM, Vázquez-Barquero A, García-Porrero JA, Duffau H (2011) Cortex-sparing fiber dissection: an improved method for the study of white matter anatomy in the human brain. J Anat 219(4):531–541. https://doi.org/10.1111/j.1469-7580.2011.01414.x

McRobbie DW, Moore EA, Graves MJ, Prince MR (2008) MRI from picture to proton (2nd ed.). AJNR Am J Neuroradiol 29(6):e50. https://doi.org/10.3174/ajnr.A0980

Nazparveen LA, Phirke DS (2017) Cranial ultrasound in critically ill neonates. J Med Res 3(6):290–293

Neha LS, Dhingra R (2013) Plastinated knee specimens: a novel educational tool. J Clin Diagn Res 7(1):1–5. https://doi.org/10.7860/JCDR/2012/5142.2657

Newton MR, Berkovic SF, Austin MC, Rowe CC, McKay WJ, Bladin PF (1995) SPECT in the localisation of extratemporal and temporal seizure foci. J Neurol Neurosurg Psychiatry 59(1):26–30. https://doi.org/10.1136/jnnp.59.1.26

Nichols DA, Meyer FB, Piepgras DG, Smith PL (1994) Endovascular treatment of intracranial aneurysms. Mayo Clin Proc 69(3):272–285. https://doi.org/10.1016/s0025-6196(12)61068-4

Niedermeyer E (2011) Niedermeyer's electroencephalography: basic principles, clinical applications, and related fields. Lippincott Williams & Wilkins

Nongena P, Ederies A, Azzopardi DV, Edwards AD (2010) Confidence in the prediction of neurodevelopmental outcome by cranial ultrasound and MRI in preterm infants. Arch Dis Child Fetal and Neonatal Ed 95(6):F388–F390. https://doi.org/10.1136/adc.2009.168997

Ogawa S, Lee TM, Kay AR, Tank DW (1990) Brain magnetic resonance imaging with contrast dependent on blood oxygenation. Proc Natl Acad Sci 87(24): 9868–9872

Pasternak O, Kelly S, Sydnor VJ, Shenton ME (2018) Advances in microstructural diffusion neuroimaging for psychiatric disorders. NeuroImage 182:259–282. https://doi.org/10.1016/j.neuroimage.2018.04.051

Pavel D, Jobe T, Devore-Best S, Davis G, Epstein P, Sinha S (2006) Viewing the functional consequences of traumatic brain injury by using brain SPECT. Brain Cogn 60:211–213

Pinto-Coelho L (2023) How artificial intelligence is shaping medical imaging technology: a survey of innovations and applications. Bioengineering (Basel) 10(12):1435. https://doi.org/10.3390/bioengineering10121435

Placzek MS, Zhao W, Wey HY, Morin TM, Hooker JM (2016) PET neurochemical imaging modes. Semin Nucl Med 46(1):20–27. https://doi.org/10.1053/j.semnuclmed.2015.09.001

Ramón C (1909) Histology of the nervous system of man and vertebrates. A. Maloine, Paris

Schröder H, Moser N, Huggenberger S (2020) Basic neurohistology. In: Neuroanatomy of the mouse. Springer, pp 1–24. https://doi.org/10.1007/978-3-030-19898-5_2

Schulz RA, Stein JA, Pelc NJ (2021) How CT happened: the early development of medical computed tomography. J Med Imaging (Bellingham) 8(5):052110. https://doi.org/10.1117/1.JMI.8.5.052110

Sheridan CJ, Matuz T, Draganova R, Eswaran H, Preissl H (2010) Fetal magnetoencephalography – achievements and challenges in the study of prenatal and early postnatal brain responses: A review. Infant Child Dev 19(1):80–93. https://doi.org/10.1002/icd.657

Srisuchinnawong A, Wang B, Shao D et al (2021) Modular neural control for gait adaptation and obstacle avoidance of a tailless gecko robot. J Intell Robot Syst 101:27. https://doi.org/10.1007/s10846-020-01285-y

Standring S (2020) Gray's anatomy: the anatomical basis of clinical practice, 42nd edn. Elsevier

Stiller W (2018) Basics of iterative reconstruction methods in computed tomography: a vendor-independent overview. Eur J Radiol 109:147–154. https://doi.org/10.1016/j.ejrad.2018.10.025

Tan Z, Wei H, Song X, Mai W, Yan J, Ye W et al (2022) Positron emission tomography in the neuroimaging of autism spectrum disorder: a review. Front Neurosci 16:806876. https://doi.org/10.3389/fnins.2022.806876

Ter-Pogossian MM (1977) Basic principles of computed axial tomography. Semin Nucl Med 7(2):109–127. https://doi.org/10.1016/s0001-2998(77)80013-5

Van Wezel-Meijler G, Steggerda SJ, Leijser LM (2010) Cranial ultrasonography in neonates: role and limitations. Semin Perinatol 34(1):28–38. https://doi.org/10.1053/j.semperi.2009.10.002

Verger A, Kas A, Darcourt J, Guedj E (2022) PET imaging in neuro-oncology: an update and overview of a rapidly growing area. Cancers (Basel) 14(5):1103. https://doi.org/10.3390/cancers14051103

von Hagens G, Tiedemann K, Kriz W (1987) The current potential of plastination. Anat Embryol (Berl) 175(4):411–421. https://doi.org/10.1007/BF00309677

Xue W, Bowman FD, Pileggi AV, Mayer AR (2015) A multimodal approach for determining brain networks by jointly modeling functional and structural connectivity. Front Comput Neurosci 9:22. https://doi.org/10.3389/fncom.2015.00022

Yeh FC, Irimia A, Bastos DCA, Golby AJ (2021) Tractography methods and findings in brain tumors and traumatic brain injury. NeuroImage 245:118651. https://doi.org/10.1016/j.neuroimage.2021.118651

3 Advanced Imaging Techniques in Neurovisualisation

Dibakar Borthakur and Rima Dada

Abstract

Structural and functional neuroimaging techniques have their own unique advantages. A common neuroimaging technique is computed tomography (CT) scanning. Magnetic resonance imaging (MRI) is known for its ability to produce detailed, high-resolution images without employing ionising radiation. An advanced version of MRI imaging is functional MRI (fMRI), for studying and correlating functions of the brain. With the addition of newer neuroimaging techniques such as positron emission tomography (PET), magnetic resonance spectroscopy (MRS), etc., studying the physiological, biochemical and molecular aspects along with the neuroanatomy has become possible. Apart from studying the normal anatomy and physiology, these techniques are immensely useful for diagnosis and monitoring of neurological diseases. They are believed to be game-changers and have revolutionised disease management in clinical neuroscience. This chapter discusses some of the commonly employed neuroimaging techniques and recent advances in them as they are evolving and their utility in neuroscience research and in clinics.

D. Borthakur · R. Dada (✉)
Department of Anatomy, All India Institute of Medical Sciences, New Delhi, India

Keywords

Neuroimaging · Structural imaging · Functional imaging · Advanced neuroimaging

3.1 Introduction

Neuroimaging has evolved significantly over the past few decades, providing unprecedented insights into the structure and function of the brain. Advanced imaging techniques have enabled researchers and clinicians to visualise neural activity, structural organisation and neurochemical processes with remarkable precision (Gilmore et al. 2018). Various types of structural as well as functional imaging techniques have been used to study and understand the anatomy of the nervous system including the brain. Structural imaging is an essential aspect of neuroanatomical studies which allows visualisation as well as analysis of the anatomical features (Uysal 2023). In these techniques, the anatomy of the brain is studied without physically penetrating the skull. Structural imaging techniques in combination with functional imaging techniques can provide additional information about the neural activity in specific anatomical regions correlating with behavioural or cognitive functions (Bijukumar et al. 2015). Functional imaging studies yield much more precise information about the neural pathway than what had been obtained from the ablation and stimulation studies (John

K. S. Ravi et al. (eds.), *Visualization in Neuroanatomical Sciences*, Biomedical Visualization 11,
https://doi.org/10.1007/978-3-032-07380-8_3

Kiernan 2013). Structural imaging techniques are also used for the evaluation of the anatomical changes that occur over a period of time, e.g. alteration of brain mass with diseases and with ageing. These techniques are routinely utilised in clinical neuroscience for the diagnosis of organic, psychiatric and vascular diseases of the nervous system.

Various parts and regions of the brain are different in microstructure and biochemical composition. This causes various regions of the brain to appear differently and distinctly in sections when examined with the naked eye. The regions composed of neuronal cell bodies appear as grey matter, and regions where axons and fibre tracts are aggregated appear as white matter. Similar contrast is also discernible in images of the brain obtained by structural imaging techniques that help to differentiate one anatomical region distinctly from the other. This means the basic goal of any structural imaging techniques is to generate images in which one can appreciate the different areas of the brain based on the carbohydrate, protein, fat, water and other mineral salt composition. The functional imaging techniques also consider region-specific biochemical as well as tissue composition to generate an image which tells more information about functions. The biochemical composition of the different anatomical parts of the nervous system is different. Neuronal cell bodies contain biomolecules such as proteins, carbohydrates, fat, etc. Similarly, axons and nerve fibre tracts are mainly composed of fat components due to the presence of myelin sheaths in myelinated nerve fibres. Similarly, the cerebrospinal fluid (CSF) of the brain ventricles and the subarachnoid spaces imparts their characteristic appearance in the obtained images.

Until the early 1970s, when X-rays were the only available imaging technology, taking a well-contrasted and well-differentiated image of the brain was barely possible (Fischl et al. 2002). The motivation and continued search for better technology of brain imaging led to the advent of today's modern-day techniques of structural brain imaging. Actually all the modern-day imaging techniques and tools have been constantly evolving over the last 30–40 years. As a result, we have better technologies with us now, which can provide good contrast and differentiation among various soft tissues of the brain and that of surrounding structures.

Various modalities of structural brain imaging have their defined role and purpose in modern neuroscience, providing researchers and clinicians with valuable insights into the anatomical organisation of the human brain. Relatively older and more routinely utilised techniques such as computed tomography (CT) and magnetic resonance imaging (MRI) scans are also capable of providing good-quality images of the brain with varying levels of resolution and specificity (Manasmita 2022). CT scanning is a quicker imaging modality, a valuable tool for acute trauma assessment due to their speed and availability in most places, but with the notable disadvantage of having radiation exposure (Fred 2004). MRI provides high-quality soft tissue contrast, making it ideal for studying the more detailed structural anatomy of the brain, cortical thickness and white matter integrity (Fischl 2012). MRI is often used for the diagnosis of structural abnormalities like tumours, strokes and traumatic brain injuries (Ashburner and Friston 2000). These imaging tools not only play crucial roles in studying the anatomy but proved very useful in research and clinical practice. This chapter explores the advanced imaging techniques used in neurosciences with their latest advancements.

3.1.1 Computerised Tomography

CT scan stands as a cornerstone technology among various neuroimaging modalities. CT scan is also known as the computed axial tomography scan or 'CAT scan'. The final image or a 'slice' or 'tomogram' is a reconstructed image after processing the data from multiple X-ray films. CT scans of the brain particularly provide excellent images that are not only useful for studying the normal neuroanatomy but also useful for detection of tumour, calcification and collection of fluids such as blood in various regions of the head. Thus, they are very useful for rapid evaluation of head injury patients in road traffic accidents.

Principles It utilises the basic principle of X-ray beam application. While performing a CT scan of the brain, a subject lies with their head positioned within the centre of a cylindrical apparatus. The CT scanner revolves around the patient's head, emitting X-ray beams that are detected by sensors positioned opposite the source. As multiple X-ray films in the axial planes are taken in a sequence, these provide much more information than a single X-ray film. The information obtained from the individual X-ray films taken from different angles is further processed with the aid of computer algorithms. The detectors in turn quantify the attenuation of X-rays reaching the detector by passing through different tissues of the brain. A central processing unit (CPU) then gathers the information and processes the scans to generate detailed, high-resolution images of the brain in axial, coronal and sagittal planes. In most of the routinely performed CT scans, eight to ten axial slices are usually obtained. A good-quality CT scan depends on the specifications of the instrument used which include width of the X-ray beam, sensitivity of the detector and the computer algorithm. The narrower the X-ray beam, the better the result.

Recent Advances CT technology has advanced significantly from its initial stage to modern multi-slice or spiral CT scanners which can yield very high spatial resolution images. The new generation of *multi-slice* or *spiral CT scanners* can acquire multiple slices of data simultaneously, thereby reducing the scanning time but with increased resolution and precision. The introduction of dual-energy CT and spectral imaging techniques enabled better tissue characterisation. This has been augmented further by the concomitant developments in the software used for acquiring and processing images.

Applications in Neuroscience An accurate CT scan reveals the anatomy of the brain in axial, coronal and sagittal sectional views. CT films are excellent means of teaching neuroanatomy to medical students. Various congenital malformations of the nervous system such as hydrocephalus are nicely revealed in CT films, and thus the atypical anatomy of the brain can also be studied in CT images (Mayer 2011).

Application in Clinics A CT scan of the brain is widely employed for the rapid evaluation of head trauma patients, stroke, tumour, vascular anomalies, etc. The application and usage of brain and spinal cord CT imaging has revolutionised patient care. As it is noninvasive and cost-effective and provides the desired information in a rapid manner, it is very much popular among clinicians. The use of ionising radiation has obvious disadvantages, and it is absolutely contraindicated for use in pregnant women and to be used with caution in young children.

Despite the disadvantages of ionising radiation, newer advanced technology is emerging that involves lower radiation exposure which is deemed safer. Newer CT scan modalities provide improved quality of images, and with the addition of artificial intelligence and machine learning (AI and ML), interfaces help to rapidly analyse and interpret images. The role of CT in neuroimaging is expected to expand further, promising even greater precision and efficiency in modern neuroimaging.

3.1.2 Magnetic Resonance Imaging (MRI)

MRI is a revolutionary neuroimaging technology in the sense that it not only provides very detailed and high-resolution images but at the same time is noninvasive and is not radio-hazardous. MRI technology can be considered as a pinnacle in medical imaging technology, particularly in the field of neurology.

Principles MRI technology relies on the principle of interaction between magnetic fields and radio waves with the body's hydrogen atoms. The subject's body is placed within a strong magnetic field which makes the hydrogen atoms or protons of the human body to align with the externally applied magnetic field (Segebarth and Decorps 2010). The protons then release radiofrequency signals which are detected and processed by a processing unit for image generation. Later the images are reconstructed to visualise sectional images in any of the desired planes, viz. axial, coronal and sagittal.

Recent Advances MRI is relatively newer in comparison to the CT scan. Similar to CT scans, MRI technology has revolutionised clinical neurology in the evaluation of several structural neurological disorders and has a role in associated vascular anatomy (Reichenbach et al. 1997). Since its first clinical use, the MRI technology has been continuously progressing and advanced by means of developing stronger magnets such as superconducting magnets. Strong magnetic fields beyond 1.5 Tesla to 3.0 Tesla MRI scanners have been developed which has improved the spatial resolution and signal-to-noise ratio significantly (Filippi et al. 2023). The more advanced versions of MRI, such as diffusion-weighted imaging (DWI) or diffusion tensor imaging (DTI), perfusion-weighted imaging (PWI), functional MRI (fMRI), magnetic resonance spectroscopy (MRS), etc., have enabled for even more detailed assessment of the anatomy including microstructure and biochemical composition assessment.

Application in Neuroscience and in Clinics Apart from studying the normal anatomy of the brain, the use of MRI is indispensable across various disciplines including its game-changing role in clinical neurology. For diagnosing and monitoring a wide range of neurological disorders, including vascular events, MRI is crucial. MRI is an excellent tool for detailed assessment of traumatic brain injuries which can identify contusions, haemorrhages, diffuse axonal injuries, etc., that may not be as readily visible in other imaging modalities. Preoperative MRI is also increasingly being utilised for delineating tumour boundaries, assessing proximity to critical structures, etc., which can help achieve better surgical planning without any major neurological deficit post-surgery (Ashburner and Friston 2000).

MRI can offer unparalleled insight into the soft tissue structures with very good contrast and detailed tissue characterisation without any radiation exposure. However, there are contraindications for usage of MRI in certain individuals such as having a pacemaker. As an MRI does not involve ionising radiation, it is particularly suitable in those scenarios where multiple scanning may be required, e.g. longitudinal follow-up of a neurological disorder and in cases of paediatric neurological imaging.

3.1.3 Functional Magnetic Resonance Imaging (fMRI)

Functional magnetic resonance imaging (fMRI) is an advanced MRI technology to study the brain connectivity and functions. Much of the information about the functioning of various areas of the brain had been obtained from ablation studies or observations made in patients of neurological disorders. Since the introduction of fMRI, the precise functions of various areas of the brain could be studied with more precision and accuracy.

Principles In addition to the basic principles of MRI, fMRI utilises the strong magnetic fields and radiofrequency pulses for the detection of changes of oxygenation and blood flow in the brain (Ogawa et al. 1990). fMRI is based on the principle of 'blood-oxygen-level-dependent (BOLD) contrast' mechanism. A more active brain region consumes more oxygen, and to deliver more oxygen, blood flow is increased (Logothetis 2008). That means the region has more concentration of oxygenated haemoglobin and less concentration of deoxygenated haemoglobin. The presence of paramagnetic deoxygenated haemoglobin in a brain region results in signal loss in fMRI, whereas more active brain regions with high concentration of oxygenated haemoglobin, but lower concentration of deoxygenated haemoglobin result in intense signals (Fig. 3.1). Thus, it creates the BOLD contrast between more and less oxygen-consuming areas, and that is then detected and visualised in fMRI. Whenever a particular anatomical region of the brain is involved in neural activity, that requires more oxygen and in turn more cerebral blood flow (Raichle and Snyder 2007). The increased cerebral blood flow is detected in fMRI as signal intensity and is an indirect method of visualisation of the brain activity at that functional state (Fox and Greicius 2010).

Recent Advances Since its introduction, there have been significant developments in fMRI technology. Now it uses a higher magnetic field range of MRI machines capable of generating 3–7 Tesla magnetic fields. The high Tesla scanners with strong magnetic fields are capable of even minute changes in the neural activity. With the addition of newer imaging sequences, such as *echo planar imaging (EPI)*, rapid and dynamic acquisition of images is possible. High-resolution and high-field fMRI can differentiate activity within the layers of the cortex and small nuclei of the brain. Multimodal imaging integrates fMRI data with other neuroimaging modalities, such as electroencephalography (EEG) and magnetoencephalography (MEG), which enhances knowledge about brain dynamics (Yeo et al. 2011). Neuropsychiatric disorders have clinical heterogeneity and a long latent phase of clinical presentation, which limits accurate diagnosis at times, thereby hindering the management strategies. Multidimensional insight towards the pathophysiology of neuropsychiatric illness is now possible with the aid of multimodal neuroimaging techniques. For example, T1-weighted (T1w) MRI provides high-resolution images with excellent contrast which is much preferred for visualising the grey matter, white matter and cerebrospinal fluid and also for quantifying regional brain volumes. So, T1w MRI can detect abnormal cortical growth and pattern of atrophy of grey matter that give valuable clues for clinical diagnosis of neuropsychiatric disorders. T2-weighted (T2w) MRI and fluid-attenuated inversion recovery (FLAIR) MRI can efficiently detect hyperintense lesions that indicate neuroinflammation, demyelination,

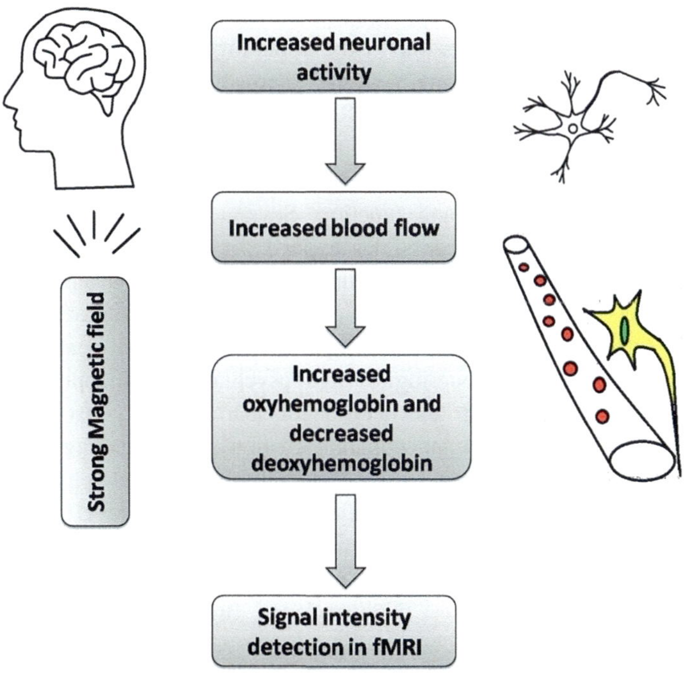

Fig. 3.1 Schematic representation of the basic principles of the functional magnetic resonance imaging (fMRI)

ischaemia, etc., pointing towards clinical conditions such as multiple sclerosis (MS). Unlike traditional MRI, the diffusion-weighted MRI (dMRI) gathers information about the microstructural integrity by quantifying the diffusion of water molecules within tissue. The technique of dMRI enables tractographic mapping of white matter pathways which provides crucial information about white matter alterations and network disruptions which can be the earliest signs associated with a neuropsychiatric disorder. Similarly, EEG and MEG when combined with these imaging techniques can directly measure neural activity via electrical and magnetic signals and provide other related information (Banihashemi et al. 2023; Luo et al. 2024; Micha 2017; Huang and Shu 2025).

Integration of AI and ML algorithms further aids to identify complex neural pathways and connectivity patterns which have the potential to be used for prediction and diagnosis of several neurological diseases. AI and ML has transformed traditional neuroimaging to an arena with significant breakthroughs towards a rapidly evolving field that incorporates the use of several neuroimaging techniques and AI and ML algorithms to identify various brain disorders with much more accuracy, speed and efficiency. This has been made possible by the technological advancements and computational innovations associated with AI and ML. Researchers and clinicians now have a powerful tool through which novel approaches can be made to explore the brain's structure and function. Several AI models have been successfully applied for the analysis and interpretation of neuroimaging data. These AI algorithms aid clinicians and researchers in diagnosing, treating and monitoring patients with neurological and psychiatric disorders. AI and ML and deep neural learning has the multipronged and unparalleled potential for biomarker discovery, disease prediction and personalised treatment strategies in several neuropsychiatric disorders (Sanchez et al. 2023).

Applications in Neuroscience Broad applications of fMRI in both clinical and research settings are to map brain functions related to motor, sensory, language and cognitive processes (Fox and Greicius 2010). It also helps in pre-surgical planning by localisation of regions responsible for vital functions and avoiding post-operative deficits. In addition to that, fMRI has been able to identify normal and aberrant circuitry and therefore helps in the understanding of complex processes such as cognition. *Resting-state fMRI (rs-fMRI)* measures brain activity when an individual is not engaged in any particular task (Van Essen et al. 2013). fMRI can localise regions of the brain specific to a particular function, e.g. during language processing, certain areas of the brain such as Broca's area and Wernicke's areas are activated (Fox and Raichle 2007). Similarly, tasks based on memory lead to change in vascularity in the hippocampus. It also addresses the complex questions related to neural plasticity, effect of ageing on the brain, etc.

Application in Clinics Identification of function-specific regions by fMRI helps clinicians to better deal with neurological disorders. By understanding the functional aspects of the normal and diseased brain and elucidating neural circuitry, understating the mechanisms of causation of some of the complex neurological diseases and their treatment has become easier. In clinical neurosciences too, fMRI has a significant role to play. It can detect and localise epileptogenic foci in the brain, and the same can guide a neurosurgeon in surgical interventions done in refractory cases (Abou-Khalil 2007). In addition to that, fMRI is used for longitudinal follow-up of the chronic neurodegenerative diseases like multiple sclerosis (Poldrack et al. 2011). Altered functional connectivity patterns in several psychiatric disorders like schizophrenia, bipolar affective disorder and anxiety disorders have been identified, a deep understanding of which is expected to decipher the pathophysiology of these diseases (van den Heuvel and Hulshoff Pol 2010; Howes and Kapur 2009).

Challenges Interpretation of fMRI data requires expertise in the field and calibration, and optimisation of instrument and data analysis algorithms is also essential to minimise false positive results in neural connectivity and brain mapping analysis. Nevertheless, fMRI promises to uncover new layers of complexity in brain function and connectivity in the coming days.

3.1.4 Diffusion Tensor Imaging (DTI)

Diffusion tensor imaging (DTI) is an MRI-based technique by which the microstructural properties of white matter in the brain can be studied. It provides valuable information about the integrity of a neural pathway.

Principles The basic principle of DTI is diffusion of water molecules in the brain. In the brain the diffusion of water molecules is anisotropic due to the presence of abundant myelinated axons (Mukherjee et al. 2008). *Diffusion tensor* is measured by the technique that represents the three-dimensional diffusion process. There are several parameters in the diffusion tensor. These are fractional anisotropy, mean diffusivity, axial diffusivity, radial diffusivity, etc. (Alexander et al. 2007). From these parameters measured in the diffusion tensor, the integrity of the white fibre tract can be determined.

Recent Advances Some of the recent advancements in DTI are *high-resolution DTI* which uses advanced apparatus for image acquisition with upgraded software and allows more detailed white matter tract visualisation. *Multi-shell DTI* is another advanced approach involving acquisition of image data in multiple diffusion weightings. It also serves to provide much more detailed information about the white matter microstructure and integrity. Other techniques include *neurite orientation dispersion and density imaging (NODDI)* and *diffusion kurtosis imaging (DKI)* with the ability to probe into the microstructural properties of the white matter beyond the conventional DTI (Zhang et al. 2012; Mitchell et al. 2019).

Applications in Neuroscience The white matter pathways and connectivity pattern can be studied which is called tractography. DTI can be useful for studying the normal developmental pattern of the white matter tracts in the early years of life (Hermoye et al. 2006; Karlsgodt et al. 2008).

Application in Clinics DTI has an essential role in clinical neuroscience aiding the diagnosis and monitoring of several neurological disorders. It is of much value in traumatic brain injuries, as it can detect microstructural injuries (Thomalla et al. 2004). Disease progression of multiple sclerosis, Alzheimer's disease, Parkinson's disease, etc., reveals characteristic changes in DTI (Douaud et al. 2013). In attention-deficit hyperactivity disorder (ADHD), abnormalities are detected in the pathways in the frontal lobe, basal nuclei, brain stem and cerebellum. Abnormalities of two structurally and functionally different neural pathways, the uncinate fasciculus (UF) and the cingulate bundle (CB), are observed in schizophrenia. DWI with fluid-attenuated inversion recovery (FLAIR) sequence is employed for fibre tractography of human corpus callosum (Bailey et al. 2013; Soni et al. 2017).

3.1.5 Functional Near-Infrared Spectroscopy (fNIRS)

Functional near-infrared spectroscopy (fNIRS) is an optical imaging technique employing infrared rays and measuring neural activity by determining the changes in haemodynamic and blood oxygenation status (Ferrari and Quaresima 2012). It has gained much popularity in recent times on account of its portability, safety and cost-effectiveness.

Principles fNIRS utilises near-infrared light which can penetrate biological tissues and gets absorbed by *chromophores*, which in the human body is mainly the haemoglobin. It also estimates the levels of changes in oxyhaemoglobin and deoxyhaemoglobin concentration from the absorption pattern of near-infrared light (Ferrari and Quaresima 2012; Boas et al. 2014). The difference in absorption pattern of these two forms of haemoglobin varies according to the neural activity states and thus provides an indirect way of assessment of brain activity.

Recent Advances High-density fNIRS increases the spatial resolution using high-density *optode* arrays. It can be combined with other techniques in multimodal imaging modalities (Sato et al. 2013). Development of wearable fNIRS devices facilitated studying the brain activity during real-world interactions and ambulatory activities (Piper et al. 2014).

Applications in Neuroscience In cognitive neuroscience, fNIRS has become an indispensable tool for studying the brain functions. The mechanisms of language processing and its lateralisation can be studied which provides meaningful insight into the language processing problems. Execution of higher mental functions such as memory, attention, decision-making, empathy and social interactions can be studied (Quaresima and Ferrari 2019; Aslin and Mehler 2005). Neurodevelopmental processes can be studied safely with the fNIRS technique, especially in infants and children because it is easy to administer and free from ionising radiation (Lloyd-Fox et al. 2014).

Applications in Clinics fNIRS has shown promising prospects in various neurodevelopmental disorders such as autism spectrum disorder (ASD), ADHD, etc. (Lloyd-Fox et al. 2014). The monitoring of a patient during the therapeutic interventions for neurorehabilitation is also guided by fNIRS (Obrig 2014). Neural correlates of several psychiatric disorders—depression, anxiety and schizophrenia—have been obtained through fNIRS. The application of fNIRS has important implications for optimising various training strategies for neurorehabilitation in children with cerebral palsy and other developmental disorders. The human brain is a complex system, where various regions perform definite assigned tasks which exhibit different activation states in accordance with performance of the particular task. The designated region executes a task by interacting with other brain regions, but the precise functional mechanism involving the neural activation pathways of several diseases is not fully deciphered. Understanding the functional neural mechanisms and their alterations during diseased states is essential for a thorough understanding of the pathophysiology of these neurological disorders and developing effective strategies to deal with them. Neurorehabilitation of some of the common neurological disorders involves cortical activation and neural remodelling. Studying the living human brain by invasive means is not possible to look for the changes in the brain during performance of a task, and a vast majority of information is obtained through animal experimentations. Nevertheless, information obtained from in vitro studies cannot be exactly similar to how the brain functions in humans. This has led to the emergence of functional neural imaging modalities which can assess various aspects of brain functions like cognition, memory and behaviour in real time. Knowledge of such functional neural imaging has tremendously changed the present focus of brain research. fNIRS and fMRI are two such modalities which are commonly employed for functional brain research. fNIRS is more advantageous because it is more rapid and cost-effective. It is important to understand the principle behind the functional imaging technology fNIRS. When we intend to do something or we think about a task and plan accordingly, a specific region of the brain gets activated and undergoes chemical reactions. When we think, feel and act, neurons in the brain undergo chemical reactions and exhibit energy transformation. These changes are reflected by the changes in the blood flow to the particular region. The changes in the blood flow and hence blood oxygen level are used for the indirect assessment of the neural activity in a particular region. The alteration of energy transformations can be reflected by changes in blood flow. Thus, changes in blood oxygen levels are used to assess neuronal activity in different regions of the brain. fNIRS is easy to administer to patients, well tolerated, cost-effective and sensitive and at the same time has high resolution. Another advantage of the technique is that it can be safely applied to children and elderly. It has been increasingly becoming popular and being applied in basic neuroscience research involving animal studies and in clinical research. Furthermore, its efficacy in capturing functional brain activity during the performance of a task has contributed to a growing adoption in clinical settings (Huo et al. 2021; Zhang et al. 2024).

Being a noninvasive imaging modality to study the changes in the cerebral haemodynamics with changing activity of the cerebral cortex, the fNIRS has gained great interest among researchers and clinicians. One of the unique advantages of the fNIRS is that it allows neuro-investigation in a more realistic and clinical environment. Its application in stroke rehabilitation is worth mentioning. Due to the acceptability, safety and high spatial and temporal resolutions compared with EEG and fMRI, fNIRS is now increasingly being used for studying a wide range of neuropsychiatric disorders such as affective disorders, schizophrenia, organic brain disorders, neurodevelopment disorders, etc. The technique has been able to demonstrate its broad utility in understanding the traumatic brain injuries, Alzheimer's disease and related dementias (ADRD). Studies have consistently proven and demonstrated that fNIRS may provide valuable biomarkers for apathy and depression in individuals with ADRD and chronic osteoarthritic pain, with differential patterns based on cognitive function. It suggests that there can be different symptomatologies of various neuropsychiatric disorders based on the cognitive status of the individual. In the coming days, fNIRS is expected to expand its application horizon by increasing data collection capacity, reducing interference and errors and simultaneously achieving higher spatiotemporal resolution (Chang et al. 2021; Huff et al. 2025; Huo et al. 2021; Zhang et al. 2024).

3.1.6 Optical Coherence Tomography (OCT)

Optical coherence tomography (OCT) provides high-resolution cross-sectional images of biological tissues using light rays. The technique is extensively used for the neural retina visualisation.

Principles The principle of OCT is the use of low-coherence interferometry for capturing three-dimensional images in micrometre resolution from an optical scattering media such as biological tissue in the human body (Huang et al. 1991). The technique directs a beam of near-infrared light at the biological tissue sample and in turn measures the echo time delay and intensity of backscattered light. The OCT apparatus compares the tissue scattered light with a reference light beam and constructs an image of the biological tissue (Kashani et al. 2017).

Recent Advances *Swept-source OCT (SS-OCT)* is a new technology utilising a tunable laser source, for deeper penetration and fast acquisition of images. *Optical coherence tomography angiography (OCTA)* studies the retinal blood flow in a noninvasive manner and the changes in retinal vasculature in various disease conditions (Spaide et al. 2018) (Jonnal et al. 2016). The use of portable OCT devices in recent times could facilitate bedside assessment of ocular and neurological conditions.

Applications in Neuroscience Detailed high-resolution image of the retina can be obtained with OCT which guides therapeutic monitoring of several ophthalmologic diseases such as age-related macular degeneration (AMD), diabetic retinopathy, glaucoma, etc. (Grzybowski and Barboni 2016; Doustar et al. 2017). The technique also helps to diagnose conditions of papilloedema and optic neuritis and monitor optic nerve health (Jonnal et al. 2016). As the retina can be considered as a window to the brain, therefore by employing OCT, the complex retina brain connections can be explored, and the impact of the neurological conditions via the retina assessed. The neuropsychiatric diseases commonly afflicted in such cases include schizophrenia, bipolar affective disorders, major depressive episodes, etc. (ElShaarawi et al. 2022; Levi et al. 2024; Xu et al. 2024; van de Kreeke et al. 2019).

Applications in Clinics OCT has been extensively used for understanding the complex retinal structure and relationship of the neural retina with several disease conditions of the nervous system.

3.1.7 Positron Emission Tomography (PET)

Positron emission tomography (PET) is a noninvasive imaging technology capable of probing into the functional aspects of the brain. PET utilises radiolabelled tracer materials for the visualisation and measurement of metabolic activity, blood flow and neurotransmitter dynamics (Cherry et al. 2018).

Principles During a PET scan, a biologically active molecule that is labelled with a radionuclide capable of emitting positron is injected into the body. Fluorine-18 (^{18}F), carbon-11 (^{11}C) and oxygen-15 (^{15}O) are the usual molecules or tracers that are usually used (Phelps 2000). These radiolabelled molecules accumulate in the target tissues during biological metabolism. The accumulated radiolabelled molecules, while decaying, start emitting positrons which are detected by the PET scanner. Computer software and predesigned algorithms reconstruct the PET scanner data into a 3D image, wherein regions of concentrated radiolabelled isotope tracer molecules can be identified.

Recent Advances To extract more precise information, PET scanning has been utilised with new tracer molecules and by introducing hybrid PET scanning, i.e. integrating PET scanning with other imaging modalities like PET with MRI (Bailey et al. 2016). The hybrid imaging technology with further integration with AI and ML is expected to probe into anatomical details with more precision and resolution. Another recent development in PET scanning is the total-body PET which may open new avenues and opportunities in patient care and research in the coming days. Overall, PET scanning is the new hope of modern neuroscience, providing critical neuroanatomical details with simultaneous insights into the functional and molecular aspects of a wide range of neurological disorders.

3.1.8 Electroencephalogram (EEG)

Electrical activity of the brain can be recorded by a noninvasive method known as the electroencephalography (EEG). It has been in use in clinical neuroscience since the early twentieth century for the diagnosis of epilepsy, but the technique has been constantly advancing, and now it has not only been used for diagnosis of seizure disorders but equally in several areas of functional neuroscience research (Fisher et al. 2005).

Principles Neurons generate axon potentials during excitation and transmission of information. Axon potentials of the neurons are electrical signals which are recorded in an EEG (Young 2000). Postsynaptic potentials are recorded by putting electrodes over specific regions of the scalp. These electrical signals once recorded are amplified and visualised and depicted in a graphical form. Various signals of the EEG are represented in frequency unit hertz (Hz). The different frequency bands are named as delta (0.5–4 Hz), theta (4–8 Hz), alpha (8–13 Hz), beta (13–30 Hz) and gamma (30–100 Hz), and the state of the electrical activity of the brain can be recognised by identifying individual frequency band. For example, beta waves are produced during active thinking, and alpha waves are generated during relaxed state (Michel et al. 2004).

Recent Advances *High-density EEG* is an advanced form which uses more electrodes (e.g. 128 or 256 channels). As the technique captures precise localisation of the electrical activity of the brain, it allows studying the more complex neural networks. Wearable EEG devices are another recent development in which constant monitoring of the electrical activity of the brain in real time is possible outside the laboratory or clinic setting (Casson 2019). The technique has multiple applications such as in identification and recording of epileptiform discharge which is quite difficult to capture in a short-duration EEG, for example, in cases of absence seizure. Similarly, the electrical activity of the neurons can be studied while performing a particular task such as playing a musical instrument (Craik et al. 2019). In more recent times, EEG data has been integrated with AI which has opened the door for the development of more precise diagnostic and research tools (Zhang et al. 2016).

Application in Neuroscience and Clinics
EEG is a routinely used technique in clinics for the diagnosis of various types of seizure disorders. Abnormal electrical activities in various forms of seizures, known as epileptiform discharge, are distinctly identified in an EEG and utilised for accurate diagnosis of seizure disorders. As different waveforms of EEG are generated during normal functioning of the brain and during various stages of sleep, EEG is often employed for the study of sleep disorders also. EEG is a core component in *polysomnography*, a comprehensive sleep study procedure. EEG can provide definite signs of brain death by recording no electrical activity. EEG is also helpful for assessing the brain functioning in patients in severe coma (Young 2000). Cognitive neuroscience also utilises EEG to understand higher mental activity involving complex cognitive processes such as perception, attention, memory, decision-making, etc. (Craik et al. 2019). The principles of EEG are integrated in the brain-computer interface (BCI) (Wolpaw and Wolpaw 2012). Brain-computer interface (BCI) is an advanced process by which the brain can communicate with external devices. The importance of BCI can be easily understood from the fact that BCI is widely used these days for the control of artificial limbs or control of other apparatus connected to the brain. This has a promising role for functional rehabilitation in persons with disabilities.

3.1.9 Cerebral Angiography and CT Angiography (CTA)

Cerebral angiography basically deals with the visualisation of cerebral vessels. In this procedure, a radiopaque dye is injected into the cerebral circulation that delivers it into the blood vessels of the brain. Some dye also gets into the brain tissues. After a predefined amount of time has passed, a radiograph is taken which provides a good contrast image of the cerebral vessels. The radiopaque substance absorbs X-rays more than the surrounding soft tissues and makes the dye in circulation appear radiopaque, thereby yielding better contrast and differentiation with brain parenchyma. Thus, in a cerebral angiogram, the cerebral circulation is primarily prominently visualised. Any disorders of the brain vasculature such as aneurysm or blockage can be identified readily. Additionally, as the contrast of the brain parenchyma from the surrounding tissue is also enhanced, therefore, the angiogram can also provide a better image of the brain parenchyma than a conventional plain radiograph.

Cerebral CT angiography (CTA) is a noninvasive technique to visualise the cerebral vasculature. An iodine-based contrast medium is administered into the circulation followed by rapid acquisition of CT scan images (Kumamaru et al. 2010). It is useful for diagnosis and management of various cerebral vascular conditions including stenosis, aneurysms and arteriovenous malformations (AVMs).

Technique
The technique involves the administration of contrast agents which enhances the vascular structures, thereby allowing clear visualisation in the CT-acquired images. Later a three-dimensional representation of the entire cerebral circulation can be reconstructed and studied from the acquired images (Kumamaru et al. 2010).

Recent Advances Nowadays, CT scanners, including the multidetector CC scan (MDCT), perform the procedure rapidly following the contrast medium administration. This facilitates capturing high-resolution images within seconds and reducing any procedural artefacts.

Application in Neuroscience and Clinics
Cerebral CTA is indicated in a variety of diseases of the cerebral blood vessels. The technique is very helpful for aneurysm detection, stroke

evaluation, AVM, etc. (Schmidt et al. 2021) (Singh et al. 2019). It can also detect vasculitis of the cerebral vessels (Ghorishi et al. 2023). The technique can accurately identify and quantify the occlusion or stenosis of blood vessels and guide the treatment or intervention in acute ischaemic stroke (Dias et al. 2021). Due to its capacity to rapidly identify vessel occlusions, thrombolytic or endovascular therapy can be instituted promptly which improves patient outcomes (Dias et al. 2021).

Challenges Radiation exposure is a notable drawback of the technique. The contrast agent administration is usually contraindicated in patients with nephropathy. Similarly, contrast agent-induced allergic reactions are also not very uncommon (Sabarudin et al. 2014). Despite these limitations, it is considered a very useful tool. The diagnostic utility and safety of the technique is expected to improve with continued advancements in the CT technology and contrast medium.

3.1.10 Perfusion CT (PCT)

Perfusion computed tomography (PCT) is a technique for rapid qualitative and quantitative evaluation of cerebral perfusion by parameters such as cerebral blood flow (CBF), cerebral blood volume (CBV) and mean transit time (MTT). Intravenous iodine is used as a contrast medium in the procedure, and the instrument used is a 64 slice, CT scanner machine. Forty millilitres of iodine (contrast medium) is mixed with 25 ml of saline which is then injected at a controlled rate into the circulation of the subject (Schramm et al. 2004). As intravenous contrast medium is used, normal kidney function test and normal creatinine clearance is a pre-required condition prior to the procedure. A cerebral PCT is often used for quickly identifying ischaemic penumbra and for distinction between a penumbra and cerebral infarct (Sabarudin et al. 2014). In a cerebral infarct, both MTT and CBV values are low, but in the case of a penumbra, CBV is preserved even if MTT is increased. PCT has been used in combination with CTA for monitoring in patients with subarachnoid haemorrhage.

3.2 Summary and Conclusion

Various structural and functional brain imaging techniques offer their unique advantages in visualising various aspects of brain anatomy. The commonly employed neuroimaging technique is CT scanning, which employs X-ray radiation for obtaining images. As CT-based scans are faster, cheaper and more readily available than MRI-based scans, they are often utilised as the first line of investigation in acute emergency conditions such as trauma. MRI-based scanning techniques are known for their ability to produce detailed, high-resolution images without ionising radiation. MRI works on the principles of nuclear magnetic resonance, where protons of the body align with an externally applied magnetic field and emit signals when excited by radiofrequency pulses. An image of the brain tissue is then reconstructed from these signals (Ashburner and Friston 2000). The other advanced versions employing MRI technology are fMRI, MRS, etc., by which brain activity can be mapped during resting state or activity (Logothetis 2008). fMRI has broadened the arena of MRI technology to make it utilisable in many more allied fields of medicine. From the anatomical study point of view, MRI technology is instrumental in mapping functional brain regions through techniques like cortical surface reconstruction and voxel-based morphometry (VBM) (Fig. 3.2). The study of the functional aspects can help analyse differences in brain anatomy across individuals and populations and may facilitate investigations into how structural variations relate to higher mental functions like cognitive functions, behaviour, etc. (Fischl 2012). The recent introduction of optical imaging and multimodal imaging has broadened the arena of neuroimaging. Despite the many advantages, the structural and functional neuroimaging techniques have the challenges of image variability, interpreter's

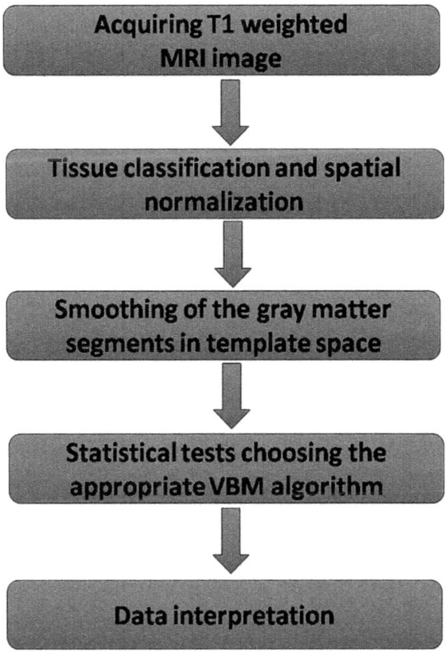

Fig. 3.2 Schematic representation of the workflow of voxel-based morphometry (VBM)

variability, artefacts and lack of standardised protocols across all centres.

Looking ahead, ongoing research aims to refine imaging techniques with the integration of artificial intelligence for greater sensitivity and specificity and also for automated data analysis which can yield not only prompt but very accurate data. In conclusion, advanced neuroimaging techniques represent a cornerstone in modern neuroscience research as well as in clinical neurology, with the capability to noninvasively look into the complex human brain and other parts of the nervous system. These advanced imaging tools strive to enrich neuroanatomical knowledge in times ahead.

References

Abou-Khalil B (2007) An update on determination of language dominance in screening for epilepsy surgery: the Wada test and newer noninvasive alternatives. Epilepsia 48(3):442–455

Alexander AL, Lee JE, Lazar M, Field AS (2007) Diffusion tensor imaging of the brain. Neurotherapeutics 4(3):316–329. https://doi.org/10.1016/j.nurt.2007.05.011

Ashburner J, Friston KJ (2000) Voxel-based morphometry – the methods. NeuroImage 11(6 I):805–821. https://doi.org/10.1006/nimg.2000.0582

Aslin RN, Mehler J (2005) Near-infrared spectroscopy for functional studies of brain activity in human infants: promise, prospects, and challenges. J Biomed Opt. https://doi.org/10.1117/1.1854672

Bailey J, Oliveri A, Levin E (2013) 基因的改变NIH public access. Bone 23(1):1–7. https://doi.org/10.1016/j.schres.2009.01.014.Uncinate

Bailey DL, Pichler BJ, Gückel B, Barthel H, Beer AJ, Botnar R et al (2016). Combined PET/MRI: from status Quo to status Go. Summary report of the fifth international workshop on PET/MR imaging; February 15–19, 2016; . Molecular Imaging and Biology. United States, Tübingen. https://doi.org/10.1007/s11307-016-0993-2

Banihashemi L, Lv J, Wu M, Zhan L (2023) Editorial: current advances in multimodal human brain imaging and analysis across the lifespan: from mapping to state prediction. Front Neurosci 17:2021–2023. https://doi.org/10.3389/fnins.2023.1153035

Bijukumar DR, McGeehan C, Mathew MT (2015) Using structural and functional brain imaging to uncover how the brain adapts to blindness Gabriella. ANN Neurosci Psychol 176(1):139–148

Boas DA, Elwell CE, Ferrari M, Taga G (2014) Twenty years of functional near-infrared spectroscopy: introduction for the special issue. NeuroImage. https://doi.org/10.1016/j.neuroimage.2013.11.033

Casson AJ (2019) Wearable EEG and beyond. Biomed Eng Lett 9(1):53–71. https://doi.org/10.1007/s13534-018-00093-6

Chang F, Li H, Zhang S, Chen C, Liu C, Cai W (2021) Research progress of functional near-infrared spectroscopy in patients with psychiatric disorders. Forensic Sci Res 6(2):141–147. https://doi.org/10.1080/20961790.2020.1720901

Cherry SR, Jones T, Karp JS, Qi J, Moses WW, Badawi RD (2018) Total-body PET: maximizing sensitivity to create new opportunities for clinical research and patient care. J Nucl Med 59(1):3–12. https://doi.org/10.2967/jnumed.116.184028

Craik A, He Y, Contreras-Vidal JL (2019) Deep learning for electroencephalogram (EEG) classification tasks: a review. J Neural Eng 16(3):31001. https://doi.org/10.1088/1741-2552/ab0ab5

Dias BA, Bezerra KB, Bezerra ASDA, Santana VG, Borges RR, Reinaux JCDF et al (2021) Importance of computed tomography angiography in acute/hyperacute ischemic stroke. Radiol Bras 54(6):360–366. https://doi.org/10.1590/0100-3984.2020.0168

Douaud G, Menke RAL, Gass A, Monsch AU, Rao A, Whitcher B et al (2013) Brain microstructure reveals early abnormalities more than two years prior to clinical progression from mild cognitive impairment to

Alzheimer's disease. J Neurosci 33(5):2147–2155. https://doi.org/10.1523/JNEUROSCI.4437-12.2013
Doustar J, Torbati T, Black KL, Koronyo Y, Koronyo-Hamaoui M (2017) Optical coherence tomography in Alzheimer's disease and other neurodegenerative diseases. Front Neurol 8(DEC):1–13. https://doi.org/10.3389/fneur.2017.00701
ElShaarawi MS, Gaafar AA, Shalaby HS, Abdelgawad RHA (2022) Optical coherence tomography changes in Egyptian patients with major depressive disorder. J Egypt Ophthalmol Soc 115(4):169–174. https://doi.org/10.4103/ejos.ejos_53_22
Ferrari M, Quaresima V (2012) A brief review on the history of human functional near-infrared spectroscopy (fNIRS) development and fields of application. NeuroImage 63(2):921–935. https://doi.org/10.1016/j.neuroimage.2012.03.049
Filippi M, Preziosa P, Arnold DL, Barkhof F, Harrison DM, Maggi P et al (2023) Present and future of the diagnostic work-up of multiple sclerosis: the imaging perspective. J Neurol 270(3):1286–1299. https://doi.org/10.1007/s00415-022-11488-y
Fischl B (2012) FreeSurfer. NeuroImage 62(2):774–781. https://doi.org/10.1016/j.neuroimage.2012.01.021
Fischl B, Salat DH, Busa E, Albert M, Dieterich M, Haselgrove C et al (2002) Whole brain segmentation: automated labeling of neuroanatomical structures in the human brain. Neuron 33(3):341–355. https://doi.org/10.1016/S0896-6273(02)00569-X
Fisher RS, van Emde Boas W, Blume W, Elger C, Genton P, Lee P, Engel JJ (2005) Epileptic seizures and epilepsy: definitions proposed by the International League Against Epilepsy (ILAE) and the International Bureau for Epilepsy (IBE). Epilepsia 46(4):470–472. https://doi.org/10.1111/j.0013-9580.2005.66104.x
Fox MD, Greicius M (2010) Clinical applications of resting state functional connectivity. Front Syst Neurosci 4 (June). https://doi.org/10.3389/fnsys.2010.00019
Fox MD, Raichle ME (2007) Spontaneous fluctuations in brain activity observed with functional magnetic resonance imaging. Nat Rev Neurosci 8(9):700–711. https://doi.org/10.1038/nrn2201
Fred HL (2004) Drawbacks and limitations of computed tomography: views from a medical educator. Tex Heart Inst J 31(4):345–348
Ghorishi A, Alayon A, Ghaddar T, Kandah M, Amundson PK (2023) MR and CT angiography in the diagnosis of vasculitides. BJR Open 5(1):20220020. https://doi.org/10.1259/bjro.20220020
Gilmore JH, Knickmeyer RC, Gao W (2018) Imaging structural and functional brain development in early childhood. Nat Rev Neurosci 19(3):123–137. https://doi.org/10.1038/nrn.2018.1
Grzybowski A, Barboni P (2016) OCT in central nervous system diseases: the eye as a window to the brain. In: OCT in central nervous system diseases: the eye as a window to the brain. Springer, pp 1–342. https://doi.org/10.1007/978-3-319-24085-5
Hermoye L, Saint-Martin C, Cosnard G, Lee S-K, Kim J, Nassogne M-C et al (2006) Pediatric diffusion tensor imaging: normal database and observation of the white matter maturation in early childhood. NeuroImage 29(2):493–504. https://doi.org/10.1016/j.neuroimage.2005.08.017
Howes OD, Kapur S (2009) The dopamine hypothesis of schizophrenia: version III–the final common pathway. Schizophr Bull 35(3):549–562. https://doi.org/10.1093/schbul/sbp006
Huang W, Shu N (2025) AI-powered integration of multimodal imaging in precision medicine for neuropsychiatric disorders. Cell Rep Med 6(5):102132. https://doi.org/10.1016/j.xcrm.2025.102132
Huang D, Swanson EA, Lin CP, Schuman JS, Stinson WG, Chang W et al (1991) Optical coherence tomography. Science (New York, NY) 254(5035): 1178–1181. https://doi.org/10.1126/science.1957169
Huff AJ, Park J, Montero-Hernandez S, Park L, Lee C, Pollonini L, Ahn H (2025) Functional near-infrared spectroscopy (fNIRS) detects brain changes for apathy and pain in patients with Alzheimer's disease and related dementias: an exploratory study. NeuroImage Rep 5(3):100266. https://doi.org/10.1016/j.ynirp.2025.100266
Huo C, Xu G, Li W, Xie H, Zhang T, Liu Y, Li Z (2021) A review on functional near-infrared spectroscopy and application in stroke rehabilitation. Med Novel Technol Devices 11(January):100064. https://doi.org/10.1016/j.medntd.2021.100064
John Kiernan RR (2013) Topography of the cerebral hemispheres. In: Barr's the human nervous system: an anatomical viewpoint. Wolters Kluwer
Jonnal RS, Kocaoglu OP, Zawadzki RJ, Liu Z, Miller DT, Werner JS (2016) A review of adaptive optics optical coherence tomography: technical advances, scientific applications, and the future. Investig Ophthalmol Vis Sci 57(9):OCT51–OCT68. https://doi.org/10.1167/iovs.16-19103
Karlsgodt KH, van Erp TGM, Poldrack RA, Bearden CE, Nuechterlein KH, Cannon TD (2008) Diffusion tensor imaging of the superior longitudinal fasciculus and working memory in recent-onset schizophrenia. Biol Psychiatry 63(5):512–518. https://doi.org/10.1016/j.biopsych.2007.06.017
Kashani AH, Chen C-L, Gahm JK, Zheng F, Richter GM, Rosenfeld PJ et al (2017) Optical coherence tomography angiography: a comprehensive review of current methods and clinical applications. Prog Retin Eye Res 60:66–100. https://doi.org/10.1016/j.preteyeres.2017.07.002
Kumamaru KK, Hoppel BE, Mather RT, Rybicki FJ (2010) CT angiography: current technology and clinical use. Radiol Clin North Am 48(2):213–235., vii. https://doi.org/10.1016/j.rcl.2010.02.006
Levi AY, Cherninkova S, Milanova V, Haykin VM, Oscar AH (2024) Application of optical coherence tomography in neurodegenerative diseases: a focus on

Parkinson's, Alzheimer's, and Schizophrenia in Bulgarian patients. Biotechnol Biotechnol Equip 38(1). https://doi.org/10.1080/13102818.2024.2419440

Lloyd-Fox S, Richards JE, Blasi A, Murphy DGM, Elwell CE, Johnson MH (2014) Coregistering functional near-infrared spectroscopy with underlying cortical areas in infants. Neurophotonics 1(2):25006. https://doi.org/10.1117/1.NPh.1.2.025006

Logothetis NK (2008) What we can do and what we cannot do with fMRI. Nature 453(7197):869–878. https://doi.org/10.1038/nature06976

Luo N, Shi W, Yang Z, Song M, Jiang T (2024) Multi-modal fusion of brain imaging data: methods and applications. Mach Intell Res 21(1):136–152. https://doi.org/10.1007/s11633-023-1442-8

Manasmita D (2022) Neuroimaging techniques and what a brain image can tell us. Neurosci News Res:3–21

Mayer J (2011) Basic approach to head ct interpretation. Eur J Paediatr Neurol 2344(234):38–41

Micha R (2017) 乳鼠心肌提取 HHS public access. Physiol Behav 176(1):100–106. https://doi.org/10.1177/0022146515594631.Marriage

Michel CM, Murray MM, Lantz G, Gonzalez S, Spinelli L, Grave de Peralta R (2004) EEG source imaging. Clin Neurophysiol 115(10):2195–2222. https://doi.org/10.1016/j.clinph.2004.06.001

Mitchell T, Archer DB, Chu WT, Coombes SA, Lai S, Wilkes BJ et al (2019) Neurite orientation dispersion and density imaging (NODDI) and free-water imaging in Parkinsonism. Hum Brain Mapp 40(17):5094–5107. https://doi.org/10.1002/hbm.24760

Mukherjee P, Berman JI, Chung SW, Hess CP, Henry RG (2008) Diffusion tensor MR imaging and fiber tractography: theoretic underpinnings. AJNR Am J Neuroradiol 29(4):632–641. https://doi.org/10.3174/ajnr.A1051

Obrig H (2014) NIRS in clinical neurology – a "promising" tool? NeuroImage 85(Pt 1):535–546. https://doi.org/10.1016/j.neuroimage.2013.03.045

Ogawa S, Lee TM, Nayak AS, Glynn P (1990) Oxygenation-sensitive contrast in magnetic resonance image of rodent brain at high magnetic fields. Magn Reson Med 14(1):68–78. https://doi.org/10.1002/mrm.1910140108

Phelps EM (2000) The merging of biology and imaging into molecular imaging. J Nucl Med 41:661–681

Piper SK, Krueger A, Koch SP, Mehnert J, Habermehl C, Steinbrink J et al (2014) A wearable multi-channel fNIRS system for brain imaging in freely moving subjects. NeuroImage 85(1):64–71. https://doi.org/10.1016/j.neuroimage.2013.06.062

Poldrack RA, Mumford JA, Nichols TE (2011) Handbook of functional MRI data analysis. Cambridge University Press

Quaresima V, Ferrari M (2019) Functional Near-Infrared Spectroscopy (fNIRS) for assessing cerebral cortex function during human behavior in natural/social situations: a concise review. Organ Res Methods 22(1):46–68. https://doi.org/10.1177/1094428116658959

Raichle ME, Snyder AZ (2007) A default mode of brain function: a brief history of an evolving idea. NeuroImage 37(4):1083–1089. https://doi.org/10.1016/j.neuroimage.2007.02.041

Reichenbach JR, Venkatesan R, Schillinger DJ, Kido DK, Haacke EM (1997) Small vessels in the human brain: MR venography with deoxyhemoglobin as an intrinsic contrast agent. Radiology 204(1):272–277. https://doi.org/10.1148/radiology.204.1.9205259

Sabarudin A, Subramaniam C, Sun Z (2014) Cerebral CT angiography and CT perfusion in acute stroke detection : a systematic review of diagnostic value. 4(4):282–290. https://doi.org/10.3978/j.issn.2223-4292.2014.07.10

Sanchez B, Duran D, Martinez K, Viera W (2023) Advances of artificial intelligence in aeronautics. Athenea:34–42. https://doi.org/10.47460/athenea.v4i12.56

Sato H, Yahata N, Funane T, Takizawa R, Katura T, Atsumori H et al (2013) A NIRS-fMRI investigation of prefrontal cortex activity during a working memory task. NeuroImage 83:158–173. https://doi.org/10.1016/j.neuroimage.2013.06.043

Schmidt VF, Masthoff M, Czihal M, Cucuruz B, Häberle B, Brill R et al (2021) Imaging of peripheral vascular malformations – current concepts and future perspectives. Mol Cell Pediatr 8(1):19. https://doi.org/10.1186/s40348-021-00132-w

Schramm P, Schellinger PD, Klotz E, Kallenberg K, Fiebach JB, Ku S et al (2004) Comparison of perfusion computed tomography and computed tomography angiography source images with imaging in patients with acute stroke of less than 6 hours ' duration. 1652–1658. https://doi.org/10.1161/01.STR.0000131271.54098.22

Segebarth C, Décorps M (2010) Functional magnetic resonance imaging. In: Tomography. Wiley, pp 377–391. https://doi.org/10.1002/9780470611784.ch15

Singh V, Vignesh S, Neyaz Z, Phadke RV, Mehrotra A, Mishra P (2019) Detection and evaluation of intracranial aneurysms in the posterior fossa by multidetector computed tomography angiography – comparison with digital subtraction angiography. Asian J Neurosurg 14(2):491–498. https://doi.org/10.4103/ajns.AJNS_290_18

Soni N, Mehrotra A, Behari S, Kumar S, Gupta N (2017) Diffusion-tensor imaging and tractography application in pre-operative planning of intra-axial brain lesions. Cureus 9(10):3–7. https://doi.org/10.7759/cureus.1739

Spaide RF, Fujimoto JG, Waheed NK, Sadda SR, Staurenghi G (2018) Optical coherence tomography angiography. Prog Retin Eye Res 64:1–55. https://doi.org/10.1016/j.preteyeres.2017.11.003

Thomalla G, Glauche V, Koch MA, Beaulieu C, Weiller C, Röther J (2004) Diffusion tensor imaging detects early Wallerian degeneration of the pyramidal tract after ischemic stroke. NeuroImage 22(4): 1767–1774. https://doi.org/10.1016/j.neuroimage.2004.03.041

Uysal S (2023) Brain imaging: Jacqueline C. Junn and Suzan Uysal. In: Functional neuroanatomy and clinical neuroscience: Foundations for Understanding Disorders of Cognition and Behavior. Oxford University Press. https://doi.org/10.1093/oso/9780190943608.003.0029

van de Kreeke JA, Nguyen H-T, den Haan J, Konijnenberg E, Tomassen J, den Braber A et al (2019) Retinal layer thickness in preclinical Alzheimer's disease. Acta Ophthalmol 97(8): 798–804. https://doi.org/10.1111/aos.14121

van den Heuvel MP, Hulshoff Pol HE (2010) Exploring the brain network: a review on resting-state fMRI functional connectivity. Eur Neuropsychopharmacol 20(8):519–534. https://doi.org/10.1016/j.euroneuro.2010.03.008

Van Essen DC, Smith SM, Barch DM, Behrens TEJ, Yacoub E, Ugurbil K (2013) The WU-Minn human connectome project: an overview. NeuroImage 80:62–79. https://doi.org/10.1016/j.neuroimage.2013.05.041

Wolpaw J, Wolpaw EW (2012) Brain–computer interfaces: principles and practice. Oxford University Press. https://doi.org/10.1093/acprof:oso/9780195388855.001.0001

Xu B, Li F, Zhang Z, Xiao Q (2024) The use of optical coherence tomography (OCT) and OCT angiography in borderline personality disorder compared to health control subjects. CNS Neurosci Ther 30(3):1–9. https://doi.org/10.1111/cns.14699

Yeo BTT, Krienen FM, Sepulcre J, Sabuncu MR, Lashkari D, Hollinshead M et al (2011) The organization of the human cerebral cortex estimated by intrinsic functional connectivity. J Neurophysiol 106(3): 1125–1165. https://doi.org/10.1152/jn.00338.2011

Young GB (2000) The EEG in coma. J Clin Neurophysiol 17(5):473–485. https://doi.org/10.1097/00004691-200009000-00006

Zhang H, Schneider T, Wheeler-Kingshott CA, Alexander DC (2012) NODDI: practical in vivo neurite orientation dispersion and density imaging of the human brain. NeuroImage 61(4):1000–1016. https://doi.org/10.1016/j.neuroimage.2012.03.072

Zhang Z, Luo D, Rasim Y, Li Y, Meng G, Xu J, Wang C (2016) A vehicle active safety model: vehicle speed control based on driver vigilance detection using wearable EEG and sparse representation. Sensors 16(2). https://doi.org/10.3390/s16020242

Zhang J, Yu T, Wang M, Zhang Y, Li H, Chen H et al (2024) Clinical applications of functional near-infrared spectroscopy in the past decade: a bibliometric study. Appl Spectrosc Rev 59(7):908–934. https://doi.org/10.1080/05704928.2023.2268416

Neuroimaging: A Computational Lens on the Brain

4

Sankara Reddy Thamma, N. B. Prajwala, N. B. Pushpa, and Apurba Patra

Abstract

Advancement in neuroimaging has improved our understanding of the human brain by allowing nonintrusive visualization of its structure and function. As imaging technologies progress, the incorporation of computational methods varies from signal processing and machine learning to advanced statistical modeling and has become necessary in extracting meaningful information from complex brain data. This chapter explores neuroimaging from the perspective of computational analysis, highlighting how algorithms enhance image acquisition, noise reduction, segmentation, and interpretation. Further, it covers the role of computational tools in functional biomarker identification, connectivity mapping, and neurological disorder diagnosis. The combination of neuroscience and data science and computational neuroimaging provides deep understanding of brain mechanisms and enables the development of actual and predictive models of brain health.

Keywords

Neuroscience · Deep brain stimulation · Gene therapy · Brain-computer interfaces · Psychiatric therapies · Cognitive enhancement · Translational neuroscience · Neurotechnology governance

4.1 Introduction

Biomedical visualization is a combination of science, arts, and technology to create visual representation of biological data and biological process useful for research, communication, and education. It uses several technologies like image processing, graphics, and other media to make complex biological process simpler and more understandable, which can be further used for patients' education, surgical training, and exploring biological systems. It acts as a bridge between scientific knowledge and visual understanding.

Through detailed analysis and a strong foundation in present research, we aim to present a roadmap for the future of neurovisual intelligence, a field where machines help us not just observe but truly understand the human brain.

Neurovisualization is a combination of multiple fields of neuroscience, computational

S. R. Thamma
Technical Delivery Master II, Deloitte Consulting LLP, Charlotte, NC, USA

N. B. Prajwala (✉)
Associate Software Engineer, The Boeing Company, Bengaluru, Karnataka, India

N. B. Pushpa
Department of Anatomy, JSS Medical College, JSSAHER, Mysore, Karnataka, India

A. Patra
Department of Anatomy, All India Institute of Medical Sciences, Bathinda, Punjab, India

K. S. Ravi et al. (eds.), *Visualization in Neuroanatomical Sciences*, Biomedical Visualization 11,
https://doi.org/10.1007/978-3-032-07380-8_4

modeling, and artificial intelligence, which offers high-resolution visual analytics to analyze complex brain functions. Since 2021, quick growth in multimodal neuroimaging, deep learning, and graph-theoretic approaches have changed our view of brain activity across spatial and temporal domains (Brown et al. 2022; Li and Zhang 2022). Key points of neuroimaging that varies from CT scans and X-ray to diffusion tensor imaging (DTI), functional MRI (fMRI), and PET/MR hybrids are examined alongside their computational foundations (Khan et al. 2022). Critical mathematical and algorithmic models like Hodgkin–Huxley simulations, graph neural networks (GNNs), and transformer-based data fusion techniques (Singh et al. 2023; Mehta et al. 2022) are also explored. Techniques like Grad-CAM and SHAP aid in explainability, while ethical concerns related to privacy, bias, and neurorights are addressed through emerging regulatory frameworks (Martínez-Pérez et al. 2022). The focus is on segmentation workflows, preprocessing standards, and modeling pipelines driven by practices like BIDS and containerized reproducibility (Wang and Ahmed 2023). Neurovisualization is extended to cloud platforms, patient-specific digital twins, and immersive VR tools (Kaur et al. 2024). By integrating computational theory, clinical relevance, and ethical foresight, this chapter gives an overall insight on neurovisualization.

4.1.1 Magnetic Resonance Imaging and Its Expansions

MRI changed neuroimaging because it gives great soft tissue detail without using harmful radiation. Developed in the late 1970s, MRI exploits the magnetic properties of hydrogen atoms to generate detailed images. Pulse sequences such as T1-weighted, T2-weighted, FLAIR, and DWI enable clinicians to visualize various brain tissues and pathologies, including edema, gliosis, and demyelination (Li and Zhang 2022).

Advances such as susceptibility-weighted imaging (SWI), quantitative susceptibility mapping (QSM), and magnetization transfer imaging (MTI) have refined tissue specificity. With parallel imaging and compressed sensing, acquisition speeds have increased, while motion correction techniques reduce artifacts. MRI continues to serve as the anatomical reference for aligning functional and diffusion data, making it central to both clinical and research workflows.

4.1.2 Functional MRI and the BOLD Signal Revolution

The introduction of BOLD (blood-oxygen-level-dependent) contrast in the early 1990s, pioneered by Ogawa and colleagues, marked a turning point in functional brain imaging. By measuring changes in deoxyhemoglobin concentration, fMRI enabled researchers to map neural activity indirectly but noninvasively. This innovation transformed cognitive neuroscience, offering millimeter-scale spatial and seconds-scale temporal resolution.

Contemporary fMRI research employs varied paradigms—block design, event-related, and naturalistic—and preprocessing standards have matured through tools like fMRIPrep. Analytical methods now include GLM, MVPA, and dynamic causal modeling.

4.1.3 High-Field MRI, MEG, and fNIRS

Ultrahigh-field MRI (>7 Tesla) enables submillimeter imaging and laminar resolution, opening new avenues in cognitive neuroscience. However, its clinical adoption remains limited due to safety constraints and artifact sensitivity. Magnetoencephalography (MEG), by contrast, offers millisecond temporal resolution, making it ideal for tracking fast neural oscillations. When combined with source localization techniques and fMRI, MEG provides rich spatiotemporal mappings (Zhou et al. 2023).

Functional near-infrared spectroscopy (fNIRS) presents a portable, cost-effective alternative for populations not suited for MRI, such as infants or bed-bound patients. Wearable fNIRS and MEG systems are emerging for ambulatory monitoring and real-time brain–computer interfaces (BCIs) (Kaur et al. 2024).

From structural to functional and molecular domains, the field of neurovisualization has evolved in tandem with imaging and computational innovations. These developments have laid the groundwork for today's integrated, multi-scale neurovisual platforms.

4.1.4 Molecular Imaging with PET and SPECT

Positron emission tomography (PET) and single-photon emission computed tomography (SPECT) enable the visualization of metabolic and molecular brain processes. The most commonly used tracer, ^18F-FDG, highlights regions of altered glucose metabolism—providing diagnostic value in Alzheimer's, epilepsy, and various forms of dementia. Other tracers, such as ^11C-raclopride and ^11C-PiB, target dopaminergic and amyloid systems, respectively.

Recent innovations include kinetic modeling methods (e.g., Logan plots) and partial volume corrections, improving quantification. PET/MRI hybrid systems are becoming more prevalent, offering concurrent anatomical and molecular data acquisition (Wang and Ahmed 2023). Figure 4.1 represents the evolution of neurovisual timeline; PET-derived features are increasingly integrated into machine learning models for early disease detection and therapy monitoring.

4.2 Image Processing Insight for Neurovisualization

Biomedical visualization is a rapidly growing field which integrates various techniques like artificial intelligence (AI), machine learning (ML), and advanced image processing to improve our understanding of complex biological system. This is very important for transferring data into useful visual representation which can be used in diagnosis, medical education, and treatment planning.

4.2.1 Components of Image Processing and Analysis in Neural Biomedical Visualization

Before preprocessing, a large dataset of brain MRI images has to be collected which is crucial process, and these images are then analyzed to get desired results. These raw images are not suitable for direct analysis since it contains various disturbances like patient movement during imaging which would result in inconsistent results. To get accurate results, the dataset must be prepared properly, and this involves several important steps. Figure 4.2 represents the components and workflow of image processing and analysis in neural biomedical visualization:

4.2.1.1 Cleaning the Images

The MRI images may contain several artifacts and noise which are raised from factors like pat-

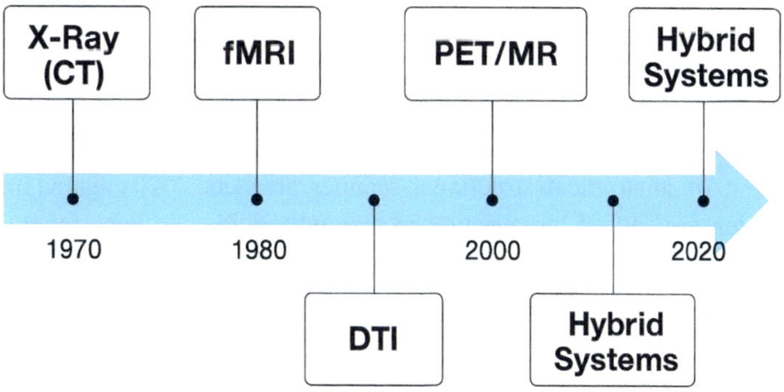

Fig. 4.1 Timeline of neurovisual evolution

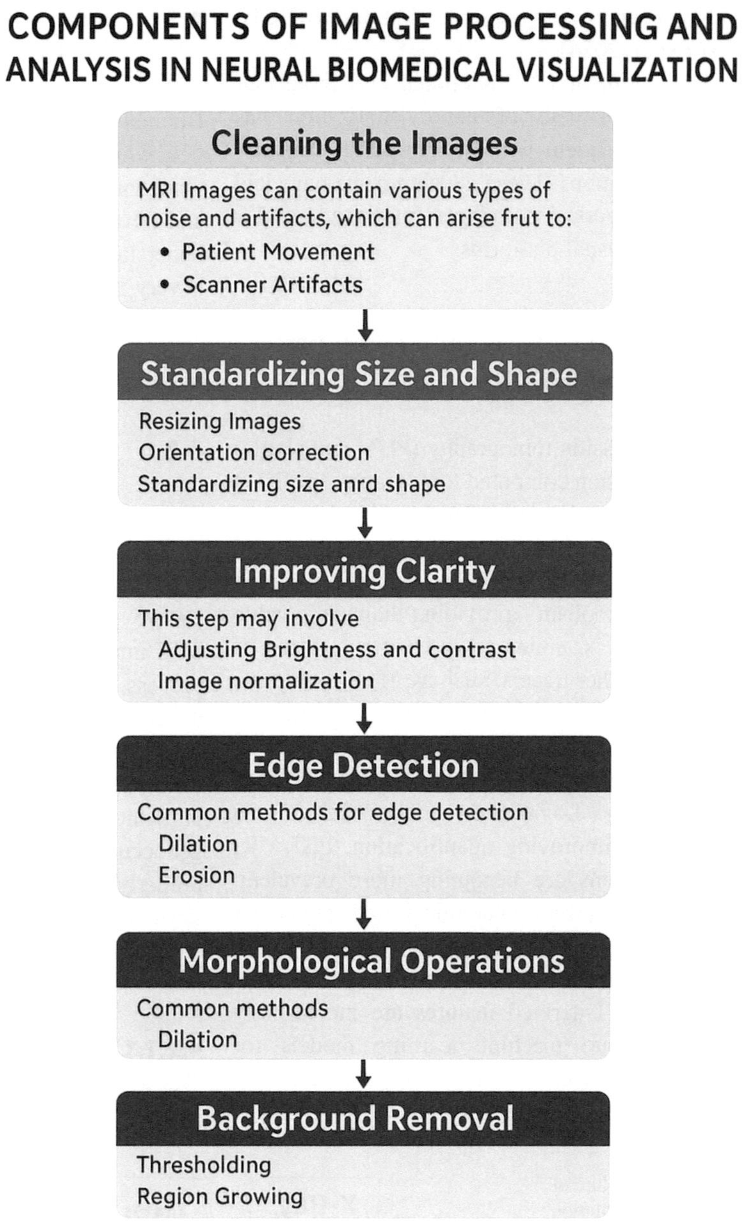

Fig. 4.2 Components of image processing and analysis in neural biomedical visualization

ent movement which leads to blurred or misalignment of anatomical structures, scanner artifacts which are caused by variation in magnetic fields or technical issues with the scanner. These issues can be addressed while preserving the features of the images using denoising algorithms like Gaussian filtering or wavelet transforms.

4.2.1.2 Standardizing Size and Shape

All cleaned images need to be in standard size and shapes for further processing; this is required for effective comparison and analysis of images when using machine learning algorithms. MRI images significantly vary in dimensions, which makes analysis complicated. Standardization involves resizing images where all images are resized to a common dimension to ensure uniformity across the dataset. Orientation correction is necessary because images in the dataset may have different orientations due to variations in the patient's position during scanning. Aligning the images ensures anatomical consistency across the dataset.

4.2.1.3 Improving Clarity

Improving MRI image clarity is essential to ensure that important features are easily identifiable. This may involve steps like adjusting brightness and contrast, which helps highlight key brain structures and makes them more visible for analysis. Image normalization is also important, as variations in brightness and intensity can affect the accuracy of analysis. Therefore, intensity values should be normalized across the dataset, especially when comparing images from different patients or scan sessions.

4.2.1.4 Edge Detection

Edge detection helps in identifying the boundaries of structures within the MRI images. This is required for isolating regions such as tumors or other abnormalities. A common edge detection method is Canny edge detection; this algorithm uses multistage process to identify a wide range of edges in images. It is widely used in medical image processing and known for its accuracy. Sobel edge detection: this method calculates image intensity gradient to highlight regions of high spatial frequency, effectively identifying edges. Overall edge detection improves the visibility of abnormalities and anatomical structures.

4.2.1.5 Morphological Operations

Morphological operations are very useful while processing images based on their shapes. It significantly improves the quality of segmented images. Common morphological operations are dilation; here pixels get added to the boundaries of the objects in an image, which in turn closes the small gaps and connects missing points in the structure. In erosion, here pixels get removed from the boundaries of objects helping to eliminate small noise and refine the shapes of larger structures.

4.2.1.6 Background Removal

This step of preprocessing helps to separate brain structure from surrounding noise and irrelevant data. This process can involve thresholding, where the image is segmented by setting a specific intensity threshold, which allows for the separation of brain tissue from the background. In region growing, this starts with seed points and grows regions based on similar pixel intensity values, which in turn isolates the brain from the background, which helps to focus on relevant structure.

4.2.1.7 Shape Identification

Shape identification helps in recognizing and categorizing different structures and shapes like tumor and abnormalities within brain, and this may involve contour detection; this algorithm is used to recognize the outlines of shape within the image, allowing to recognize specific structures. In feature extraction, this step involves analyzing the identified shapes to extract relevant features like size, shape, and volume.

4.3 Graph-Theoretic Representations of the Connectome

Key graph metrics consist of:

- **Degree/node strength:** Shows how central a node is in a network
- **Path length and efficiency:** Show how easily information flows through the network
- **Clustering coefficient and modularity**: Capture local segregation and community structures
- **Rich-club coefficient**: Highlights highly interconnected hubs critical for cognitive resilience

Graph neural networks (GNNs) are based on graph theory and let you learn by looking at the attributes of nodes and edges. These models support tasks such as disease classification and biomarker identification using richly annotated connectivity matrices (Zhang et al. 2023).

Variability across individuals, scanner protocols, and thresholding methods pose challenges in graph construction. Solutions such as individualized parcellations, ensemble graph modeling, and bootstrapped thresholding aim to enhance robustness and generalizability (Sharma et al. 2022).

4.3.1 Data-Driven Statistical and AI Models

While biophysical and graph-theoretic models prioritize interpretability, data-driven approaches focus on scalability and automation—both critical for modern neurovisual pipelines that deal with massive, high-dimensional datasets.

4.3.2 Classical Approaches

Machine learning methods such as SVMs, random forests, and logistic regression are widely used in neuroimaging. These rely on engineered features—volume measurements, shape descriptors, and radiomic patterns—and perform well on structured classification tasks (Martínez-García et al. 2021).

4.3.3 Preprocessing and Quality Control

Preprocessing is primary step to any neuroimaging analysis—it mitigates artifacts, standardizes data across the subjects, and prepares input for stable downstream modeling. For structural MRI, key steps include **skull stripping** (e.g., using BET or HD-BET), **bias field correction** (commonly with N4ITK), and **intensity normalization**.

In functional MRI workflows, preprocessing pipelines typically include:

- **Slice timing correction**
- **Motion correction and realignment** (e.g., ANTs, MCFLIRT)
- **Spatial normalization** to standard templates like MNI
- **Temporal filtering** to remove physiological noise

Physiological confounds—such as cardiac and respiratory signals—are regressed out using methods like RETROICOR and CompCor (Behzadi et al. 2007). Tools such as **MRIQC** assist in automated quality assessment, flagging outlier volumes using metrics like framewise displacement (FD) and DVARS (Power et al. 2012).

For diffusion imaging, **QSIPrep** streamlines denoising, eddy-current correction, and susceptibility distortion correction. The use of containerized tools (Docker, Singularity) ensures that preprocessing environments remain consistent across systems, enhancing reproducibility and auditability (Esteban et al. 2019).

4.4 Segmentation and Registration

Accurate segmentation and image registration are pivotal for anatomical and functional alignment. Traditional tools like **FreeSurfer** rely on probabilistic atlases and Bayesian classification to segment brain structures. Deep learning models such as **FastSurfer** and **nnU-Net** now surpass classical methods in both accuracy and speed, achieving high Dice scores across various imaging modalities (Isensee et al. 2021).

Registration techniques vary based on the analysis goal:

- **Rigid and affine transformations** provide coarse alignment.
- **Nonlinear registration** (e.g., ANTs SyN) offers fine-grained warping for cross-subject comparisons.
- **Boundary-based registration (BBR)** is used for aligning functional data to structural volumes.
- **Multimodal registration** aligns data across modalities (e.g., T1-weighted MRI and DTI) by optimizing multichannel similarity metrics.

Longitudinal studies use tailored alignment techniques that preserve within-subject structural consistency over time, critical for studies tracking disease progression.

4.4.1 Feature Extraction and Dimensionality Reduction

The extraction of meaningful features from raw images is essential for downstream modeling and statistical inference. From structural scans, features like **cortical thickness**, **gray matter volume**, and **surface curvature** are commonly derived. Diffusion MRI offers indices such as **FA**, **MD**, and **fiber orientation distributions**, while tractography provides connectomic data in the form of **streamline counts** and **adjacency matrices**.

Functional imaging generates rich time-series data, from which researchers compute:

- **ALFF** (amplitude of low-frequency fluctuations)
- **ReHo** (regional homogeneity)
- **Dynamic functional connectivity** measures

These are often supplemented with **graph-theoretic features** that characterize brain networks.

For high-dimensional data, dimensionality reduction is essential. Techniques such as **PCA**, **t-SNE**, and **UMAP** aid in noise reduction and visualization. Advanced nonlinear embeddings like **diffusion maps** or **PHATE** preserve topological structure, enabling better separation of cognitive states or disease phenotypes (Moon et al. 2022). Figure 4.3 explains the steps involved in neurovisualization pipeline.

4.4.2 Modeling Workflows and Integration

Once features are extracted, the pipeline moves into the modeling phase. Here, machine learning and deep learning algorithms transform numerical descriptors into diagnostic or predictive outcomes.

Classical workflows use models such as:

- **Support vector machines**
- **Logistic regression**
- **Random forests**

These are often evaluated with cross-validation and tuned via grid or Bayesian search.

AutoML solutions, such as **AutoPrognosis**, automate hyperparameter tuning, performance evaluation, and even interpretability visualization, accelerating translational impact (Lee et al. 2023).

4.4.3 Containerized and Cloud-Optimized Workflows

Scalable deployment is crucial for large-scale neuroimaging studies, especially those involving consortium-based research or clinical integration. **Containerization technologies** like Docker and Singularity encapsulate the runtime environment, ensuring consistent execution across platforms.

Workflow orchestration tools like **Nextflow** and **Snakemake** automate complex multistage pipelines, enabling distributed processing on HPC clusters or cloud platforms like **AWS**, **GCP**, and **Azure**.

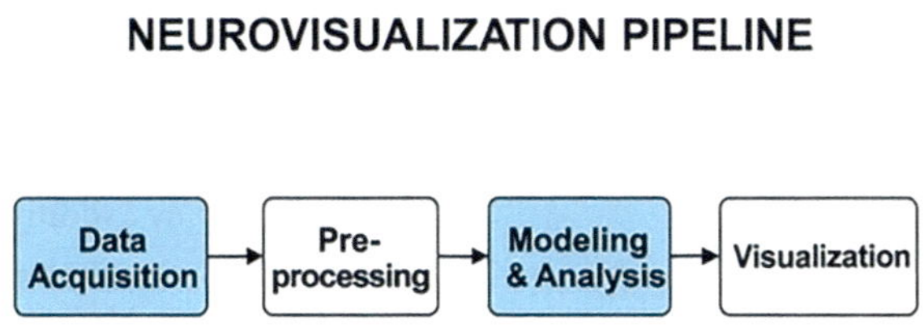

Fig. 4.3 Neurovisualization pipeline

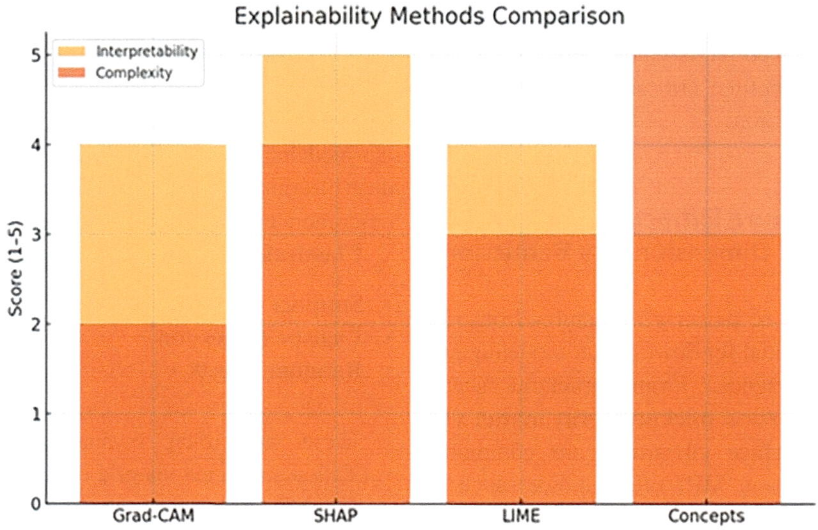

Fig. 4.4 Explainability method comparison

Emerging ecosystems such as **Neurodesk** and **Flywheel.io** offer pre-configured environments and intuitive dashboards, lowering the barrier for neuroimaging analysis. Meanwhile, **OpenNeuro** promotes open science by hosting BIDS-compliant datasets and cloud-executable pipelines.

To manage resource demands and costs, researchers are adopting:

- **GPU acceleration** for deep learning workflows
- **Preemptible instances** to reduce cloud runtime costs
- **Edge computing** and **serverless architectures** for real-time applications

As neurovisualization moves beyond static workstations to global, scalable ecosystems, these cloud-native strategies represent the next phase of innovation in brain data analytics.

4.5 Interpretability and Explainable AI in Neurovisualization

The increasing reliance on deep learning models in neuroimaging has yielded impressive gains in accuracy and automation. However, this progress has come with a critical trade-off: reduced transparency. In clinical settings, where decisions can impact patient outcomes, it is no longer sufficient for a model to perform well—it must also be interpretable. Figure 4.4 represents the interpretability and complexity.

4.5.1 Perturbation-Based Attribution Methods: Insights from LIME and SHAP

A subset of XAI approaches evaluates model interpretability by introducing controlled changes to input features and observing the resultant variations in output.

Lime (local interpretable model-agnostic explanations) builds simple surrogate models around individual predictions. By randomly masking input segments—such as brain voxels or regions—LIME identifies which areas most affect the final classification (Ribeiro et al. 2016).

SHAP (SHapley Additive exPlanations) goes a step further by leveraging concepts from cooperative game theory. It computes the marginal contribution of each feature across all possible feature subsets, resulting in more theoretically grounded and consistent attributions (Lundberg and Lee 2017). In neuroimaging, SHAP has been applied to:

- Rank the importance of brain regions in disease diagnosis
- Detect outlier predictions
- Improve cohort-level transparency

These methods are particularly valuable when explaining **tabular fusion models** that integrate imaging data with clinical or genetic features.

However, their utility in voxel-wise models remains constrained by computational demands and the challenges of defining meaningful perturbations in spatially correlated inputs. Scalable variants like **Kernel SHAP** and **Deep SHAP** are helping address these concerns.

4.5.2 Model-Agnostic Strategies: Surrogates, Counterfactuals, and Concepts

Beyond visual overlays and perturbation analyses, newer interpretability tools aim to offer **human-readable logic** behind complex models.

Surrogate modeling involves training transparent, rule-based models—such as decision trees or logistic regressions—on the predictions of a black-box model. While these surrogates sacrifice granularity, they provide a high-level view of how inputs relate to outputs, often used for regulatory auditing or clinician education (Molnar 2022).

Counterfactual explanations go further by identifying minimal, plausible changes to an input that would flip the model's decision.

Emerging Trends in Interpretability

- Human-in-the-loop (HITL) systems refer to interactive platforms that empower clinicians to evaluate, modify, or override algorithmic decisions. By integrating expert feedback into the model refinement process, these systems not only enhance user trust but also contribute to improved generalization performance through iterative retraining cycles.
- **Multimodal explanation fusion**: Integrating saliency maps from MRI with SHAP values from clinical tables offers a holistic interpretive picture.
- **Fairness-aware interpretability**: Recent studies incorporate demographic overlays to identify biases in model attention—e.g., focusing disproportionately on features correlated with ethnicity or age.

4.6 Immersive Visualization: VR/AR Ecosystems

Virtual reality (VR) and augmented reality (AR) have transformed neurovisualization. These tools facilitate spatially immersive experiences, enabling users to effortlessly engage with intricate 3D structures, including fiber tracts, dynamic activation patterns, and complex cortices. Several trends are expected to influence neurovisualization in the forthcoming years:

- Neuroscience has adapted gaming technologies like Unity3D and Unreal Engine, which are often used to make games. They let you see brain volumes, segmented regions, and statistical data overlays in real time. You can use NIfTI or DICOM files to put these together.
- **NeuroCave, OpenVR-Neuro, and BrainSpace VR**: Pre-built VR environments that enable manipulation of brain networks in 3D space, interactive querying of nodes, and immersive hypothesis testing.
- **Microsoft HoloLens**: In AR settings, platforms allow overlaying brain models onto surgical fields or physical mannequins. This is particularly useful in **neurosurgical planning**, enhancing spatial reasoning and reducing intraoperative uncertainty (Zhao et al. 2023).

Recent developments pair immersive systems with real-time inference: for instance, brain–computer interfaces (BCIs) feeding live EEG or fNIRS signals into VR platforms to drive neurofeedback, motor rehabilitation, or cognitive training protocols. Evaluations now consider not just usability but **cognitive load**, **spatial**

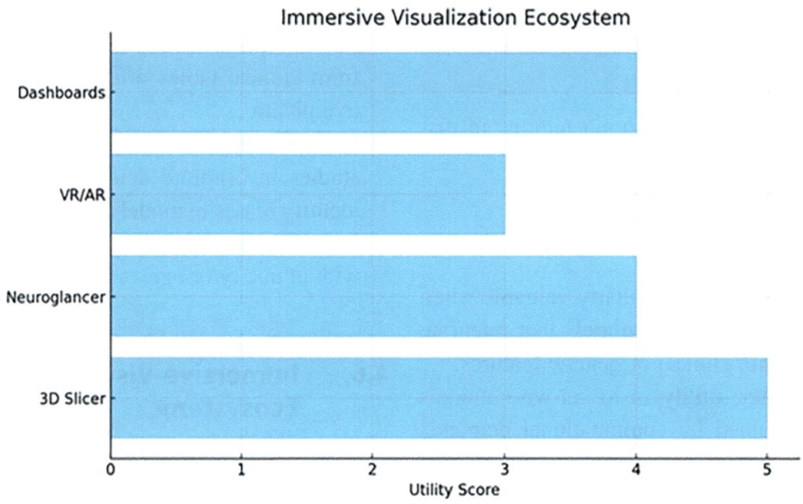

Fig. 4.5 Immersive visualization ecosystem

accuracy, and **therapeutic efficacy** (Fairclough et al. 2021). Figure 4.5 indicates immersive visualization ecosystem.

Emerging Interfaces and Novel Modalities

- **Biosignal overlays**: Platforms now support real-time visualization of physiological signals—such as heart rate, galvanic skin response, or eye-tracking—alongside neural data. This integration helps link cognitive states with autonomic responses, enriching behavioral and emotional interpretations.
- **3D printing:** Neuroscience often uses game engines like Unity3D and Unreal Engine. These engines let researchers create real-time visuals of brain scans, segmented regions, and statistical data using NIfTI or DICOM files. Plus, 3D printing can turn detailed imaging data into physical models for designing prosthetics, training surgeons, or teaching patients. Exporting STL files allows for a direct way to replicate anatomical structures.
- **AI-guided hand shaking:** Few visualization systems have AI agents that adaptively highlight abnormalities, suggest annotations, or respond to natural language questions—enhancing human–AI collaboration in exploratory analysis.

In conclusion, neurovisualization is rapidly progressing beyond mere image viewing. Collaborative dashboards and simulations are transforming how we comprehend, share, and apply brain data. As these technologies become more interactive, intelligent, and tailored, they will become indispensable in neuroscience and clinical neurotechnology.

4.7 Epilepsy Focus Localization

For patients with drug-resistant epilepsy, accurate identification of the **epileptogenic zone** is critical for surgical success. Neurovisualization plays a central role in presurgical planning, often combining EEG with fMRI to localize seizure foci. These **EEG-fMRI fusion** techniques detect interictal discharges alongside hemodynamic changes, often achieving concordance with invasive sEEG recordings (Huang et al. 2022).

Machine learning models—particularly **temporal convolutional networks** with attention mechanisms—help classify seizure types and localize abnormal activity. Functional connectivity analysis reveals disrupted hub dynamics in areas like the **posterior cingulate** and **thalamus**, aiding in differential diagnosis.

White matter tractography, using diffusion data, maps altered connectivity patterns and supports **resection planning**. Emerging platforms now integrate **VR-based simulations** of surgical pathways, allowing neurosurgeons to visualize trajectories while preserving eloquent cortex.

4.7.1 Psychiatric Disorder Biomarkers

Psychiatric conditions such as schizophrenia, bipolar disorder, and major depressive disorder often lack definitive biological markers. Neurovisualization is increasingly used to **subtype patients**, predict treatment response, and support early intervention strategies.

Studies employing **structural MRI**, **diffusion imaging**, and **resting-state fMRI** have revealed alterations in global efficiency and modularity within the **default mode**, **executive control**, and **salience networks**.

4.7.2 Brain–Computer Interfaces and Neurorehabilitation

Neurovisualization is advancing quickly, changing how we understand and use brain data through interactive tools like collaborative dashboards and simulations. These tools are becoming key in neuroscience and clinical neurotechnology as they get smarter and more user-friendly.

Brain–computer interfaces (BCIs), a prime example, translate brain signals into actions, aiding communication and movement for those with motor issues. Real-time neurofeedback, using fMRI or EEG, helps in rehabilitation, focus training, and PTSD treatment. Deep learning finds patterns in brain activity to control devices or navigate interfaces. Tracking functional connectivity helps guide training by observing brain changes.

In clinical settings, hybrid systems that combine **EEG with fNIRS** allow for **wearable neurofeedback** platforms, extending therapy to home environments (Liu et al. 2023).

Immersive VR combined with neurovisual analytics has shown promise in guiding **motor imagery exercises**, providing real-time feedback, and enhancing user engagement. Explainable AI modules monitor signal integrity and model confidence, improving safety and personalization.

4.7.3 Neuromarketing and Ethical Considerations

Outside the clinical sphere, neurovisualization has entered the realm of **consumer neuroscience**, where tools like fMRI and EEG are used to measure **emotional engagement**, **brand recall**, and **decision-making under cognitive load**. Activation in regions such as the **nucleus accumbens** and **orbitofrontal cortex** is often correlated with reward anticipation and purchasing intent.

Predictive models have demonstrated superiority over traditional surveys in forecasting consumer behavior. However, this capability introduces ethical concerns, particularly regarding **cognitive privacy** and **subconscious manipulation**.

In response, some systems now incorporate explainability mechanisms that allow consumers to understand how neural features inform behavioral predictions. These efforts align with emerging discussions around **neurorights** and the **responsible use of commercial neurotechnology** (Ienca and Andorno 2023).

From diagnosing Alzheimer's to decoding consumer preference, the applications of neurovisualization span a wide spectrum. What unites them is a common reliance on high-quality imaging, robust modeling, and interpretability tools that enable users to move from data to decisions. As case studies multiply, they not only validate the science but also shape the ethical and operational norms for this rapidly advancing field.

4.8 Challenges and Future Directions

Neurovisualization has advanced significantly; yet, substantial challenges remain regarding technology, ethics, available resources, and regulations. As we expand, the enhancement of picture data and the increased utilization of AI for decision-making necessitate a cautious evolution of this field, grounded on sound scientific principles and a consideration of societal implications.

Five critical domains necessitate enhancement in neurovisualization: standardizing diverse data types, interpreting the significance of results rather than merely obtaining accurate answers, safeguarding data privacy and protecting neurorights, addressing the constraints of existing systems, and exploring models that will influence future developments.

4.8.1 Data Heterogeneity and Standardization Barriers

Neuroimaging data often presents considerable variations. Variability arises from different scanners, acquisition protocols, population demographics, and preprocessing techniques. This variability poses a major obstacle for generalizable AI models, especially those trained on narrow or single-center datasets.

Initiatives like the **Brain Imaging Data Structure (BIDS)** aim to create standardized directory formats, naming conventions, and metadata schemas (Gorgolewski et al. 2016). Neuroimaging data often varies a great deal. Although many people use it, consistent application is lacking, particularly with older or protected data collections.

It is well known that neuroimaging results can vary a lot. Following standards can be inconsistent, especially with older or private data collections, even though they are often used. To resolve this, techniques like ComBat harmonization, domain adaptation, and adversarial networks are kept to make the discrepancies between sites less while aiming to keep the biological information safe (Fortin et al. 2018). Future workflows must support dynamic schema validation, semiautomated data cleaning, and cloud-native pipelines that enforce standardization by design.

4.8.2 The Interpretability–Accuracy Trade-Off

Deep learning models deliver state-of-the-art accuracy across numerous neuroimaging tasks. However, their opaque decision-making limits clinical trust and regulatory acceptance. In medical contexts, where **explainability is not optional**, this trade-off remains a core tension.

Efforts to bridge the gap include:

- Embedding **interpretable constraints** (e.g., attention regularization)
- Developing **prototype-based learning** systems
- Incorporating **symbolic rules** into neural architectures

Neuro-symbolic AI, which combines rule-based reasoning with data-driven learning, shows promise in producing outputs that are both accurate and traceable (Wang et al. 2022). As explainability becomes a regulatory requirement—especially under frameworks like the EU AI Act—such hybrid models are likely to become essential.

4.8.3 Privacy, Consent, and Neurorights

Neuroimaging data offers deeply personal insights, capable of revealing emotional states, intentions, and even preferences. This makes neurodata among the most sensitive forms of biometric information, raising urgent concerns around **privacy**, **consent**, and **autonomy**.

The concept of **neurorights** has emerged to address this gap. These include the rights to:

- Mental privacy
- Cognitive liberty
- Psychological continuity
- Protection from algorithmic manipulation

In 2021, Chile became the first country to enshrine neurorights into its constitution (Ienca and Andorno 2023), setting a precedent for global regulatory conversations.

While traditional anonymization techniques like skull stripping and facial masking are standard, recent studies show that **GANs can reconstruct identifiable faces from MRI scans**, undermining privacy. Solutions such as **differential privacy**, **synthetic data generation**, and **federated learning** are being explored—but their clinical validity and legal frameworks remain under development.

Consent models are also evolving. Instead of static, one-time approvals, **dynamic consent frameworks** allow participants to grant tiered, revocable permissions. Blockchain-based ledgers have been piloted to provide transparent audit trails for data usage (Gupta et al. 2022).

4.8.4 Infrastructure Gaps and Deployment Bottlenecks

Many neurovisualization pipelines rely on **high-performance computing (HPC)**, **GPU clusters**, and advanced cloud services. This creates a digital divide—institutions in low-resource settings may lack access to the infrastructure necessary to run modern AI workflows.

Even within resource-rich environments, **integration into clinical systems** like PACS or EHRs is rarely straightforward. Issues around data governance, system compatibility, and staff training often delay adoption.

Emerging technologies offer partial solutions:

- **Containerized workflows** (Docker, Singularity) ensure reproducibility.
- **Workflow orchestration engines** (e.g., Nextflow, Snakemake) enable scalability.
- **Infrastructure-as-code** tools (Terraform, Pulumi) automate cloud deployment.

The next frontier includes **serverless neuroimaging** (e.g., AWS Lambda functions for on-demand preprocessing), **microservices for modular tool deployment**, and **edge AI** on portable MRI/EEG devices for real-time rural diagnostics.

Training, open-source access, and cross-sector partnerships will be key to democratizing these capabilities.

4.8.5 The Future: Paradigms Shaping the Next Decade

Several transformative trends are set to redefine the future of neurovisualization:

- **Federated learning**: Enables model training across decentralized datasets without moving sensitive data—enhancing both privacy and scalability (Sheller et al. 2020).
- **Neuro-symbolic integration**: Combines statistical inference with logic-based reasoning, offering interpretable models that comply with clinical guidelines and policy constraints.
- **Quantum neuroinformatics**: Though still emerging, quantum computing could accelerate similarity searches, kernel methods, and network analyses in neuroimaging (Smith et al. 2024).
- **Internet of Brain Things (IoBT)**: As wearable devices become mainstream, neurovisual systems will need to support continuous, secure, and interoperable data streaming for real-time analysis.
- **Regulatory harmonization**: Global organizations like WHO, UNESCO, and IEEE are actively developing ethical codes and compliance standards for AI in neurotechnologies. Convergence of GDPR, HIPAA, and emerging neurorights will shape how brain data is stored, shared, and governed.

While the science behind neurovisualization continues to advance rapidly, the real test lies in ensuring its responsible deployment. This work includes creating tools that are understandable, fair, and easy to get to, all based on strong systems, clear rules, and careful thought about what is right and wrong. The next 10 years should bring not only better algorithms but also better ways to manage and more chances for everyone to be part of new ideas.

4.9 Ethical, Legal, and Societal Implications (ELSI)

Infrastructure, clear regulations, and ethical planning support understandable, fair, and accessible tools. Smarter algorithms, better governance, and more inclusive ideas are expected in the future years.

Brain-scanning technology is entering daily life, posing ethical, legal, and social concerns. Brain signals can reveal mental states, measure moods, and forecast conduct, raising challenges for philosophy, law, human rights, and global policy, not just data science and medicine.

The key ethical issues include mental privacy, permission, bias, business use, and worldwide rules. These difficulties must be addressed for brain-scanning tech to be useful, inclusive, and dependable.

4.9.1 Mental Privacy and Cognitive Autonomy

Brain signals differ from other biometric data because they can show internal experiences like emotions, memories, and intentions that people might not say out loud. This makes studying the brain more intrusive and raises ethical issues. Because of these concerns, neurorights are being developed to protect a person's inner thoughts and identity.

These include:

- The right to **mental privacy** (centers on an individual's ability to control access to their brain data)
- The right to **cognitive liberty** (freedom of thought)
- The right to **psychological continuity** (preserving identity over time)
- The right to **freedom from algorithmic manipulation**

Countries like Chile have taken pioneering steps by embedding neurorights into constitutional frameworks (Ienca and Andorno 2023). However, broader legal recognition remains limited. There is a pressing need for international treaties or digital rights charters that explicitly protect mental data from misuse in surveillance, law enforcement, or political targeting.

4.9.2 Informed Consent and Data Reuse

Traditional consent models are poorly suited to the evolving landscape of neuro-AI. Participants often give permission for a specific study, unaware that their brain data may later be reused for unrelated applications—including commercial or behavioral profiling.

Dynamic consent frameworks offer a solution. These models allow participants to:

- Grant different levels of access for different use cases
- Revoke consent at any time
- Monitor who has used their data and for what purpose

Platforms like OpenNeuro and EBRAINS are experimenting with **blockchain-led consent tracking**, ensuring transparency and accountability (Gupta et al. 2022). These mechanisms not only empower individuals but also build public trust in neurotechnology.

4.9.3 Bias, Fairness, and Demographic Representation

Bias in neuro-AI systems is not hypothetical—it is measurable and consequential. Many large neuroimaging datasets are skewed toward white, Western, and neurotypical populations. As a result, models trained on these datasets often perform poorly when applied to underrepresented groups.

For example:

- Age-related brain changes vary across populations.
- Psychiatric symptom profiles differ across cultures.

- Structural brain volumes are affected by socio-economic and lifestyle factors.

Addressing these gaps requires both **technical interventions** (e.g., fairness-aware algorithms, reweighting techniques, domain adaptation) and **structural solutions**, such as building more inclusive datasets. Initiatives like the **NIH All of Us Program** and the **UK Biobank's diversity expansion** are critical steps in this direction.

Journals and funding bodies are now beginning to mandate **bias audits** and **demographic reporting** as prerequisites for publication and grant approval—a trend likely to intensify.

4.9.4 Commercialization and Cognitive Exploitation

Private companies are increasingly using neuroimaging tools to probe consumer behavior, optimize advertising, and develop brain-based authentication systems. While many of these applications claim to enhance user experience, they also carry risks of **subliminal influence**, **behavioral nudging**, and **psychological manipulation**.

A particularly troubling frontier is **neuromarketing**, where brain activity is analyzed to predict purchase intent—even when users are unaware of the stimuli being measured. In such contexts, the line between analytics and exploitation becomes dangerously thin.

To counteract these risks, ethical guidelines are being proposed that include:

- Clear disclosure of neurotech use in consumer settings
- Restrictions on predictive analytics without consent
- Bans on targeting vulnerable populations (e.g., minors, people with cognitive impairments)

Frameworks like the **OECD AI Guidelines** and **IEEE Brain Initiative** are starting to address these issues, but enforceable regulation remains inconsistent across regions.

4.9.5 Global Governance and Regulatory Harmonization

Neurovisualization intersects multiple domains—healthcare, data protection, consumer rights, and international law, yet global governance remains fragmented. While the **GDPR** in Europe provides strong protections for identifiable health data, it does not explicitly cover **inferred mental states**. In the USA, **HIPAA** governs medical records but often excludes research imaging data or direct-to-consumer neurotools.

To close these gaps, international organizations like **UNESCO**, **WHO**, and **OECD** are calling for harmonized legal frameworks. These efforts focus on:

- Establishing global standards for **data interoperability**
- Defining what constitutes **neural identity**
- Creating ethics review boards specialized in **neuro-AI technologies**

Ultimately, governance must be **multidisciplinary**, combining input from neuroscientists, ethicists, legal scholars, patient advocates, and AI developers. A "wait-and-see" approach is no longer tenable—the time to build proactive, agile, and inclusive oversight structures is now.

Summary

As neurovisualization tools become more powerful and pervasive, ethical stewardship becomes as important as scientific accuracy. Looking out for mental freedom, treating all people the same, and keeping watch over how brain data is used to make money is not just a good idea—it is key to making sure this science grows in a good way. If we want brain data to do good things, we need rules that are fair and keep everyone's rights and respect safe.

4.10 Conclusion and Future Outlook

In the coming years, here are some trends that will shape neurovisualization:

Federated learning and **edge AI** will bring analysis closer to patients while preserving privacy. **Neuro-symbolic systems** will blend logical reasoning with deep learning for more interpretable, policy-aligned diagnostics. **Quantum computing** may unlock entirely new possibilities for large-scale brain modeling and signal decoding. **Global regulatory frameworks** will increasingly emphasize neurorights, dynamic consent, and ethical transparency.

Perhaps most importantly, the future of neurovisualization will be shaped by our collective commitment to responsible innovation.

References

Behzadi Y, Restom K, Liau J, Liu TT (2007) A component-based noise correction method (CompCor) for BOLD and perfusion-based fMRI. NeuroImage 37(1):90–101

Brown T, Gupta S, Lee J (2022) Population-level neural mass models for dynamic brain oscillations. Front Syst Neurosci 16:855441

Esteban O, Markiewicz CJ et al (2019) fMRIPrep: a robust preprocessing pipeline for functional MRI. Nat Methods 16:111–116

Fairclough D, Jackson P, Harris T et al (2021) Title of the article. Journal of Physiology 56(1):456–467.

Fortin J-P, Cullen N, Sheline YI et al (2018) Harmonization of cortical thickness measurements across scanners and sites. NeuroImage 167:104–120

Gorgolewski KJ et al (2016) The brain imaging data structure, a format for organizing and describing outputs of neuroimaging experiments. Sci Data 3:160044

Gupta A, Singhal N, Kalra R (2022) Blockchain frameworks for dynamic consent in neurodata management. IEEE Trans Emerg Top Comput 10(3):455–467

Huang W, Zhao H, Xie F (2022) EEG-fMRI fusion improves focus localization in surgical planning for epilepsy. Epilepsia 63(10):2121–2133

Ienca M, Andorno R (2023) Towards new human rights in the age of neuroscience and neurotechnology. Nat Hum Behav 7(1):19–27

Isensee F, Jaeger PF et al (2021) nnU-net: a self-configuring method for deep learning-based biomedical image segmentation. Nat Methods 18:203–211

Kaur P, Singh A, Dhillon S et al (2024) Title of the article. Artificial Intelligence in Medicine 53(1):12–23.

Khan A, Smith J, Lee K et al (2022) Title of the article. Journal of Medical Imaging 34(2):123–134.

Lee C, Patel V, Yan M (2023) AutoPrognosis 2.0: a platform for automating clinical predictive modeling pipelines. J Biomed Inform 140:104309

Li Y, Zhang X (2022) Deep learning in multimodal brain imaging: trends and applications. Front Neurosci 16: 875412

Liu Y, Chen Q, Li F (2023) EEG-based BCI systems for home neurorehabilitation: a deep learning approach. Front Hum Neurosci 17:1123029

Lundberg SM, Lee S-I (2017) A unified approach to interpreting model predictions. Adv Neural Inf Proces Syst 30

Martínez-Pérez G, Rodríguez M, García F et al (2021) Title of the article. Medical Image Analysis 18(4): 456–467.

Martínez-Pérez G, Rodríguez M, García F et al (2022) Title of the article. Journal of Computational Biology 19(2):157–168.

Mehta R, Gupta D, Choudhury P et al (2022) Title of the article. Journal of Neuroscience Methods 29(3): 234–245.

Moon S, Kim D, Park J et al (2022) Title of the article. Frontiers in Neuroscience 31(6):1234–1245.

Molnar C (2022) Interpretable machine learning: a guide for making black box models explainable, 2nd edn, Leanpub

Power D, Roberts M, Wilson F et al (2012) Title of the article. NeuroImage 19(8):1123–1134.

Ribeiro MT, Singh S, Guestrin C (2016) "Why should I trust you?" explaining the predictions of any classifier. In: Proceedings of the 22nd ACM SIGKDD, pp 1135–1144

Sharma R, Gupta A, Verma P et al (2022) Title of the article. Journal of Computational Neuroscience 48(4): 678–689.

Sheller MJ, Edwards B, Reina GA (2020) Federated learning in medical imaging: enabling distributed learning while preserving patient privacy. Med Image Anal 65:101765

Singh B, Patel R, Kumar S et al (2023) Title of the article. Computers in Biology and Medicine 45(5):678–690.

Smith H, Lin C, Zhao L (2024) Quantum kernels for diffusion-weighted imaging in brain Connectomic. Front Phys 12:1100892

Wang J, Lin D, Raina P (2022) Neuro-symbolic systems in precision diagnostics: hybrid logic meets deep learning. Nat Machine Intell 4(12):1051–1063

Wang Y, Ahmed S (2023) Title of the article. Journal of Biomedical Informatics 25(3):234–245.

Zhang Y, Li L, Zhao Y et al (2023) Title of the article. NeuroImage 12(5):345–356.

Zhao T, Zhang Y, Sun D (2023) AR-assisted planning in complex brain tumor surgery: accuracy and usability

Zhou X, Liu J, Wang H et al (2023) Title of the article. IEEE Transactions on Medical Imaging 40(6): 987–998.

Emerging Technologies in Neurovisualization

5

Jyothi Thittamranahalli Kariyappa, Andre Bongers, Simone Zanoni, Lydia Tong, Triinetrra Suresh Chanuvegowda, and Ken W. S. Ashwell

Abstract

The field of neurovisualization has been significantly reshaped by cutting-edge imaging technologies, notably high-field magnetic resonance imaging (MRI) and diffusion tensor imaging (DTI). These innovations have made it possible to generate detailed connectomes through noninvasive, high-resolution mapping of neural connections. This chapter reviews key advancements in neurovisualization, focusing on MRI and DTI technologies and their applications in mapping neural architecture. A case study of the koala brain demonstrates how ultrahigh-field MRI and tractography can reveal conserved mammalian connectivity patterns, highlighting the evolutionary, biomedical, and translational relevance of these tools. We discuss methodological insights, limitations, and future directions in the context of both comparative neuroanatomy and human neuroscience.

Keywords

Neurovisualization · Tractography · Connectomes · Cerebral cortex · Evolution · Metatheria

Abbreviations list is adapted and modified from the author's previously published article (Thittamranahalli Kariyappa et al. 2022).

J. T. Kariyappa (✉)
Department of Anatomy, University of Sydney, School of Rural Health, Dubbo, NSW, Australia

A. Bongers · S. Zanoni
Biological Resources and Imaging Laboratory, The University of New South Wales, Sydney, NSW, Australia

L. Tong
Taronga Wildlife Hospital, Taronga Zoo, Taronga Conservation Society Australia, Sydney, NSW, Australia

T. S. Chanuvegowda
Sir Seewoosagur Ramgoolam Medical College, University of Mauritius, Reduit MU, Belle Rive, Mauritius

K. W. S. Ashwell
Department of Anatomy, School of Medical Sciences, The University of New South Wales, Sydney, NSW, Australia

Abbreviations

α	Alpha sulcus of the cerebral cortex
ac	Anterior commissure
aci	Anterior commissure, intrabulbar limb
Amg	Amygdala
Au1	Primary auditory cortex
CA	Cornu ammonis
Cd	Caudate nucleus
cg	Cingulum
Cg (λ)	Cingulate (l) sulcus of the cerebral cortex
Cg1	Cingulate cortex, area 1
Cg2	Cingulate cortex, area 2
Cl	Claustrum
cp	Cerebral peduncle
dcw	Deep cerebral white matter

K. S. Ravi et al. (eds.), *Visualization in Neuroanatomical Sciences*, Biomedical Visualization 11,
https://doi.org/10.1007/978-3-032-07380-8_5

DEn	Dorsal endopiriform nucleus
Ent	Entorhinal cortex
f	Fornix
fa	Fasciculus aberrans
FrA	Frontal association cortex
GI	Granular insular cortex
hc	Hippocampal commissure
Ib(ζ)	Interbrachial, parietal (z) sulcus of the cerebral cortex
Ic	Internal capsule
IC	Inferior colliculus.
IG	Indusium griseum
LEnt	Lateral entorhinal cortex
LG	Lateral geniculate nucleus
Lo	Lateral olfactory tract
LO	Lateral orbital cortex
LPO	Lateral preoptic area
LS	Lateral septal nucleus
M	Motor cortex
MB	Mammillary body
MD	Mediodorsal nucleus
MEnt	Medial entorhinal cortex
mfb	Medial forebrain bundle
MG	Medial geniculate nucleus
mlf	Medial longitudinal fasciculus
MO	Medial orbital cortex
MPA	Medial preoptic area
MS	Medial septal nucleus
ON	Olfactory nerve layer of bulb
opt	Optic tract
pa (β)	Parietal (b) sulcus of the cerebral cortex
PaS	Parasubiculum
Pir	Piriform cortex
Post	Post-subiculum
PRh	Perirhinal cortex
Pu	Putamen
py	Pyramidal tract
rf	Rhinal fissure
RSG	Retrosplenial granular.
S1	Primary somatosensory cortex
SIB	Substantia innominate B
S1FL	Primary somatosensory cortex, forelimb area
S1Md	Primary somatosensory cortex, mandible area
S1My	Primary somatosensory cortex, mystacial area
S2	Secondary somatosensory cortex
SC	Superior colliculus
scp	Superior cerebellar peduncle
st	Stria terminalis
sp5	Spinal trigeminal tract
ST	Nuclei of stria terminalis.
te(δ)	Temporal (d) sulcus of the cerebral cortex
TeA	Temporal association cortex
TeD	Dorsal temporal cortex
TeV	Ventral temporal cortex
tfp	Transverse fibers of pons
Tu	Olfactory tubercle
tz	Trapezoid body
V1	Primary visual cortex
V2L	Secondary visual cortex, lateral area
VEn	Ventral endopiriform nucleus
Vhc	Ventral hippocampal commissure
VO	Ventral orbital cortex
VP	Ventral pallidum

5.1 Introduction

Recent advances in neurovisualization have profoundly transformed our ability to observe, analyze, and interpret brain structure and function across species. Magnetic resonance imaging (MRI) and diffusion tensor imaging (DTI) are established noninvasive imaging methods that enable detailed reconstruction of the brain's structural circuitry. These technologies not only enable the identification of major fiber bundles but also facilitate connectome mapping—extensive mapping of both anatomical and functional connections across the brain.

Originally developed for clinical application in humans, DTI has expanded into the realm of comparative neuroanatomy, offering new opportunities to study evolutionary adaptations and species-specific brain architectures. High-field MRI systems, in particular, provide the spatial resolution necessary to delineate fine structural and cytoarchitectonic details in small animal brains. The integration of tractography algorithms further enhances our capacity to explore cortical-subcortical relationships, thalamocortical projections, and interhemispheric connectivity.

This chapter explores recent advances in imaging technologies and their impact on the field of neurovisualization, illustrated through a case study of the koala (*Phascolarctos cinereus*) brain. This marsupial species, characterized by a relatively low encephalization quotient and a highly specialized ecological niche, offers a unique model for investigating cortical modularity and energy-efficient brain organization. By applying ultrahigh-field MRI and advanced DTI techniques, this work demonstrates how contemporary neurovisualization tools can uncover not only the anatomical complexity of underexplored species but also broader principles of brain organization relevant to mammalian evolution and translational neuroscience.

5.2 Overview of Emerging Technologies in Neurovisualization

The field of neurovisualization has undergone a paradigm shift in recent decades, driven by technological innovations that enable high-resolution, noninvasive imaging of brain structure and connectivity. Below, we outline the core technologies that have reshaped our understanding of the brain and set the stage for studies like the one presented in this chapter.

5.2.1 High-Field Magnetic Resonance Imaging (MRI)

MRI systems operating at magnetic field strengths of 7 Tesla and above (e.g., 7 T and 9.4 T) offer superior spatial resolution, which supports the examination of fine neuroanatomical features such as cortical layers, subcortical nuclei, and intricate white matter pathways (Ugurbil 2014; Van der Zwaag et al. 2016). This is particularly advantageous in small brains, where anatomical boundaries are often compressed. These capabilities support applications in neurodevelopmental studies, brain atlases, and cytoarchitectonic mapping (Toga and Thompson 2003).

- **Applications:** Neurodevelopmental studies, brain atlases, and cytoarchitectonic mapping.
- **Benefits:** Superior contrast-to-noise ratio and signal uniformity.
- **Limitations:** Susceptibility artifacts and limited availability.

5.2.2 Diffusion Tensor Imaging (DTI)

DTI enables visualization of white matter architecture by tracking anisotropic water diffusion patterns that reflect the orientation of neural fibers (Basser et al. 1994; Le Bihan 2003). It provides crucial insights into the anisotropic properties of neuronal fibers and allows for the construction of connectivity matrices and three-dimensional tractography maps.

- **Applications:** Connectomics, neurodegenerative disease tracking, surgical planning.
- **Limitations:** Limited accuracy in areas with fiber crossings; sensitive to noise and artifacts (Tournier et al. 2011).

5.2.3 Tractography and Fiber Tracking Algorithms

Tractography methods, whether deterministic or probabilistic, translate DTI data into coherent fiber pathways that model brain connectivity (Mori and van Zijl 2002). Deterministic algorithms offer straightforward and visually intuitive outputs, though they can miss complex crossing fibers. Probabilistic models (Behrens et al. 2007) provide a more nuanced view of fiber uncertainty and complexity.

- **Recent Advances:** Orientation distribution functions (ODFs) combined with high-angular resolution diffusion imaging (HARDI) techniques are employed to resolve fiber crossings within complex white matter regions.

5.2.4 Multimodal Integration

Structural, diffusion, and functional imaging modalities are now routinely combined to enhance brain mapping accuracy. This integration is supported by platforms such as FSL (Jenkinson et al. 2012), 3D Slicer (Fedorov et al. 2012), and Advanced Normalization Tools (ANTs; Avants et al. 2009), which enable precise alignment of anatomical and functional data to support interpretation of physiological and behavioral associations.

- **Future Direction:** Real-time imaging and AI-based segmentation tools, such as deep learning models like U-Net, are being developed to enhance diagnostic precision and speed (Ronneberger et al. 2015).

5.3 Case Study: Mapping the Koala Brain Connectome

5.3.1 Technical and Methodological Insights

To illustrate how cutting-edge neurovisualization technologies can be leveraged in comparative neuroanatomy, this case study focuses on the adult koala (*Phascolarctos cinereus*). As a diprotodont marsupial with a notably low encephalization quotient, the koala brain (Fig. 5.1) presents a compelling model for examining structural brain organization in a species shaped by strong ecological constraints. The imaging methodology employed in this study including ex vivo high-field MRI, diffusion tensor imaging (DTI), and tractography pipelines follows previously validated protocols used to generate the connectome of another marsupial, the quokka (*Setonix brachyurus*), a macropod species (Thittamranahalli Kariyappa et al. 2022). This consistency in imaging and post-processing workflows enables comparative assessments of cortical organization and network topology across marsupial lineages.

The methodological pipeline applied in this study highlights the capabilities and limitations of present neurovisualization tools. Sample preparation included immersion in NaCl solution followed by suspension in Fomblin for susceptibility matching. Structural imaging protocols were optimized for tissue contrast and resolution, while diffusion sequences balanced signal strength against angular resolution and scan duration.

Using a 9.4 T high-field MRI system, we acquired high-resolution structural and diffusion-weighted images of a fixed koala brain specimen. Structural scans included 2D and 3D T1- and T2-weighted fast spin echo sequences, as well as multi-gradient echo (MGE) volumes. These scans delineated cortical layers and subcortical structures with sufficient clarity to permit cytoarchitectonic analysis (Fig. 5.2). Diffusion datasets were collected across multiple b-values using a segmented EPI sequence and processed into fractional anisotropy (FA) and colored FA (FAC) maps (Figs. 5.3, 5.4, and 5.5).

DWI DICOM files were converted to 4D NIfTI datasets. Gradient directions and b-values were extracted using the same processing script. To correct for susceptibility-related distortion artifacts, we employed the *top-up* tool from FSL, as described by Andersson et al. (2003). Correction for eddy current-induced image distortions was performed using the *eddy_correct* function available in the FSL toolkit. After applying all corrections, the data were reformatted into NRRD files compatible with the 3D Slicer (4.10.2) platform for further tractographic analysis (Fedorov et al. 2012). To generate the brain mask, we used an intensity-based thresholding approach derived from the method proposed by Otsu (1979), with implementation informed by earlier diffusion imaging work (Basser et al. 1994). Single tensor models were fit to generate FA and colored FA maps (Le Bihan 2003).

Tractography was performed using SlicerDMRI (Norton et al. 2017). Seeds included functionally significant ROIs like somatosensory, visual, auditory cortices, hippocampus, dorsal thalamus, nucleus accumbens, caudate, and putamen. ROIs were drawn primarily in coronal and occasionally sagittal views. Fiber occupation in the anterior commissure (ac) was visualized and

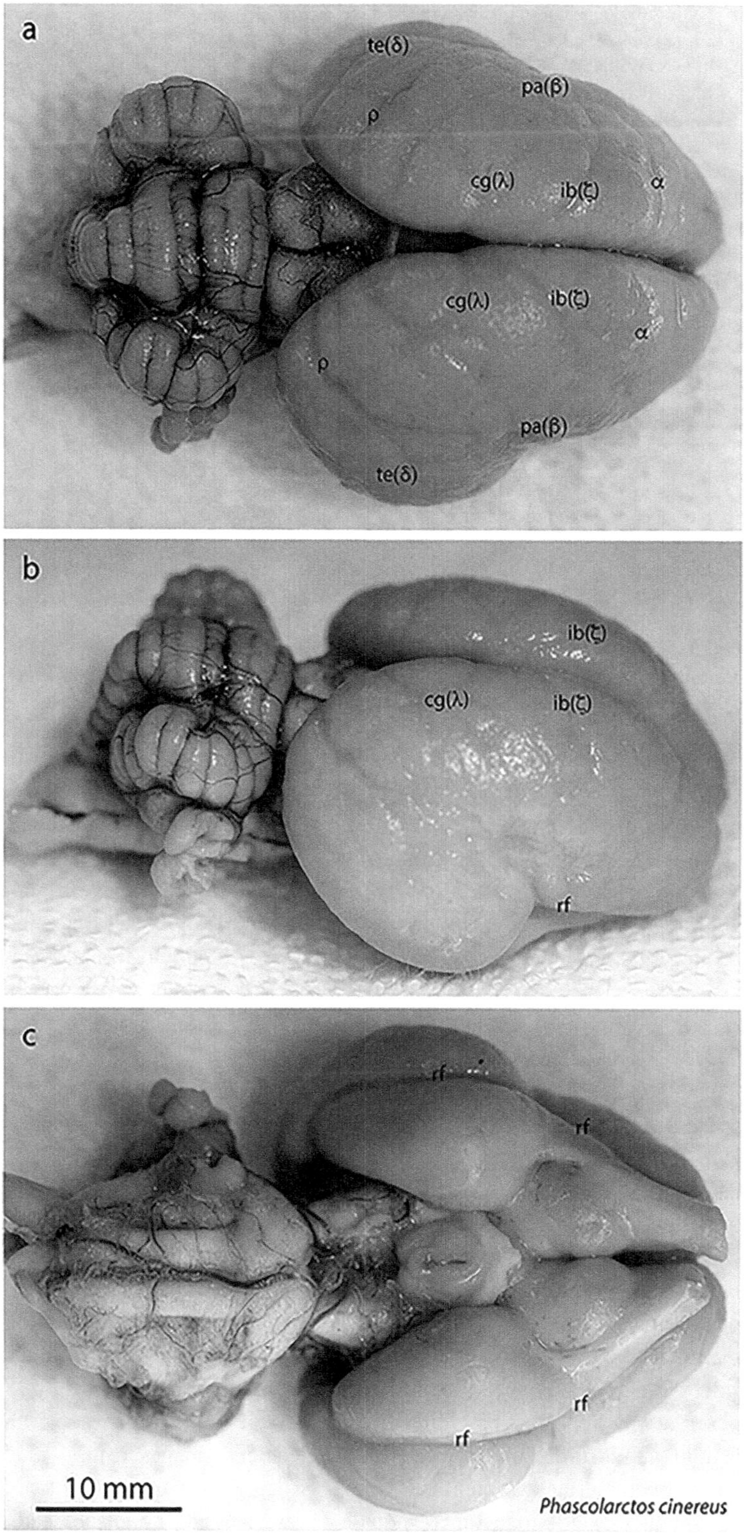

Fig. 5.1 *Phascolarctos cinereus*—sulci and major fissures of the cerebral cortex annotated with Greek letters, following the nomenclature of Rowe (1990). Full names of the labeled structures are provided in the list of abbreviations. Scale bar = 10 mm

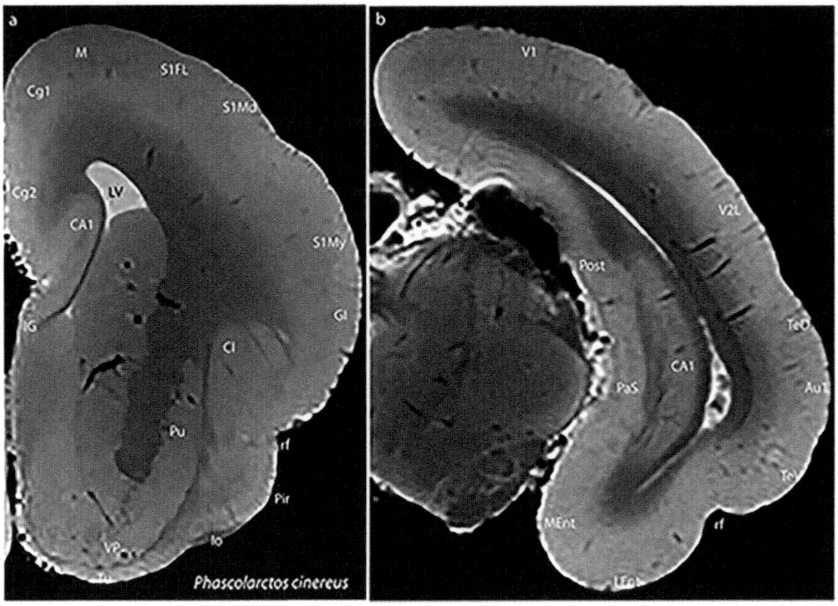

Fig. 5.2 Coronal sections of the *Phascolarctos cinereus* brain from T1-weighted MRI scans. Images are scaled to enhance cortical delineation based on laminar differentiation, with anatomical comparisons informed by the tammar wallaby brain atlas developed by Ashwell and Marotte (2010). Full names of labeled structures are provided in the list of abbreviations

measured using ImageJ (Schindelin et al. 2012). We used ANTs (Advanced Normalization Tools) for co-registering the anatomical and diffusion volumes, following the methodology outlined by Avants et al. (2009).

5.3.2 Results

5.3.2.1 Gross Anatomy and Surface Features

As depicted in Fig. 5.1, the koala brain had a mass of 17.1 grams and showed a dorsal outline that was ovoid, a result of the parieto-frontal region's lateral prominence. A slight asymmetry was observed, with the right hemisphere appearing marginally more expanded than the left. The sulci were generally shallow, aligning with prior observations reported by Johnson (1977) and Haight and Murray (1981), and included sulci aligned with surface vasculature.

Notable external features include a well-defined rhinal sulcus, serving as the anatomical divider between the neocortex and the piriform cortex. The alpha sulcus outlines the transition from the prefrontal region to the parieto-frontal cortex, whereas the beta sulcus distinguishes the parieto-frontal area from the adjacent temporal and posterior parietal cortices. The relative reduction in brain size in the koala appears primarily due to a diminished parieto-frontal cortex, a feature consistent with its low encephalization quotient and ecological specializations.

5.3.2.2 T1-Weighted MRI-Based Analysis of Koala Brain Sectional Anatomy

Magnetic resonance imaging identified the cortical areas, nuclei clusters, and principal fiber pathways within the koala brain, with T1-weighted (T1w) scans providing a clear distinction of nuclei and fiber tracts. Coronal T1w images (Fig. 5.2) served to distinguish internal structures, guide fractional anisotropy analysis, and delineate seed regions for pathway tracing. These findings were compared to the tammar

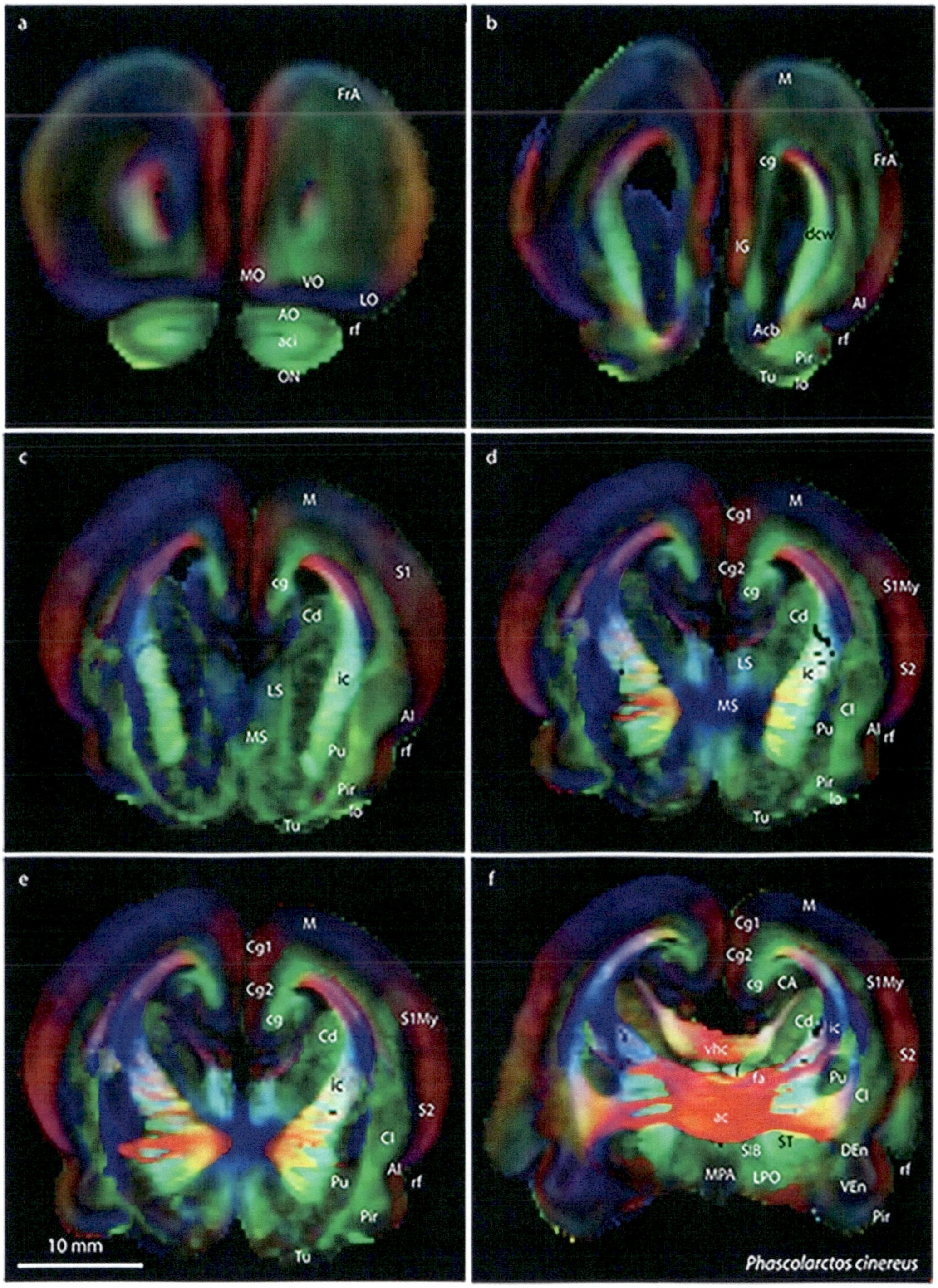

Fig. 5.3 Transverse sections of the koala brain displaying serial color FA (fractional anisotropy) maps, ranging from the olfactory bulb to the anterior commissure (**a–f**; ~15 mm range). Axes S, A, and R denote fiber orientations: superior, anterior, and right, respectively. Transverse fibers are shown in red, longitudinal fibers in green, and vertical fibers in purple/blue. See the list of abbreviations for full names of the labeled structures. Scale bar = 10 mm

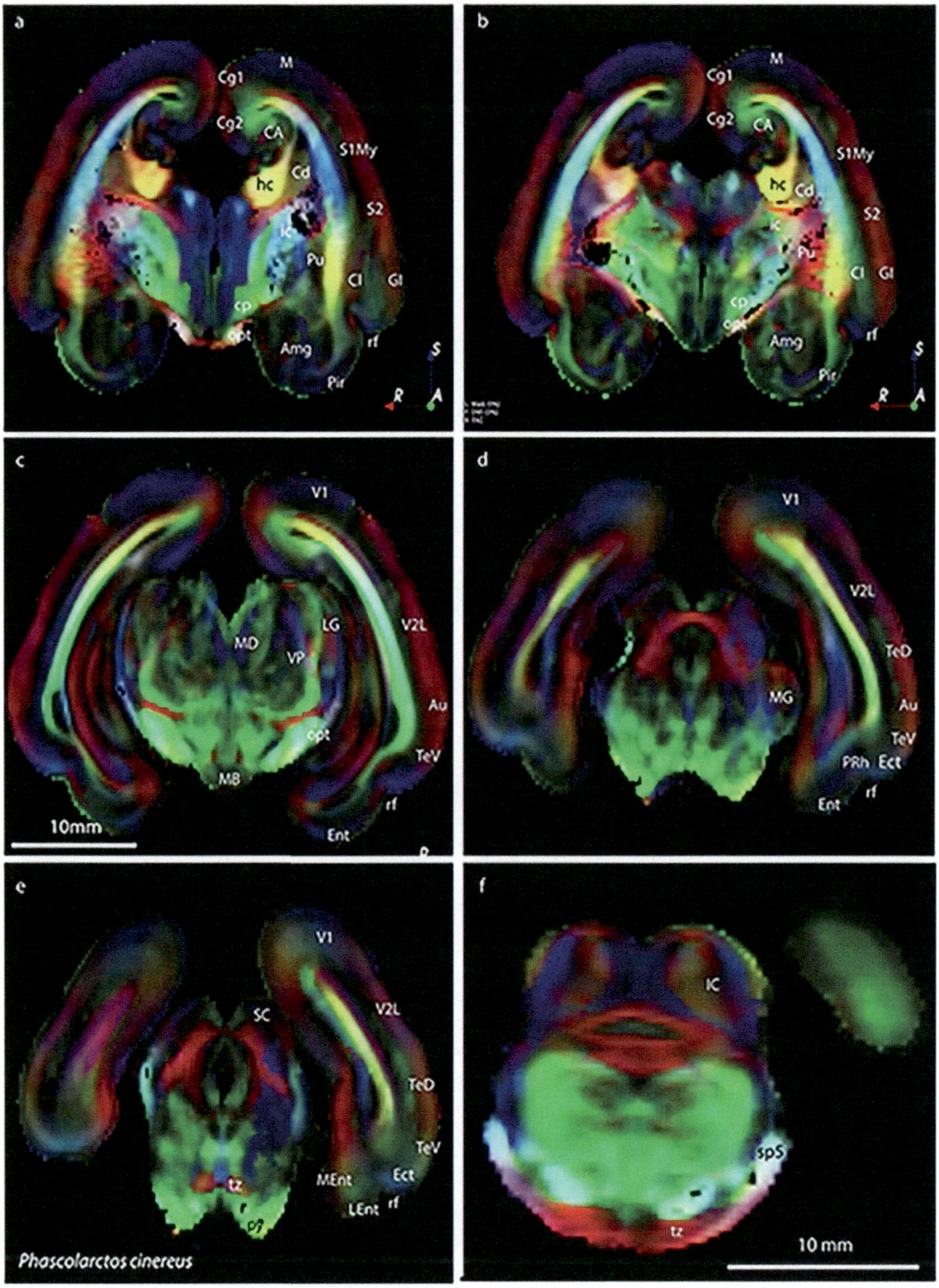

Fig. 5.4 FAC-enhanced coronal slices of the koala brain demonstrate anatomical progression from the level of the hippocampal commissure to that of the trapezoid body in the medulla. Axes S, A, and R indicate fiber orientations: superior, anterior, and right, respectively. Transverse fibers are shown in red, longitudinal fibers in green, and vertical fibers in purple/blue. Full names of labeled structures are provided in the list of abbreviations. Scale bar = 10 mm

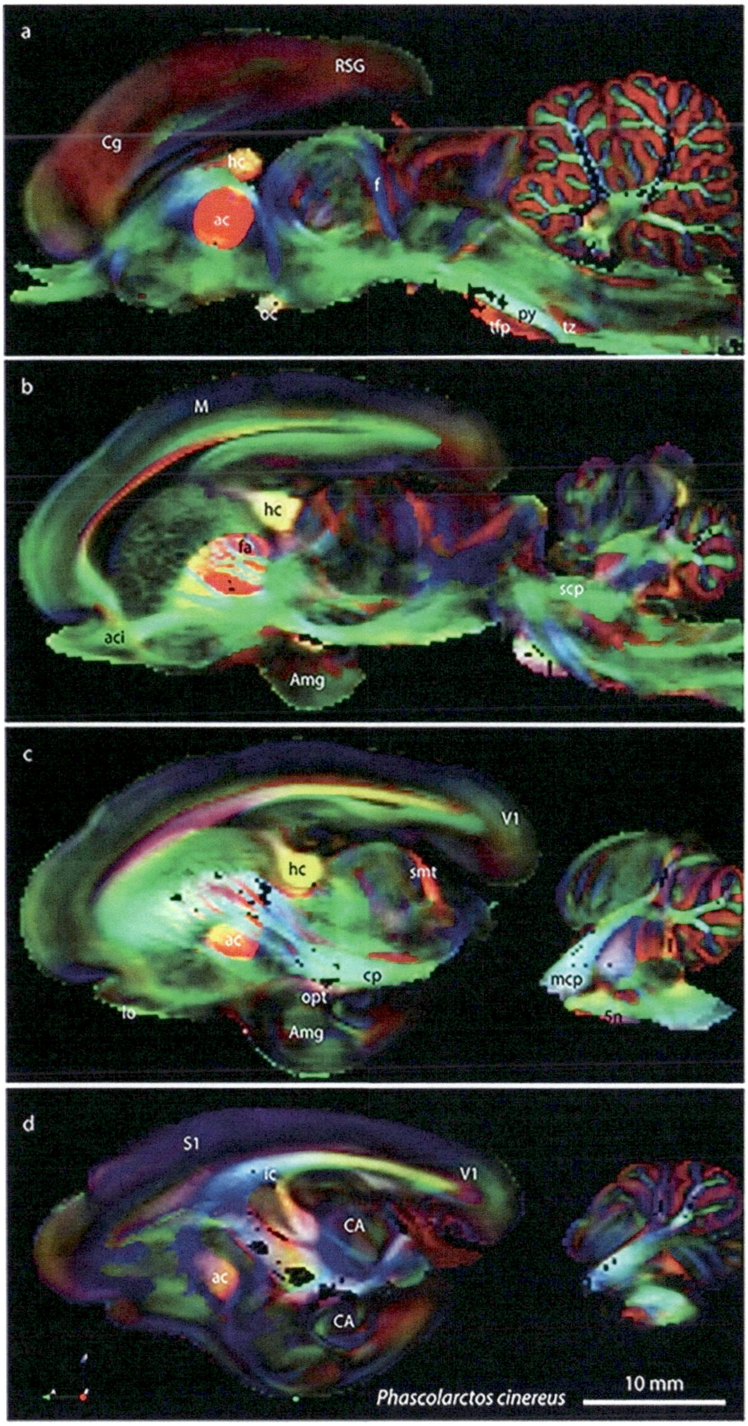

Fig. 5.5 Fractional anisotropy color (FAC) maps of the koala brain displayed in sagittal orientation. Images are presented as a mediolateral series spaced 2 mm apart. Axes S, A, and R denote fiber orientations: superior, anterior, and right, respectively. Transverse fibers are shown in red, longitudinal fibers in green, and vertical fibers in purple/blue. Full names of labeled structures are provided in the list of abbreviations

wallaby brain atlas, its closest relative (Ashwell and Marotte 2010), with higher scaling in T1w images enhancing cortical layer discrimination.

5.3.2.3 FAC-Mapped DTI of the Koala Brain

Figures 3, 5.4, and 5.5 display fractional anisotropy color (FAC) images derived from DTI analysis, where red denotes transverse fibers, blue indicates vertical projections, and green represents longitudinal pathways. Guided by diffusion-weighted imaging data, these visualizations allowed the identification of cortical territories and prominent commissural structures. Notably, connections such as the anterior and hippocampal commissures, along with the fasciculus aberrans, were distinguishable by reddish hues in Fig. 5.3f. Key features identified include:

- Lateral olfactory tract (lo) (Fig. 5.3a–d).
- Fimbriae (Fig. 5.3f).
- The medial forebrain bundle, visualized in Fig. 5.4b.
- Cerebral peduncle fibers (cp) appear prominently in Fig. 5.5c.
- The pyramidal tract can be seen in Fig. 5.5a.

A bluish-green signal characterized both the internal and external capsules, as seen across sections in Figs. 5.3a–f and 5.4a–b. In contrast, the cingulate cortex was prominently visualized in red within the same planes. Adjacent to the cingulate, the motor cortex is dark blue. S1 and S2 are identifiable by their reddish hue in the rostral sections of ascending thalamocortical projections (Figs. 5.3d–f and 5.4a, b), while V1 and V2 exhibit a comparable coloration in the caudal sections.

Scale bar = 10 mm.

5.3.2.4 Tractography of the Koala Brain

We employed deterministic fiber tracking via the 3D SlicerDMRI extension to reconstruct both global and localized connectomes. These "tracks" are mathematical interpolations of a diffusion model and are subject to uncertainties such as noise, incorrect crossings, or premature terminations (Huang et al. 2004). Fiber directions were visually distinguished using color coding: transverse pathways appeared in red, anteroposterior tracts in green, and vertically oriented fibers in blue. Oblique fibers appeared as magenta, yellow, or cyan combinations (Bell and Abdrabou 2013).

Region-of-interest (ROI) seeds were manually defined within functionally significant cortical and subcortical areas such as the primary regions for somatosensory (S1), visual (V1), auditory (Au1) processing, hippocampus, and components of the basal ganglia. This comprehensive mapping successfully captured the connectomes and yielded new insights into several major brain circuits mentioned below.

5.3.2.4.1 Whole-Brain Projections

Figure 5.6 displays projections of all brain fibers onto central slices across coronal, caudal, and sagittal planes. Longitudinal bundles were mainly ventral to the anterior commissure, with vertically oriented patterns prominent dorsally (Fig. 5.6a–d). Structures such as the stria terminalis and medial forebrain bundle were observed (Fig. 5.6a, b, d–f). Long association fibers linked ipsilateral cortical regions (Fig. 5.6a, b), including putative homologs of the superior and inferior longitudinal fasciculi.

5.3.2.4.2 Thalamocortical and Basal Ganglia Connections

ROI-based tractography from the dorsal thalamus produced fiber pathways directed toward the orbitofrontal and frontal cortices, along with somatosensory, auditory, and visual regions (Fig. 5.7a–f). Additional pathways were observed reaching the amygdala and entorhinal cortex (Fig. 5.7c–f). Basal ganglia seeds (caudate, putamen, nucleus accumbens) revealed cortico-striatal pathways connecting frontal association areas and dorsomedial isocortex (Figs. 5.8 and 5.9).

Scale bar = 10 mm.

5.3.2.4.3 Somatosensory Cortex Connections

Seeding from S1 showed ipsilateral connectivity within S1 and to rostral/caudal cortices (Fig. 5.10c–f), frontal association areas (Fig. 5.10a–d), and other sensory regions

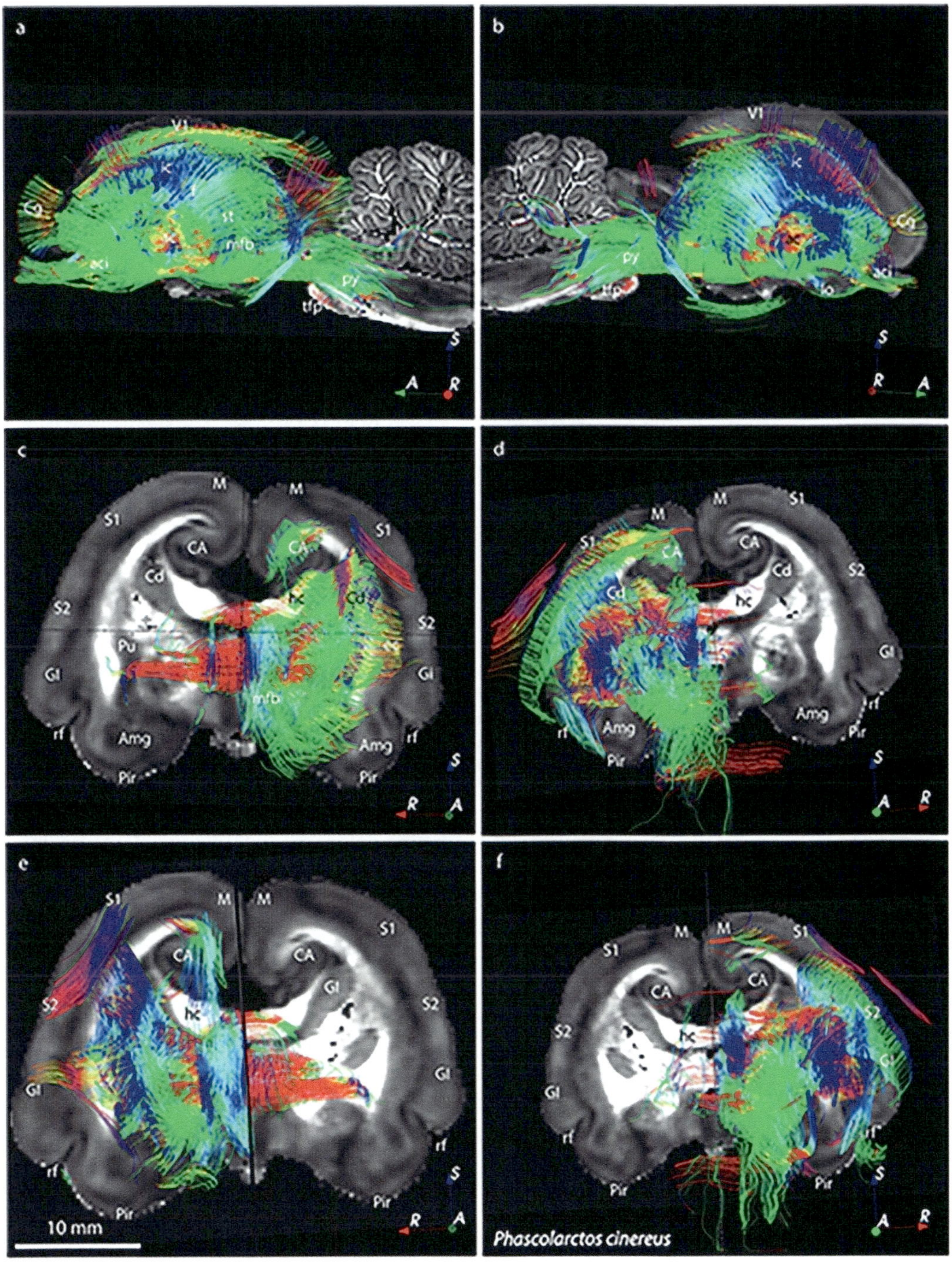

Fig. 5.6 Tractography reconstruction of the koala brain is shown from multiple perspectives, including frontal, posterior, and both sagittal views (right and left), with streamlines initiated from the right hemisphere in a rostrocaudal direction. Commissural pathways traversing to the opposite hemisphere are evident at the levels of the anterior commissure (ac) and hippocampal commissure (hc). Panels c–d and e–f represent mirrored configurations. Directionality of fibers is indicated by color: red for mediolateral, green for anteroposterior, and purple to blue for dorsoventral tracts. Abbreviations for anatomical labels are listed in the accompanying glossary. Scale bar = 10 mm

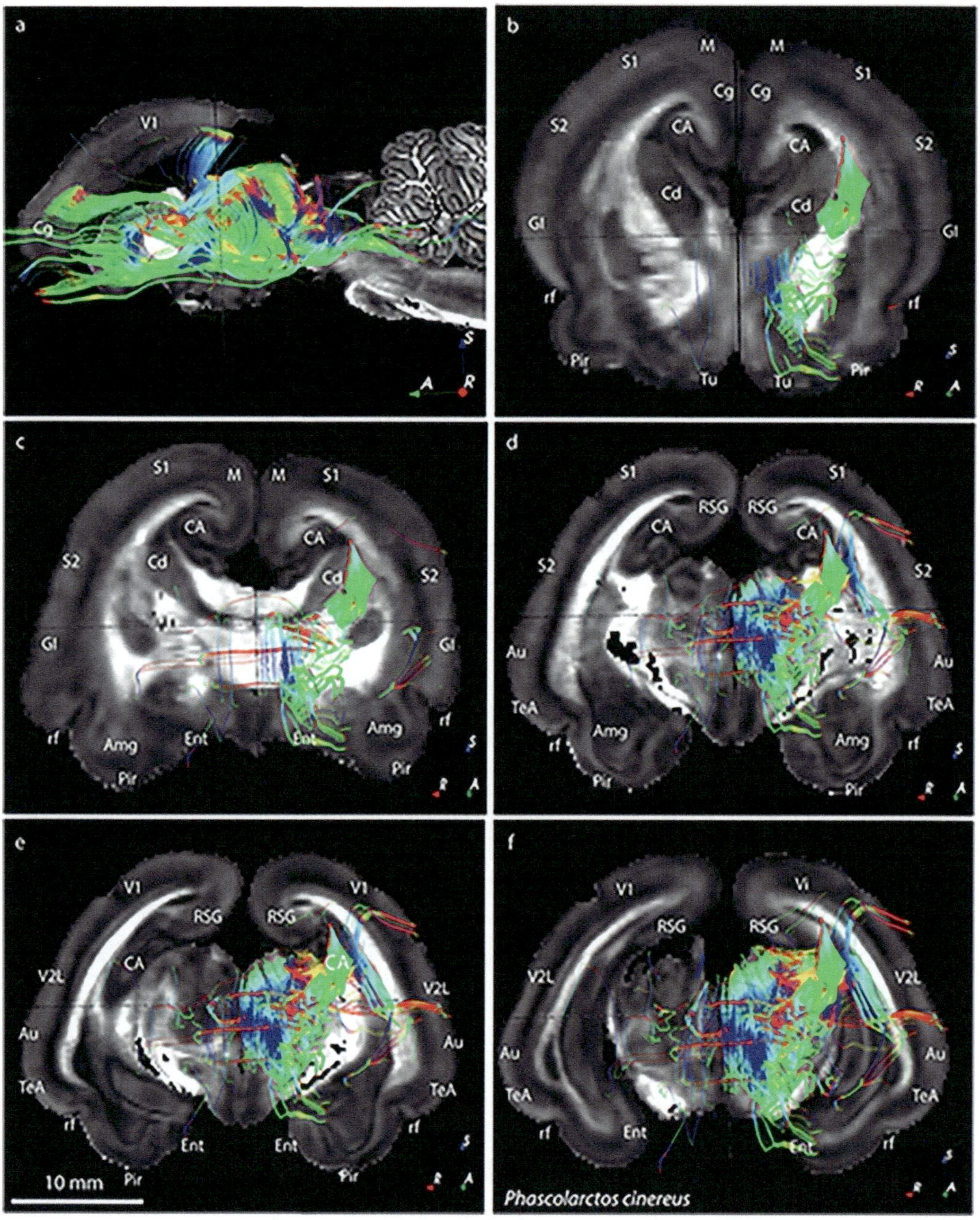

Fig. 5.7 Fiber tracking initiated in the dorsal thalamus mapped cortical connectivity patterns across the koala cerebrum. Reference axes: R (right), S (superior), and A (anterior). Fiber orientations are represented by diffusion colors: reddish yellow for commissural fibers, green for longitudinal fibers, and varying shades of blue for vertically oriented fibers. Full names of labeled structures are provided in the list of abbreviations

including auditory and visual cortices, amygdala, and entorhinal cortex (Fig. 5.10f). Thalamic projections were primarily from VP, VL, and VM nuclei (Figs. 5.10f, 5.7d). The anterior commissure carried commissural projections (Fig. 5.11a). No distinct fiber pattern was evident in the mystacial region (Figs. 5.11 and 5.12), which differs from the organization previously

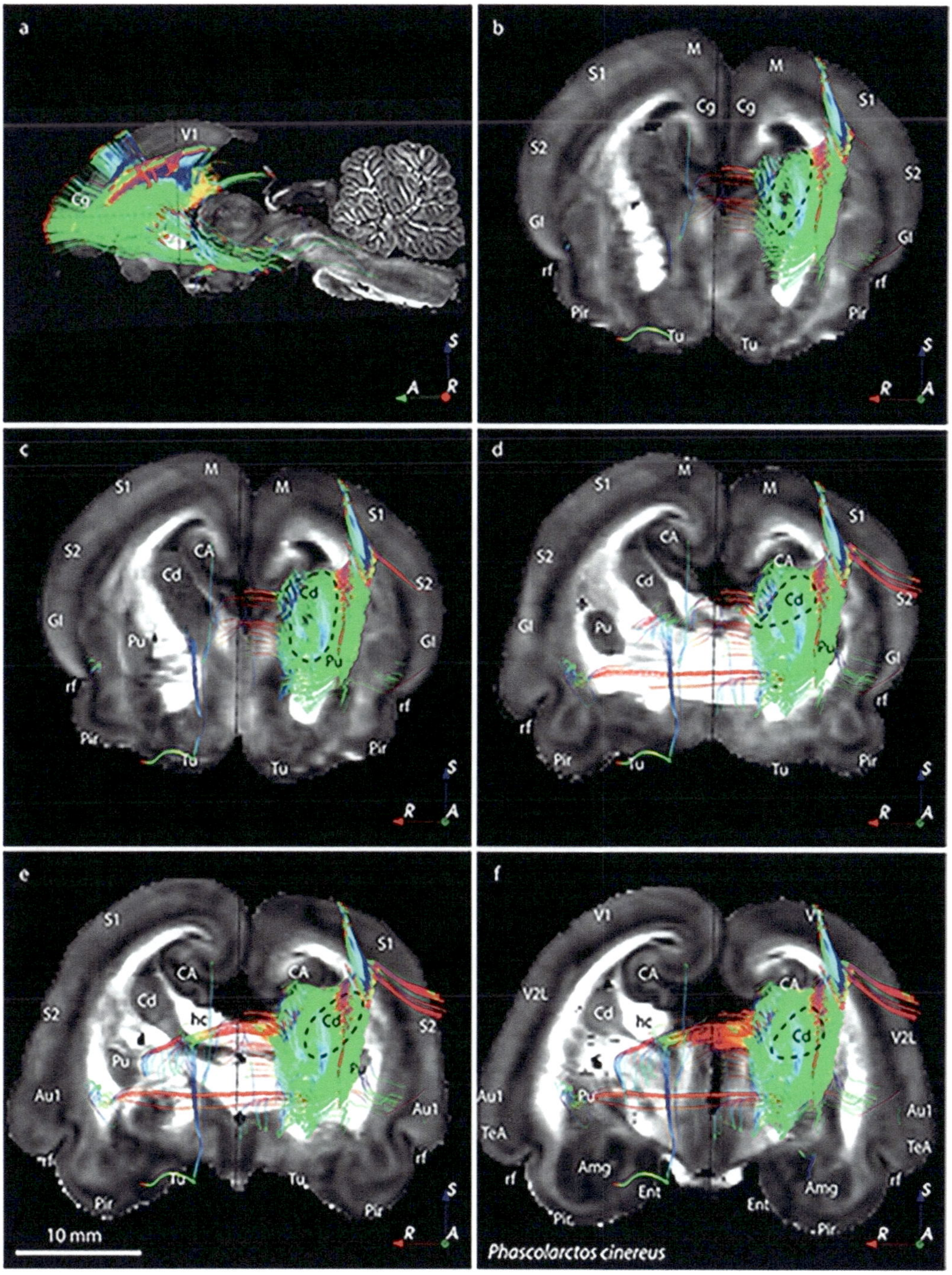

Fig. 5.8 Tractographic connections of the caudate nucleus head and body in the koala brain mapped through seed-based tractography. The black dotted line indicates the seeded region. Axes: R (right), S (superior), and A (anterior). Colors indicate fiber orientation: red, yellow for mediolateral (commissural) tracts, green for anteroposterior pathways, and blue gradients for dorsoventral projections. Full names of labeled structures are provided in the list of abbreviations. Scale bar = 10 mm

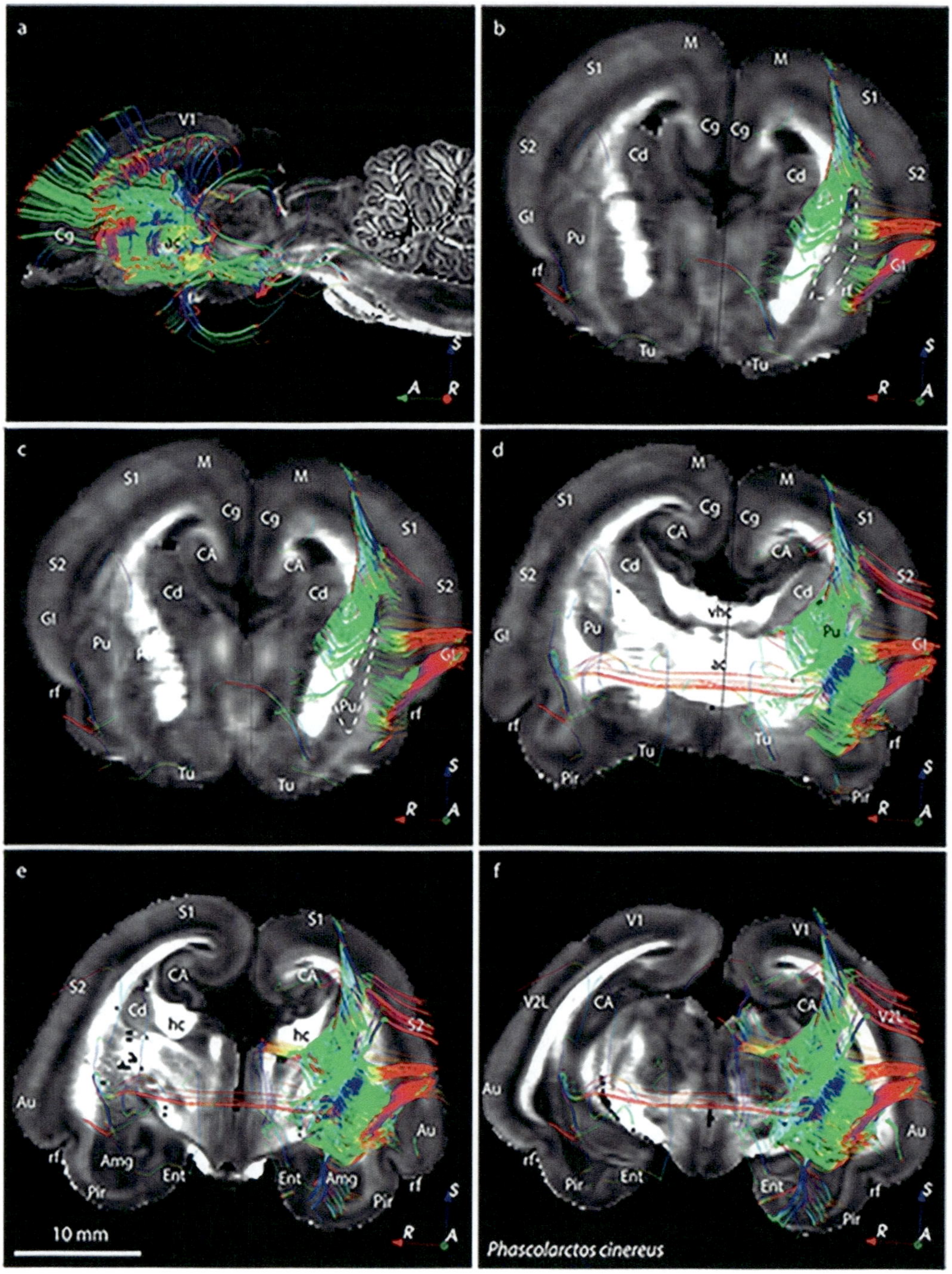

Fig. 5.9 Tractographic connections of the putamen in the koala brain identified via seed-based tracing. The white dotted line marks the seeded region. Axes: R (right), S (superior), and A (anterior). Fiber orientations are color-coded according to water diffusion direction: red, yellow for mediolateral (commissural) tracts, green for anteroposterior pathways, and blue gradients for dorso-ventral projections. Full names of labeled structures are provided in the list of abbreviations. Scale bar = 10 mm

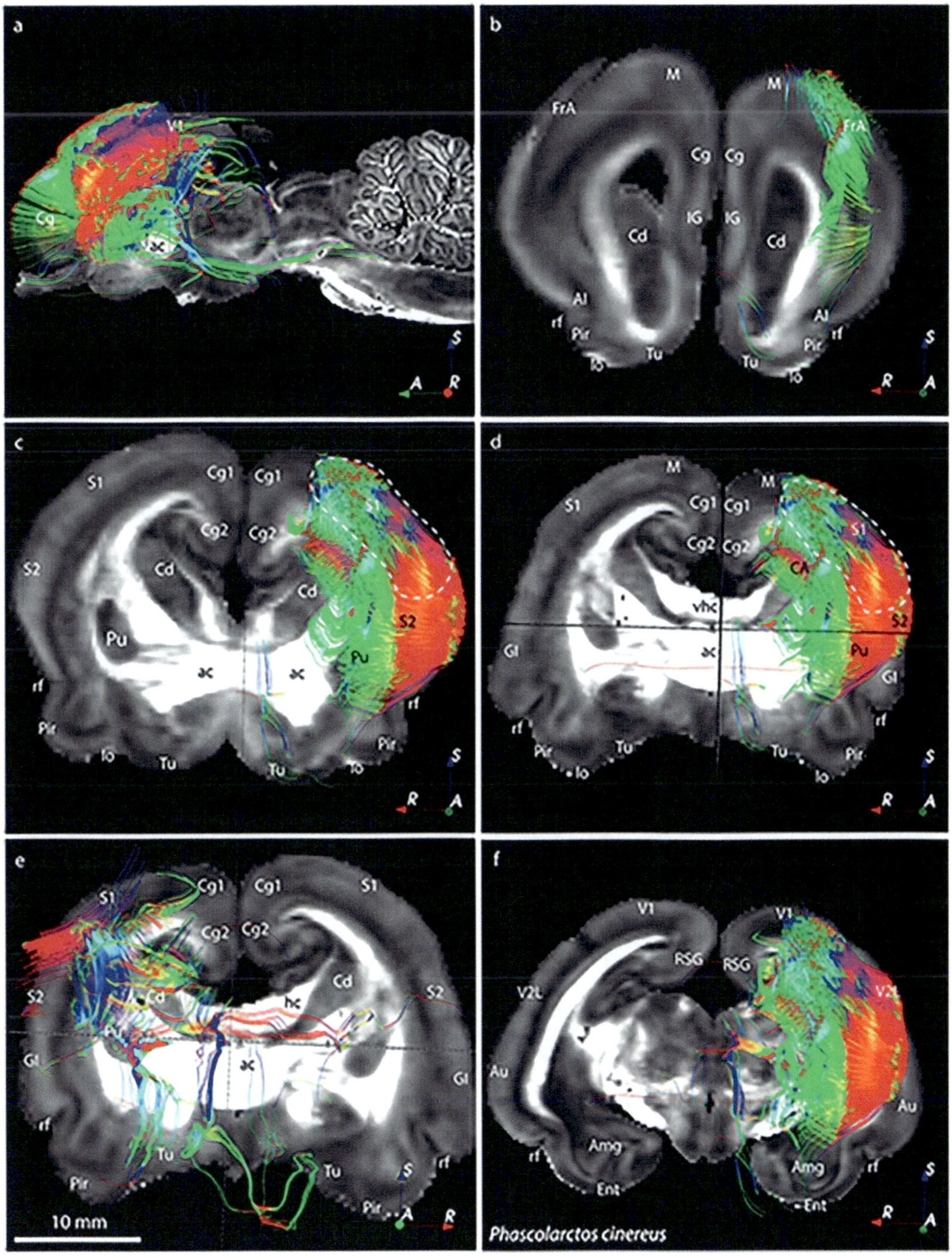

Fig. 5.10 Mapping somatosensory connectivity by tractography initiated from the primary somatosensory cortex (S1). White dotted line marks the seeded area. Identified projections include connections from S1 to the caudolateral field (S2), as well as to orbital, motor, auditory, and retrosplenial regions. Additional fibers extend to cortex rostral to S1 and distant anterior areas. Connectivity with visual cortices and commissural projections to the contralateral hemisphere are also evident. Axes: R (right), S (superior), and A (anterior). Fiber orientations are color-coded according to water diffusion direction: red, yellow for mediolateral (commissural) tracts, green for anteroposterior pathways, and blue gradients for dorsoventral projections. Full names of labeled structures are provided in the list of abbreviations. Scale bar = 10 mm

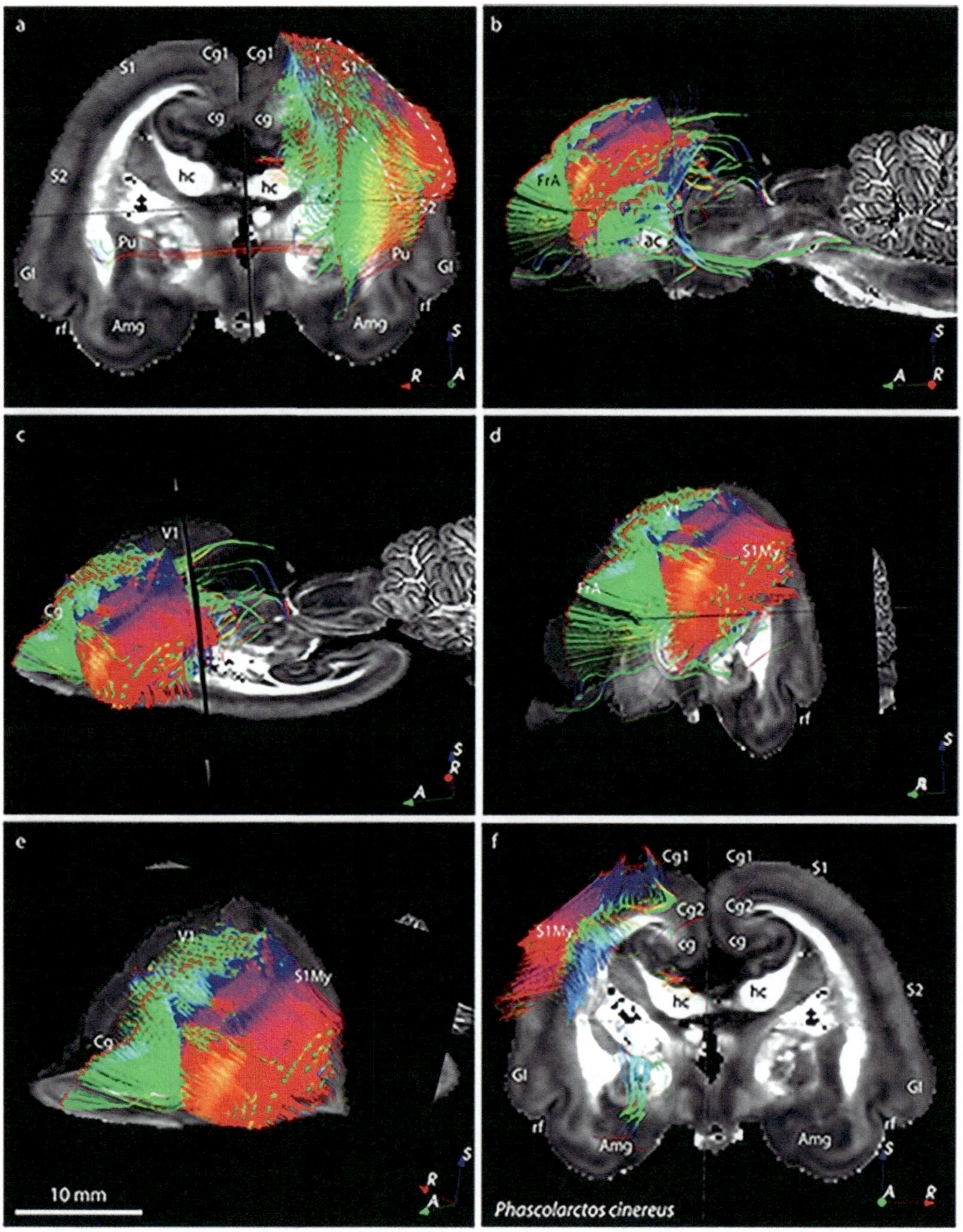

Fig. 5.11 Tractographic visualization of somatosensory connectivity initiated from the primary somatosensory cortex (S1). The white dotted line indicates the seeded area. Axonal projections are visible throughout the somatosensory cortex, including the mystacial region, which exhibits a fiber distribution pattern consistent with the rest of the somatosensory field. Axes: R (right), S (superior), and A (anterior). Fiber orientations are color-coded according to water diffusion direction: red, yellow for mediolateral (commissural) tracts, green for anteroposterior pathways, and blue gradients for dorso-ventral projections. Full names of labeled structures are provided in the list of abbreviations. Scale bar = 10 mm

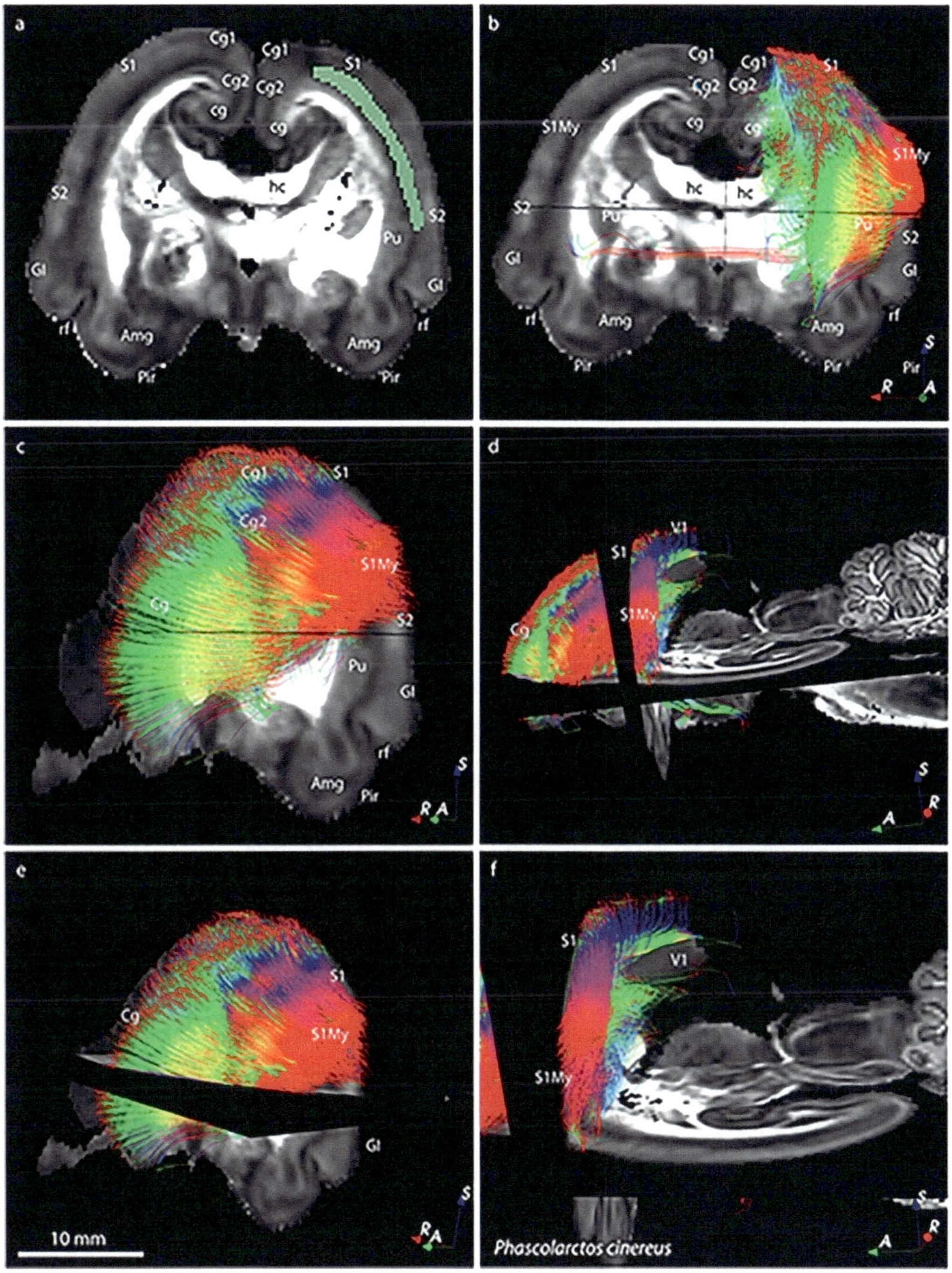

Fig. 5.12 Fiber tracking conducted with a seed placed in the S1 region of the cortex to visualize somatosensory connectivity. Figure 5.15a illustrates targeted seeding of the mystacial area to evaluate fiber organization, potentially reflecting the existence or lack of barrel-type architecture. Axonal projections in the mystacial region exhibit a distribution pattern similar to that of the broader somatosensory cortex. Axes: R (right), S (superior), and A (anterior). Fiber orientation is color-coded: longitudinal tracts appear in green, vertical pathways in purple/blue, and transverse fibers in red. Full names of all labeled structures can be found in the list of abbreviations. Scale bar = 10 mm

reported in the quokka brain (Thittamranahalli Kariyappa et al. 2022).

5.3.2.4.4 Visual Connections

V1 exhibited strong ipsilateral connections with V2 and adjacent cortices (Fig. 5.13a–f). LG was the primary thalamic input (Fig. 5.13b–e), with sparse commissural links. Additionally, projections originating from the inferior optic radiation formed Meyer's loop as they extended into the temporal lobe (Fig. 5.13b, c, d).

5.3.2.4.5 Auditory Connections

Au1 is connected to retrosplenial, temporal, frontal, and visual cortices (Fig. 5.14a–f) and to the medial geniculate nucleus (MG) (Fig. 5.14b–f). Contralateral connections were limited.

5.3.2.4.6 Limbic and Commissural Connections

The fornix was traced from hippocampal seeds, revealing pre- and postcommissural components (Figs. 5.15a–f, 5.6c, e). Fiber projections included prefrontal, retrosplenial, sensory, and limbic areas. Commissural fibers showed topographic organization (Fig. 5.16a). Interhemispheric fibers traversed the anterior commissure and hippocampal commissures (Figs. 5.16 and17). Isocortical fibers primarily used the external capsule. The fasciculus aberrans carried fibers from dorsal isocortex (Fig. 5.16b–d).

5.4 Discussion

5.4.1 Evolutionary and Biomedical Implications

The koala connectome reveals organizational principles shared across metatherian and eutherian mammals. Despite the koala's low brain-to-body size ratio, its cortical architecture displays patterns of modularity, connectivity, and hierarchical organization that align with general mammalian neurodevelopmental principles (Fig. 5.6; Karlen and Krubitzer 2007; Bullmore and Sporns 2009). The observed wiring patterns support the hypothesis that evolutionary pressures such as energy efficiency and ecological adaptation shape brain structure alongside phylogenetic constraints (Van den Heuvel and Sporns 2011).

Key findings, such as dense ipsilateral projections from S1 to S2, frontal, and auditory cortices (Fig. 5.10), mirror connectivity seen in primates (Markov et al. 2013) and underscore the role of somatosensory integration in complex behaviors (Honey et al. 2009). Visual cortex connections (Figs. 5.13 and 5.14) suggest strong intrahemispheric integration but sparse interhemispheric connectivity, reflecting both phylogenetic lineage and ecological demands (Oh et al. 2014; Gozzi and Schwarz 2016).

Such comparative analyses have implications beyond evolutionary biology. In human neuroscience, similar tractography-based methods are used to investigate neurodevelopmental and neurodegenerative conditions (Tournier et al. 2011). The koala brain serves as a valuable platform for advancing these neuroimaging tools, especially in linking anatomical structures to functional outcomes and in identifying potential biomarkers of behavioral traits (Dehay and Kennedy 2007; Neal et al. 2006).

5.4.2 Limitations and Future Directions

While DTI and tractography offer powerful tools for brain mapping, they are not without limitations. As discussed, DTI assesses fiber architecture indirectly via water diffusion, which can be influenced by nonneuronal structures. Previous work has shown that errors like fiber misrouting or abrupt termination may occur in areas of low SNR (signal-to-noise ratio) or where fibers intersect at acute angles (see Schmahmann et al. 2007; Tournier et al. 2004).

Deterministic tracking methods, though visually intuitive, lack robustness in ambiguous regions compared to probabilistic approaches. Future applications should incorporate advanced modeling techniques such as orientation distribution functions (ODFs) or constrained spherical deconvolution to overcome these limitations.

Furthermore, the lack of functional imaging in postmortem studies restricts the ability to directly link anatomical pathways with behavior or

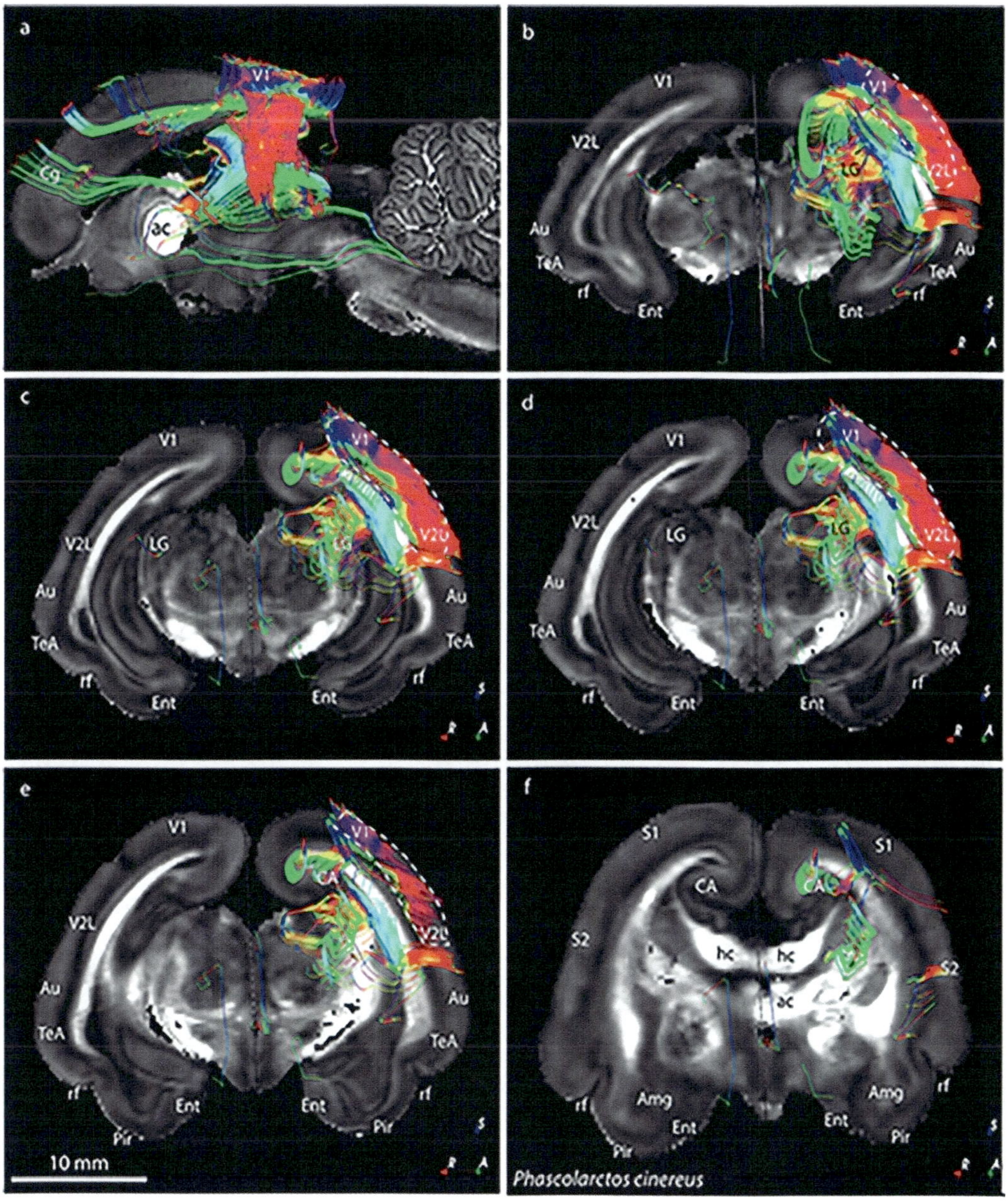

Fig. 5.13 Tractographic visualization of visual cortical connectivity in the koala brain. Connections from the visual cortex to adjacent cortical areas and thalamocortical projections targeting the visual region are evident, with relatively few commissural fibers observed. The white dotted line marks the seeded region used for tractography. Axes: R (right), S (superior), and A (anterior)**.** Fiber orientation is color-coded: longitudinal tracts appear in green, vertical pathways in purple/blue, and transverse fibers in red. Full names of all labeled structures are seen in the list of abbreviations. Scale bar = 10 mm

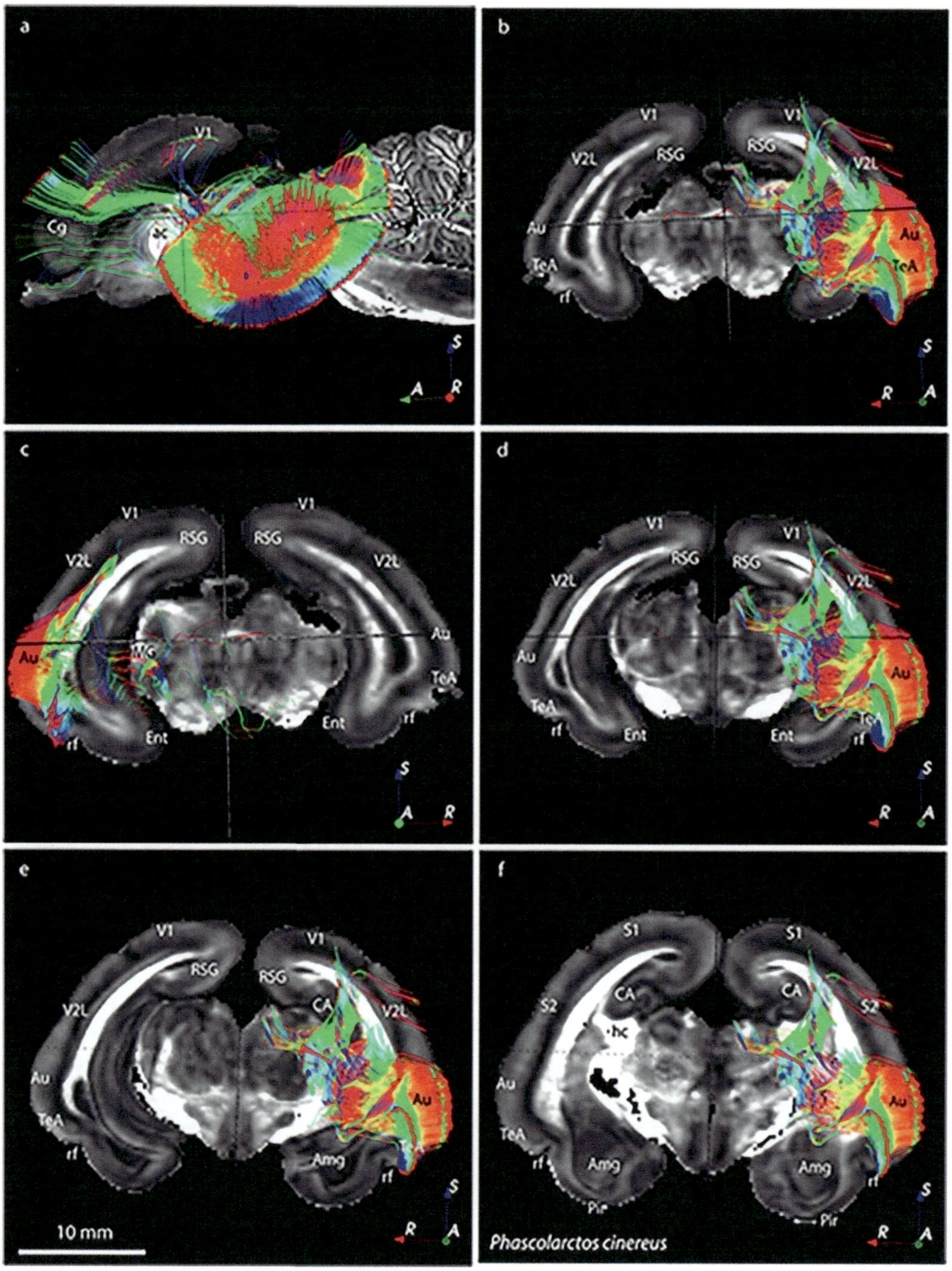

Fig. 5.14 Tractographic representation of auditory cortical connectivity in the koala brain. Projections from the auditory cortex to neighboring cortical areas are visible, with limited commissural connections observed extending to the opposite hemisphere. Directional axes are labeled as R (right), S (superior), and A (anterior). Fiber orientation is color-coded: longitudinal tracts appear in green, vertical pathways in purple/blue, and transverse fibers in red. Full names of all labeled structures are seen in the list of abbreviations. Scale bar = 10 mm

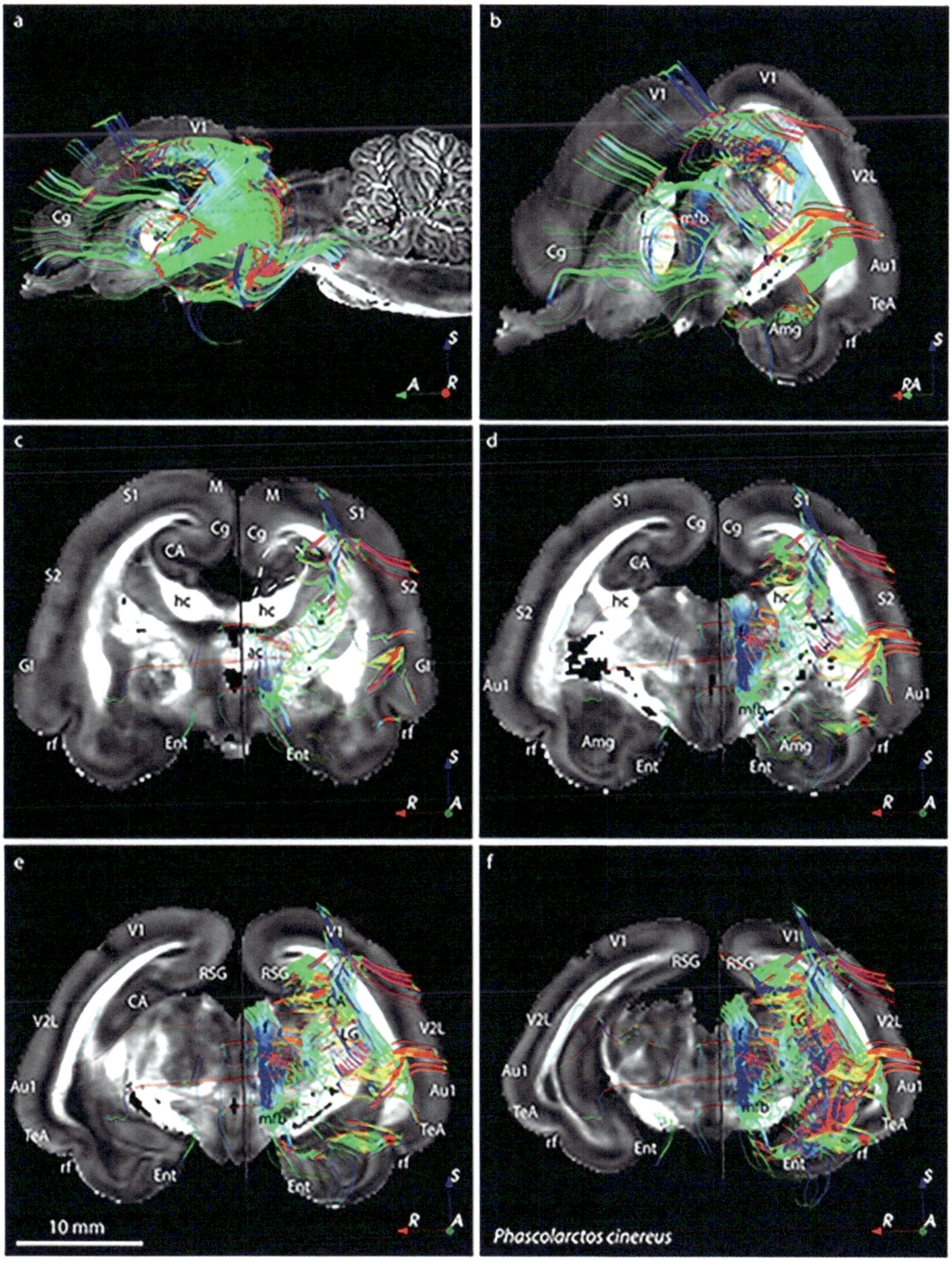

Fig. 5.15 Tractographic visualization of limbic system connectivity following hippocampal seeding in the koala brain. Seeding the hippocampus reveals projections associated with limbic pathways and related structures. Axes: R (right), S (superior), and A (anterior). Colors correspond to fiber orientation: longitudinal fibers are shown in green, vertical tracts in shades of purple and blue, and transverse pathways in red. Full names of all labeled structures are seen in the list of abbreviations. Scale bar = 10 mm

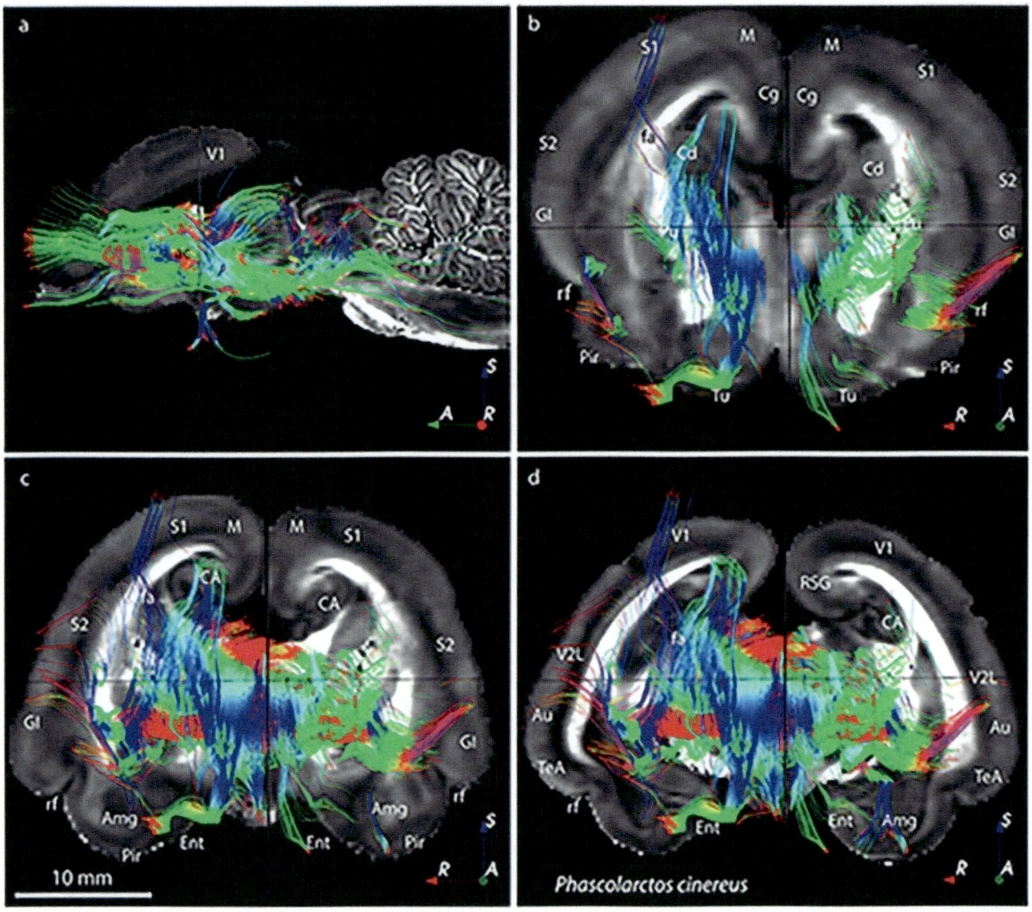

Fig. 5.16 Fiber pathways connecting the right hemisphere contralaterally. Fiber pathways originating in the right hemisphere traverse to the opposite side of the brain through key commissural structures, including the anterior commissure, hippocampal formation, and ventral telencephalic regions. Orientation axes are labeled as R (right), S (superior), and A (anterior). Colors represent fiber orientation: green for longitudinal tracts, purple and blue shades for vertical fibers, and red for transverse pathways. Full names of all labeled structures are seen in the list of abbreviations. Scale bar = 10 mm

physiology. Integrating structural and functional modalities, either in vivo or via cross-species validation, remains a critical next step.

5.5 Conclusion

This chapter illustrates how emerging neurovisualization technologies, particularly high-field MRI and DTI, enable detailed exploration of brain architecture and connectivity in non-model organisms. The koala brain connectome serves as a compelling case study, revealing modular, energy-efficient neural circuits shaped by evolutionary and ecological factors.

Beyond comparative neuroanatomy, these technologies hold translational potential in human neuroscience, offering insights into the structural basis of cognition, behavior, and brain disorders. As imaging tools continue to evolve, their integration into interdisciplinary research will be essential for decoding the complexities of brain organization across species.

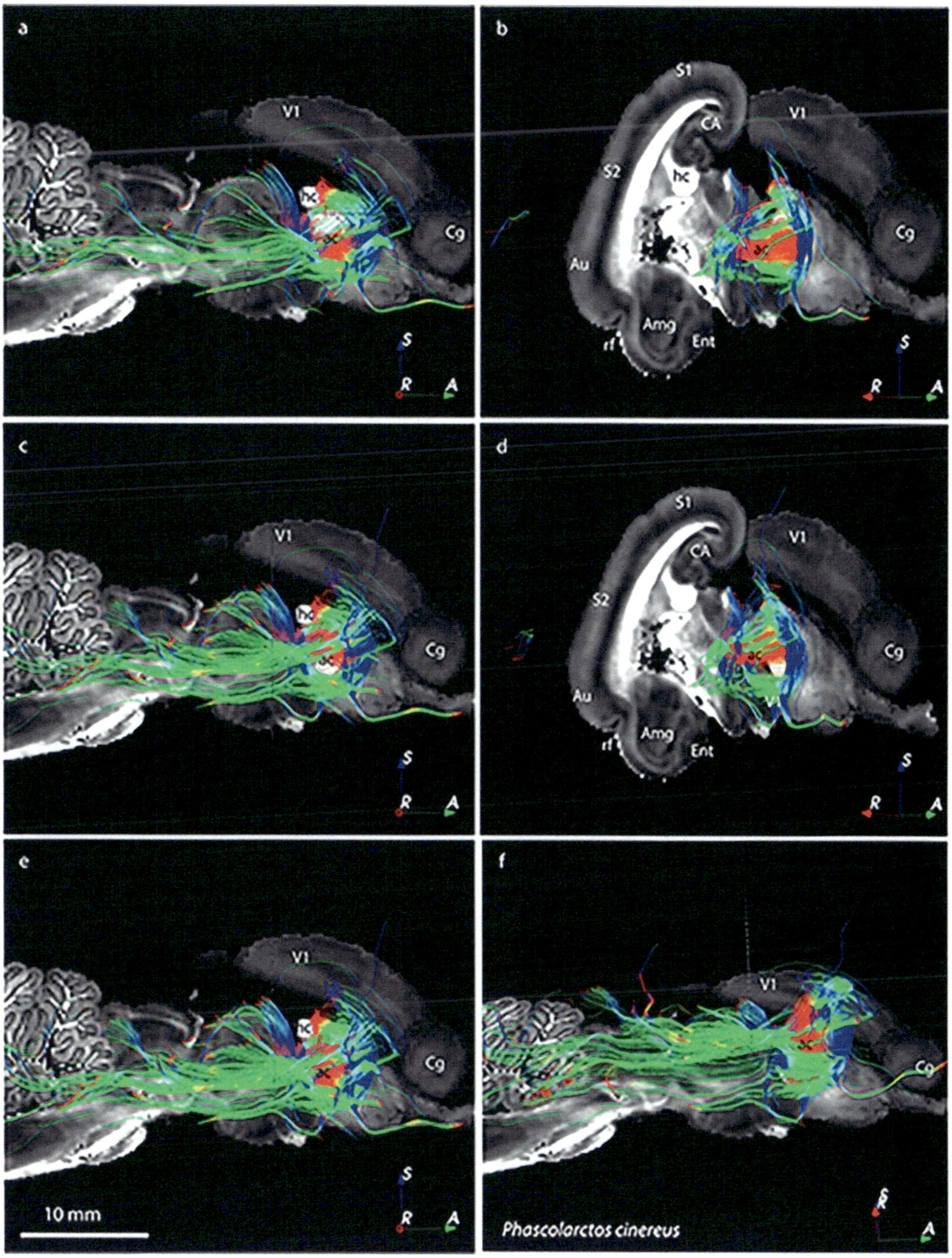

Fig. 5.17 Topographic organization of fibers within the anterior commissure. Figure 5.17a, b illustrates tractography results initiated from the inferior portion of the anterior commissure, while Fig. 5.17c and 5.17d shows outcomes from seeding its superior portion. Figure 5.17e and 5.17f presents data from seeding the entire anterior commissure. This segmented seeding strategy was employed to clarify fiber distribution percentages, as commissural fibers reaching target areas were not distinctly defined. Axes: R (right), S (superior), and A (anterior). Colors correspond to fiber orientation: transverse fibers appear red, longitudinal fibers are shown in green, and vertical fibers are represented by shades of purple and blue. Full names of all labeled structures are seen in the list of abbreviations. Scale bar = 10 mm

Acknowledgments We gratefully acknowledge the support of the National Imaging Facility and the Biological Imaging Laboratory at the UNSW Mark Wainwright Analytical Centre for access to and assistance with the high-field preclinical MRI system. We also thank Taronga Zoo, Australia, for providing the koala brain specimen used in this study. Their collaboration was essential in enabling this comparative neuroanatomy research.

Declaration by Authors Regarding Ethical Approval for the Content Used in this Chapter Ethical approval was not required for this study as the koala had died from natural causes under veterinary supervision during a routine health assessment at Taronga Zoo, Sydney. The animal was part of a geriatric population monitored under the zoo's established aged animal welfare program. No experimental procedures were conducted. The brain was collected postmortem during a standard necropsy performed by qualified zoo veterinary staff. All procedures adhered to institutional policies and national guidelines for the respectful use of deceased animals in research.

References

Andersson JLR, Skare S, Ashburner J (2003) How to correct susceptibility distortions in spin-echo echo-planar images: application to diffusion tensor imaging. NeuroImage 20(2):870–888. https://doi.org/10.1016/s1053-8119(03)00336-7

Ashwell K, Marotte L (2010) Stereotaxic atlas of the brain of the tammar wallaby. In: Ashwell KWS (ed) The neurobiology of Australian marsupials. Cambridge University Press, pp 242–244

Avants B, Tustison NJ, Song G (2009) Advanced normalization tools: V1.0. Insight J. https://doi.org/10.54294/uvnhin

Basser PJ, Mattiello J, LeBihan D (1994) Mr diffusion tensor spectroscopy and imaging. Biophys J 66(1):259–267. https://doi.org/10.1016/s0006-3495(94)80775-1

Behrens TEJ, Berg HJ, Jbabdi S, Rushworth MFS, Woolrich MW (2007) Probabilistic diffusion tractography with multiple fibre orientations: what can we gain? NeuroImage 34(1):144–155. https://doi.org/10.1016/j.neuroimage.2006.09.018

Bell D, Abdrabou A (2013) Diffusion tensor imaging and fibre tractography. Radiopaediaorg. https://doi.org/10.53347/rid-22702

Bullmore E, Sporns O (2009) Complex brain networks: graph theoretical analysis of structural and functional systems. Nat Rev Neurosci 10(3):186–198. https://doi.org/10.1038/nrn2575

Dehay C, Kennedy H (2007) Cell-cycle control and cortical development. Nat Rev Neurosci 8(6):438–450. https://doi.org/10.1038/nrn2097

Fedorov A, Beichel R, Kalpathy-Cramer J, Finet J, Fillion-Robin JC, Pujol S, Bauer C, Jennings D, Fennessy F, Sonka M, Buatti J, Aylward S, Miller JV, Pieper S, Kikinis R (2012) 3D slicer as an image computing platform for the quantitative imaging network. Magn Reson Imaging 30(9):1323–1341. https://doi.org/10.1016/j.mri.2012.05.001

Gozzi A, Schwarz AJ (2016) Large-scale functional connectivity networks in the rodent brain. NeuroImage 127:496–509. https://doi.org/10.1016/j.neuroimage.2015.12.017

Haight J, Murray P (1981) The cranial endocast of the early Miocene marsupial, *Wynyardia bassiana:* an assessment of taxonomic relationships based upon comparisons with recent forms. Brain Behav Evol 19:17–36

Honey CJ, Sporns O, Cammoun L, Gigandet X, Thiran JP, Meuli R, Hagmann P (2009) Predicting human resting-state functional connectivity from structural connectivity. Proc Natl Acad Sci 106(6):2035–2040. https://doi.org/10.1073/pnas.0811168106

Huang H, Zhang J, van Zijl PCM, Mori S (2004) Analysis of noise effects on dti-based tractography using the brute-force and multi-roi approach. Magn Reson Med 52(3):559–565. https://doi.org/10.1002/mrm.20147

Jenkinson M, Beckmann CF, Behrens TEJ, Woolrich MW, Smith SM (2012) FSL. NeuroImage 62(2):782–790. https://doi.org/10.1016/j.neuroimage.2011.09.015

Johnson JI (1977) Central nervous system of marsupials. In: Hunsaker D (ed) The biology of marsupials. Academic Press, Inc, New York, pp 157–278

Karlen SJ, Krubitzer L (2007) The functional and anatomical organization of marsupial neocortex: evidence for parallel evolution across mammals. Prog Neurobiol 82(3):122–141. https://doi.org/10.1016/j.pneurobio.2007.03.003

Le Bihan D (2003) Looking into the functional architecture of the brain with diffusion MRI. Nat Rev Neurosci 4(6):469–480. https://doi.org/10.1038/nrn1119

Markov NT, Ercsey-Ravasz M, Van Essen DC, Knoblauch K, Toroczkai Z, Kennedy H (2013) Cortical high-density Counterstream architectures. Science 342(6158). https://doi.org/10.1126/science.1238406

Mori S, van Zijl PC (2002) Fiber tracking: principles and strategies—a technical review. NMR Biomed 15(7–8):468–480. https://doi.org/10.1002/nbm.781

Neal J, Takahashi M, Silva M, Tiao G, Walsh CA, Sheen VL (2006) Insights into the gyrification of developing ferret brain by magnetic resonance imaging. J Anat 210(1):66–77. https://doi.org/10.1111/j.1469-7580.2006.00674.x

Norton I, Essayed WI, Zhang F, Pujol S, Yarmarkovich A, Golby AJ, Kindlmann G, Wassermann D, Estepar RS, Rathi Y, Pieper S, Kikinis R, Johnson HJ, Westin C-F, O'Donnell LJ (2017) Slicerdmri: open-source diffusion MRI software for brain cancer research. Cancer

Res 77(21). https://doi.org/10.1158/0008-5472.can-17-0332

Oh SW, Harris JA, Ng L, Winslow B, Cain N, Mihalas S, Wang Q, Lau C, Kuan L, Henry AM, Mortrud MT, Ouellette B, Nguyen TN, Sorensen SA, Slaughterbeck CR, Wakeman W, Li Y, Feng D, Ho A et al (2014) A mesoscale connectome of the mouse brain. Nature 508(7495):207–214. https://doi.org/10.1038/nature13186

Otsu N (1979) A threshold selection method from gray-level histograms. IEEE Trans Syst Man Cybern 9(1): 62–66. https://doi.org/10.1109/tsmc.1979.4310076

Ronneberger O, Fischer P, Brox T (2015) U-net: convolutional networks for biomedical image segmentation. In: International conference on medical image computing and computer-assisted intervention. Springer, pp 234–241. https://doi.org/10.1007/978-3-319-24574-4_28

Rowe M (1990) Organization of the cerebral cortex in monotremes and marsupials. In E. G. Jones & A. Peters (Eds.), Cerebral cortex (pp. 263– 334). Plenum Press.

Schindelin J, Arganda-Carreras I, Frise E, Kaynig V, Longair M, Pietzsch T et al (2012) Fiji: an open-source platform for biological-image analysis. Nat Methods 9(7):676–682. https://doi.org/10.1038/nmeth.2019

Schmahmann JD, Pandya DN, Wang R, Dai G, D'Arceuil HE, de Crespigny AJ, Wedeen VJ (2007) Association fibre pathways of the brain: parallel observations from diffusion Spectrum imaging and autoradiography. Brain 130(3):630–653. https://doi.org/10.1093/brain/awl359

Thittamranahalli Kariyappa J, Zanoni S, Bongers A, Tong L, Ashwell KW (2022) Magnetic resonance imaging and diffusion tensor imaging reconstruction of connectomes in a macropod, the quokka (*setonix brachyurus*). J Comp Neurol 530(12):2188–2214. https://doi.org/10.1002/cne.25328

Toga AW, Thompson PM (2003) Mapping brain asymmetry. Nature Reviews Neuroscience 4(1):37–48.https://doi.org/10.1038/nrn1009

Tournier J-D, Calamante F, Gadian DG, Connelly A (2004) Direct estimation of the fiber orientation density function from diffusion-weighted MRI data using spherical deconvolution. NeuroImage 23(3): 1176–1185. https://doi.org/10.1016/j.neuroimage.2004.07.037

Tournier J, Mori S, Leemans A (2011) Diffusion tensor imaging and beyond. Magn Reson Med 65(6): 1532–1556. https://doi.org/10.1002/mrm.22924

Ugurbil K (2014) Magnetic resonance imaging at ultrahigh fields. IEEE Trans Biomed Eng 61(5):1364–1379. https://doi.org/10.1109/TBME.2014.2313610

Van den Heuvel MP, Sporns O (2011) Rich-club organization of the human connectome. J Neurosci 31(44): 15775–15786. https://doi.org/10.1523/jneurosci.3539-11.2011

Van der Zwaag W, Schäfer A, Marques JP, Turner R, Trampel R (2016) Recent advances in ultra-high field MRI: applications to functional imaging of the human brain at 7 T. NeuroImage 168:101–118. https://doi.org/10.1016/j.neuroimage.2017.02.018

Recent Advances in Neurohistology

6

Apurba Patra, N. B. Pushpa, and Kumar Satish Ravi

Abstract

Neurohistology, microscopic examination of neural tissue, has never been advancing so quickly in contemporary times, driven by advances in technology and methods. The availability of high-resolution imaging devices like super-resolution microscopy and light-sheet fluorescence microscopy allows unprecedented imaging of neural structure. Molecular biological developments have permitted new markers to provide accurate cellular identification and functional characterization and to reveal the intricacies of the nervous system. Artificial intelligence and machine learning are transforming data analysis to improve investigation of neural connectivity and dynamics. Also, tissue clearing technologies and organoid models are facilitating investigation of three-dimensional neural architecture and disease modeling. These new technologies are explained in this chapter along with their application toward the comprehension of brain repair mechanisms, neural development, and neurodegenerative diseases. Recent advances in neurohistology using the multidisciplinary approaches are rewriting the rules of neuroscience and opening the doors of discovery and therapeutic advancement.

Keywords

Neurohistology · Histological techniques · Neurons · Immunohistochemistry · Electron microscopy · Fluorescent antibody technique

6.1 Introduction

6.1.1 Definition and Field of Neurohistology

Neurohistology is a specialized branch of histology that focuses on the microscopic structure and organization of nervous tissue (Rampelliwar et al. 2025). This branch encompasses the microscopic study of the central and peripheral nervous systems at the cellular and subcellular levels. This discipline delves into the intricate architecture of neurons—the principal signaling cells, various neuroglial cells (such as astrocytes, oligodendrocytes, and microglia), synaptic structures, myelin sheaths, and the extracellular matrix that supports and modulates neural function. The detailed microscopic analysis performed in neurohistology

A. Patra
Department of Anatomy, All India Institute of Medical Sciences, Bathinda, Punjab, India

N. B. Pushpa
Department of Anatomy, JSS Medical College, JSSAHER, Mysore, Karnataka, India

K. S. Ravi (✉)
Department of Anatomy, National Institute of Medical Sciences and Research, Jaipur, Rajasthan, India

Department of Anatomy, KMC Mangalore, MAHE, Manipal, Karnataka, India

K. S. Ravi et al. (eds.), *Visualization in Neuroanatomical Sciences*, Biomedical Visualization 11,
https://doi.org/10.1007/978-3-032-07380-8_6

provides critical insights into the spatial arrangement, morphological variations, and functional interrelationships among different neural elements. Additionally, advanced staining techniques, immunohistochemistry, and electron microscopy allow better visualization of synaptic connections, dendritic arborization, axonal integrity, and the ultrastructure of neural and glial cells (Schröder et al. 2020).

This technique serves as a cornerstone for the understanding of neural physiology and pathophysiology, indispensable in neuroscience research, clinical neurology, and diagnostic neuropathology. By helping in the identification of subtle histopathological changes, neurohistology plays a key role in diagnosing a broad spectrum of neurological disorders including neurodegenerative diseases (e.g., Alzheimer's disease, Parkinson's disease), brain and spinal cord tumors, demyelinating conditions (like multiple sclerosis), congenital malformations, and neurodevelopmental disorders such as autism spectrum disorders (Perl 2010).

In recent times, with the integration of digital pathology and molecular techniques, neurohistology is rapidly advancing toward more precise, quantitative, and high-throughput analyses. This evolution has transformed it into a dynamic field that not only aids in disease diagnosis but also in the exploration of novel therapeutic targets and biomarkers in modern neuroscience.

6.1.2 Neurohistology Is an Area that Integrates Numerous Disciplines

Cellular and molecular neuroanatomy: It is concerned with the structure and function of neurons and glial cells through highly sophisticated staining techniques, electron microscopy, and immunohistochemistry (Doyle 1978).

Neurodevelopmental studies: It helps in understanding neuronal differentiation, migration, and synaptogenesis during embryogenesis (Rakic 2003).

Neuropathology: Neurohistology is also important in the diagnosis of conditions such as Alzheimer's, Parkinson's, and multiple sclerosis based on histopathological assessment (Braak and Braak 1991).

Mapping of neuroanatomy: It is used to outline brain areas and neural circuits necessary for functional neuroimaging studies (Mesulam 2000).

As digital pathology and AI-facilitated histological examination progress, neurohistology further advances, deepening our insights into the nervous system's intricate nature.

6.1.3 Historical Perspective and Milestones

Neurohistology traces its origins to the nineteenth century with Camillo Golgi's silver staining technique (Golgi 1873), which revolutionized neuronal visualization. Ramón y Cajal, using Golgi's method, formulated the neuron doctrine (Cajal 1889), establishing neurons as individual units. Nissl staining (1894) facilitated the study of neuronal cell bodies, while myelin staining (Studnička 1900) enhanced white matter analysis. The advent of electron microscopy (1950s) provided ultrastructural insights, followed by immunohistochemistry (1970s) for molecular-level studies. Modern advancements include various newer techniques such as super resolution microscopy, confocal microscopy and Clear Lipid-exchanged Acrylamide-hybridized Rigid Imaging/ Immunostaining (CLARITY) (Epp et al. 2015). Moreover, AI-driven digital pathology facilitates rapid, automated morphological analysis, enabling the identification of subtle histopathological features that might be challenging to figure out through traditional methods. This technological progression not only enhances our understanding of neurological diseases but also contributes to the development of targeted therapies and improved patient outcomes (Jensen et al. 2024).

6.1.4 Recent Developments in Neurohistology

Recent developments in neurohistology, such as AI-based image analysis, super-resolution microscopy, clear lipid-exchanged acrylamide-hybridized rigid imaging/immunostaining/in situ hybridization-compatible tissue hydrogel (CLARITY), and multiplex immunohistochemistry, have transformed studies of neural tissues (Nabi et al. 2024). These technologies allow for high-resolution, three-dimensional visualization of neural networks, enabling improved disease diagnosis and treatment plans. AI algorithms, particularly those employing deep learning techniques, have demonstrated remarkable capabilities in analyzing histopathological images with high precision. This advancement is leading to improved diagnostic accuracy, streamlined workflows, and personalized treatment strategies in neuropathology. Such advances link microscopy and computational neuroscience, ensuring improved neuronal connectivity and pathology understanding (Bera et al. 2019; Broggi et al. 2024; Redlich et al. 2024).

6.2 Technological Advances in Neurohistology

6.2.1 Super-Resolution Imaging Techniques

The most recent trend in advanced neurohistology utilizes super-resolution microscopy, such as stimulated emission depletion microscopy (STED), structured illumination microscopy (SIM), photoactivated localization microscopy (PALM) (STED, SIM, PALM), scanning electron microscopy (SEM), transmission electron microscopy (TEM), and light-sheet fluorescence microscopy, to achieve accurate high-resolution neural imaging (Fig. 6.1). These imaging methods have nanometer-level resolution to offer true synapse, axon, and glial interface visualization essential to the comprehension of neurodegenerative diseases and connectivity between brain regions (Hell 2009).

6.2.2 Super-Resolution Microscopy

The advent of super-resolution microscopy has revolutionized the field of neurohistology by enabling imaging well below the threshold of conventional microscopy (down to tens of nanometers). Among the most powerful microscopes in this domain are STED, PALM, and SIM. These innovative technologies have pushed the boundaries of what can be visualized within the nervous system (Betzig 2015).

6.2.2.1 STED Microscopy

STED by using a depletion laser achieves super-resolution by selectively deactivating fluorophores in the periphery of the focal spot, leaving only a nanometer-sized central active area

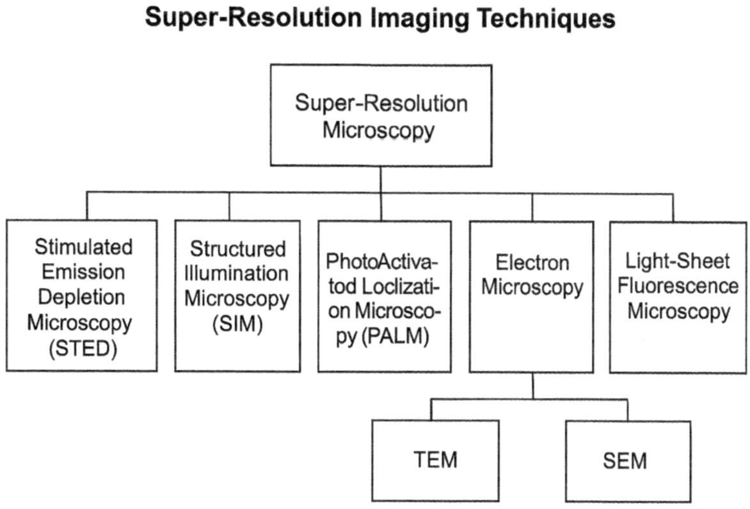

Fig. 6.1 Super-resolution imaging techniques

for imaging. This allows for continuous imaging in living cells with minimal photodamage, making it highly convenient for observing synaptic vesicles, axonal projections, and neurofilament organization in real time (Westphal et al. 2008).

6.2.2.2 PALM Microscopy

PALM utilizes photoactivatable fluorescent proteins to switch on only a small subset of fluorophores at a time. By computationally localizing each molecule with extreme precision, it reconstructs a high-resolution image from thousands of exposures. PALM is particularly beneficial for mapping individual synaptic proteins, revealing structural abnormalities in neurodegenerative diseases like Alzheimer's and Huntington's disease (Shroff et al. 2013).

6.2.2.3 SIM Microscopy

SIM projects patterned light onto the specimen and captures the interference patterns, thus enhancing resolution. Later on, these interference patterns are computationally processed to reconstruct a super-resolved image. SIM is well suited for live-cell imaging and capable of studying the neuronal cytoskeleton dynamics and synaptic remodeling under physiological and pathological conditions (Arizono et al. 2023). The ability of these super-resolution techniques to visualize fine neural structures (synaptic clefts, dendritic spines, microtubules, neurofilaments, etc.) in detail has significant implications in neurodegenerative disorders. Subtle changes in synaptic morphology and protein aggregation can now be visualized early, facilitating early diagnosis and therapeutic monitoring (Kashiwagi and Okabe 2021).

6.2.3 Light-Sheet Fluorescence Microscopy

Light-sheet fluorescence microscopy (LSFM) represents a cutting-edge, rapid, high-resolution, three-dimensional imaging technique to visualize biological specimens with minimal phototoxicity. Unlike traditional confocal microscopy, LSFM illuminates the specimen with a thin sheet of light from the side, thus restricting excitation to a single optical plane. This technique is capable of reducing photodamage and photobleaching significantly, making it an ideal method for imaging delicate neural tissues, especially in live or long-term studies (Stelzer et al. 2021).

One of the most transformative applications of LSFM lies in whole-brain imaging after optical tissue clearing. This approach preserves the 3D structure of the brain while allowing deep tissue penetration and uniform staining, thereby enabling detailed exploration of neural architecture, developmental processes, and large-scale connectomics (Hillman et al. 2019).

6.2.4 Electron Microscopy Advances

Electron microscopy (EM) remains a gold standard for ultrastructural analysis in neurohistology. It offers nanometer-scale resolution far beyond the limits of light microscopy. It is primarily used in two modes:

Transmission electron microscopy (TEM): Provides detailed two-dimensional images of internal cellular structures and allows visualization of organelles, synaptic vesicles, and mitochondrial architecture within neurons and glial cells.

Scanning electron microscopy (SEM): Produces high-resolution surface topography, helping in detailed imaging of neural membranes, axonal pathways, and extracellular matrix.

Advanced variants of SEM, such as serial block-face SEM (SBF-SEM) and focused ion beam SEM (FIB-SEM) are capable of 3D reconstructions of tissues at ultrastructural resolution. These techniques are critical in the field of connectomics, which aims to map neural circuits and synaptic connections across large brain volumes (Shi and Toga 2017). By enabling detailed visualization of neuron-to-neuron contacts and synaptic architectures, EM significantly contributes to understanding brain organization, development, and pathology at the finest level (Briggman and Bock 2012).

6.3 Molecular Advancements

6.3.1 Novel Cellular and Molecular Markers

Progress in neurohistology has brought new biomarkers such as NeuN, neuronal nuclei (neurons); Iba1, ionized calcium-binding adapter molecule 1 (microglia); and GFAP, glial fibrillary acidic protein (astrocytes) (Fig. 6.2), in addition to molecular markers for synaptic function and neuroinflammation. Multiplex immunohistochemistry and ribonucleic acid (RNA) probes increase cellular resolution for neurodegenerative and psychiatric diseases (Jessen 2004; Bennett et al. 2016).

6.3.2 Specificity in Cellular Identification

High-specificity techniques like fluorescent in situ hybridization (FISH), single-cell RNA sequencing, and multiplexed antibody staining refine cell-type classification. Advanced marker panels distinguish neuronal subtypes, astrocyte states, and microglial activation profiles, crucial for neurodevelopmental and disease research (Tasic et al. 2018; Lein et al. 2017).

6.3.3 Functional Characterization of Neurons and Glial Cells

Techniques like calcium imaging, optogenetics, and electrophysiology assess neuron-glia interactions and cellular activity. Live-cell imaging of astrocyte calcium waves and microglial responses provides functional insights into neuroimmune interactions, synaptic modulation, and brain plasticity (Fields et al. 2015; Scanziani and Häusser 2009).

6.3.4 Genetic and Proteomic Tools

RNA sequencing, mass spectrometry proteomics, and spatial transcriptomics allow for high-resolution molecular characterization of brain cells. These tools detect disease biomarkers, exposing mechanisms in neurodegeneration, synaptic dysfunction, and glial reactivity (Uhlen et al. 2015; Zeisel et al. 2018).

6.3.5 CRISPR-Based Tools for Histological Studies

Clustered regularly interspaced short palindromic repeats (CRISPR)-based methods facilitate gene expression, lineage tracing, and fluorescent

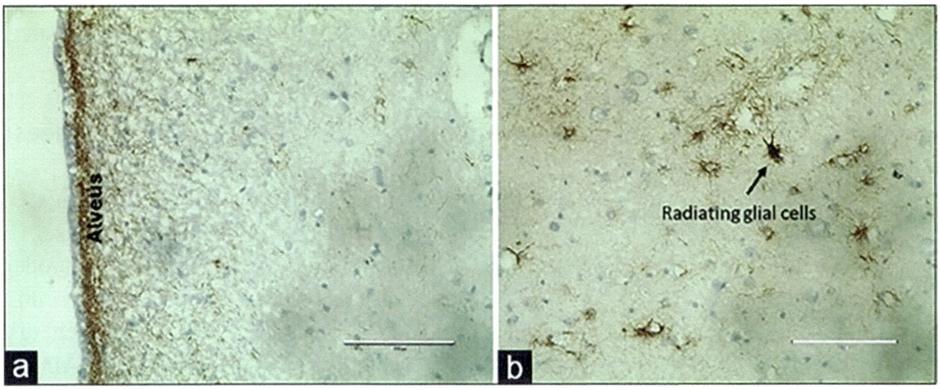

Fig. 6.2 Glial fibrillary acidic protein (GFAP) staining of cornu ammonis (CA) 3 under low magnification. (**a**) Alveus in CA3 region and (**b**) arrow indicates radiating glial cell in CA3 (Source: Cytoarchitecture of the Hippocampal Subfields CA3, CA4, and the Dentate Gyrus in Human Cerebrum: A Detailed Histological Analysis, National Journal of Clinical Anatomy14(1):35–40, Jan-Mar 2025)

labeling of neural cells. CRISPR-Cas9 knock-ins and CRISPR activation (CRISPRa) facilitate functional genetic experimentation in neuronal circuits to contribute to neurodevelopmental and neurodegenerative disease models (Shin et al. 2017; Chen et al. 2019).

6.3.6 Single-Cell Transcriptomics in Neural Research

Single-cell RNA sequencing (scRNA-seq) uncovers cellular heterogeneity, gene expression signatures, and disease-associated changes in transcription within the brain. It traces neuron-glia connections, developmental pathophysiology, and molecular features of disorders such as Alzheimer's and schizophrenia (Saunders et al. 2018; Lake et al. 2016).

6.4 Tissue Preparation and Visualization

6.4.1 Tissue Clearing Methods

6.4.1.1 CLARITY and Expansion Microscopy

CLARITY, developed by Chung et al. (2013), uses hydrogel-embedding to fix molecular structures while making tissues transparent, allowing deep, high-resolution imaging of intact neural circuits. Expansion microscopy (ExM), developed by Chen et al. (2015), physically expands tissues by isotropic expansion, permitting nanoscale visualization of subcellular structures beyond diffraction limits.

6.4.1.2 Improvements in Three-Dimensional Tissue Analysis

Recent developments in light-sheet fluorescence microscopy (LSFM) and two-photon microscopy greatly enhance 3D imaging of cleared tissues, providing high-speed, large-scale brain mapping with low photodamage (Tomer et al. 2014; Denk et al. 1990). These techniques advance neuroanatomical research, particularly for large-volume reconstructions of neural circuits.

6.4.2 New Developments in Staining and Labeling

6.4.2.1 Fluorescent Dyes and Quantum Dots

Photostability and selectivity are improved in fluorescent probes, with quantum dots (QDs) providing high brightness and multiplexing capabilities for long-term neuroimaging (Michalet et al. 2005). The fluorophores at the nanoscale can offer multiplex staining of synapses that complements high-resolution neurohistology studies (Resch-Genger et al. 2008).

6.4.2.2 Immunohistochemistry Advances

Multiplex immunohistochemistry (mIHC) techniques allow simultaneous detection of multiple proteins, increasing cellular and molecular neuroscience characterization (Tan et al. 2020). High-resolution immunohistochemical methods now include AI-supported quantification for precise neuropathological assessment (Toth et al. 2016).

6.5 Neurohistology and Disease Modeling

6.5.1 Neurodegenerative Diseases

Alzheimer's Disease (AD), Parkinson's Disease (PD), and Huntington's Disease (HD): Neurohistology forms the basis of neurodegenerative disease research by studying the cellular and molecular changes in brains of such diseases. Histopathological analysis in AD shows amyloid-beta plaques and neurofibrillary tangles as pathognomonic (Braak and Braak 1991). Dopaminergic neuron degeneration of substantia nigra defines PD, Lewy bodies (Spillantini et al. 1997). In HD huntingtin protein aggregates and striatal atrophy causes neuronal dysfunction (Vonsattel et al. 1985).

6.5.1.1 Identification of Pathological Markers

Histopathological markers such as phosphorylated tau (p-Tau) and amyloid-beta (Aβ) for AD, α-synuclein for PD, and mutant huntingtin (mHTT) for HD enable diagnosis and disease progression (Irwin et al. 2017). These markers are analyzed through immunohistochemistry (IHC), fluorescence in situ hybridization (FISH), and proteomic analysis.

6.5.2 Neurodevelopmental Disorders

6.5.2.1 Autism Spectrum Disorder (ASD) and Congenital Malformations

In ASD, histological findings reveal changes in the lamination of the cortex, microglial activation, and reduction in synaptic density, resulting in multiple connectivity and cognitive impairment (Stoner et al. 2014). In congenital malformations, such as lissencephaly and microcephaly, histology reveals disruption of neuronal migration and reduced cortex, which is typically due to genetic mutation of lissencephaly 1 (LIS1) and abnormal spindle-like microcephaly-associated (ASPM) genes (Barkovich et al. 2005).

6.5.3 Traumatic Brain Injury (TBI) and Stroke

Data on neuronal loss, glial reactivity, and vascular disease after TBI and stroke can be provided by neurohistology. Histopathological alterations of TBI include microglial activation, diffuse axonal injury (DAI), and astrocytosis (Johnson et al. 2013). Stroke pathology includes ischemic necrosis, disruption of the blood-brain barrier, and inflammatory cell infiltration, and they are investigated using Nissl staining and IHC for GFAP (Dirnagl et al. 1999).

6.6 Applications of Advanced Neurohistology

6.6.1 Functional Mapping of Neural Circuits

Sophisticated neurohistological techniques, including super-resolution microscopy, optogenetics, and viral tracers, enable precise mapping of neural circuits. Synaptic connections are investigated with rabies virus and anterograde tracers (Callaway and Luo 2015). High-resolution tissue clearing technologies, including CLARITY, facilitate 3D imaging of intact circuits (Chung et al. 2013). Calcium imaging and genetically encoded voltage indicators reveal activity-dependent circuit dynamics (Knöpfel 2012).

6.6.2 Insights into Neuroplasticity and Regeneration

Histological investigations illustrate molecular and structural changes involved in neuroplasticity and regeneration. In vivo two-photon microscopy illustrates dendritic spine plasticity following learning and injury (Trachtenberg et al. 2002). Single-cell transcriptomics is used to emphasize gene expression alterations involved in synaptic plasticity (Habib et al. 2020). The activation of glial cells and axonal sprouting after injury point toward potential targets for neural repair (Silver et al. 2015).

6.6.3 Neural Interface Technologies

Neurohistology is essential for assessing brain-machine interfaces (BMIs), neural implants, and bioelectronic prostheses. Histopathological examination of electrode-tissue interfaces evaluates glial scarring, inflammation, and neuronal integration (Polikov et al. 2005). Optogenetic modulation and fiber photometry allow for real-time modulation of neural circuits in BMI use (Boyden et al. 2005). Graphene-based and flexible bioelectronic device advances enhance biocompatibility and signal fidelity in neural interfaces (Khodagholy et al. 2013).

6.7 Ethical and Practical Considerations

6.7.1 Challenges of Adopting New Technology

Adoption of advanced neurohistological technologies has several challenges, including technical complexity, cost, and limited access. High-resolution technologies like super-resolution microscopy and clearing tissue require specialized equipment and technical expertise, thus limiting their widespread use (Zhu et al. 2020). The cost of high-end imaging instrumentation and the time-consuming nature of tissue preparation may act as a barrier for bulk usage in a research or clinical setting (Vickery et al. 2019). Also, issues like sample preparation protocols cross-compatible across a broad range of species are yet to be perfected (Kim et al. 2017).

6.7.2 The Ethical Consequences of Modern Neurohistological Science

Advanced neurohistology, especially in areas like gene manipulation, brain organoid research, and neuroengineering, is deeply ethically nuanced. For example, the application of human pluripotent stem cells to generate brain organoids for disease modeling is ethically nuanced in terms of personhood and whether and how such entities may become neural activity that simulates consciousness (Bonnicksen et al. 2020). CRISPR is also a gene editing technology with risks of germline modification that has implications for future generations (Liao et al. 2020). Privacy and consent exist as well in brain-machine interface and neuroprosthetics, especially in patients with compromised cognition (Harris 2019).

6.7.3 Standardization and Reproducibility Issues

The use of sophisticated neurohistological methods is not standardized and is a matter of reproducibility. Differences in tissue preparation, imaging procedures, and staining influence reliability and accuracy of findings (Schneider et al. 2018). The issue is more acute in multi-lab collaborative or extensive studies where reproducibility is crucial in verifying observations (Baker 2016). As a remedy, there are attempts to create standard operating procedures (SOPs), computerized data-analysis systems, and open-source programs to enhance reproducibility between laboratories (McNutt 2014).

6.8 Future Directions

6.8.1 Multidisciplinary Integration

The way forward for neurohistology is multidisciplinary integration, bringing neuroscience, engineering, artificial intelligence (AI), and biotechnology together. The development of AI-supported image analysis, especially through machine learning algorithm-based tissue segmentation and feature extraction, is capable of boosting processing power and precision of data in neuroanatomical research (Esteva et al. 2019). Besides, the blending of biomaterials and neural engineering to form functional tissue scaffolds for healing damaged neural circuits will boost histological studies, favoring both therapeutic and diagnostic applications (Zhao et al. 2020). Interdisciplinary collaboration among neuroscientists, bioengineers, and data scientists will lead to more consolidated models of brain function and disease, which will facilitate improvements in disease modeling and neurotherapy (Hochgeschwender et al. 2019).

6.8.2 Prospects for Personalized Medicine Opportunities

The opportunity to study the individual-specific neuroanatomy by employing state-of-the-art neurohistological methodologies, such as single-cell RNA sequencing, brain organoids, and genomic editing, holds untold prospects for personalized medicine. With the analysis of patient-derived organoids, scientists can model patient-specific pathogenesis of disease and screen drugs with a known molecular or genetic signature (Fischer et al. 2020). At the same time, biomarkers and neural interfaces can be used to develop personalized therapy according to the patient's unique neurological characteristics, which will improve clinical results as well as reduce side effects (Khan et al. 2019).

6.8.3 New Trends and Technologies

One of the new trends in neurohistology is multi-omics, wherein genomics, transcriptomics, and proteomics are integrated with histological examination to provide an image of brain function and disease as a whole (Fig. 6.3) (Tognini et al. 2020). The technique of 3D bioprinting makes it possible to generate personalized, functional neural tissue, and optogenetics and gene editing provide the precision of controlling neural activity in intact tissue (Liu et al. 2018). In addition, nanotechnology breakthroughs are creating new possibilities for in vivo imaging, improving depth and resolution of neurohistological research (Mazzolai et al. 2020). These breakthroughs will transform our knowledge of brain function, disease, and possible treatments.

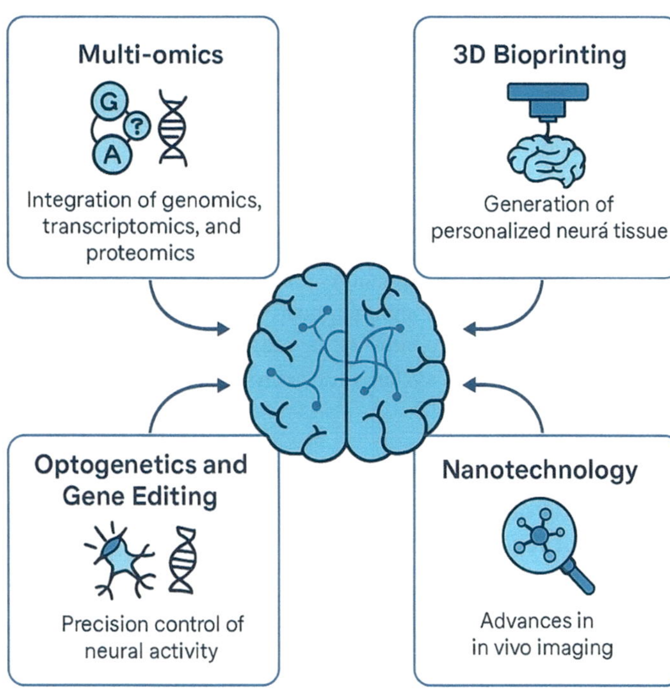

Fig. 6.3 Newer trends in neurohistology

6.9 Conclusion

6.9.1 Summary of Key Advances

Neurohistology has come a long way with the development of high-resolution imaging techniques such as super-resolution microscopy and tissue clearing methods like CLARITY that have enabled 3D imaging of neural circuits. The overlap between genetic and molecular tools, such as CRISPR/Cas9 gene editing, has transformed our knowledge on neurodevelopment and disease. The development of neural tissue engineering, i.e., brain organoids (Knoblich. 2013) and biodegradable scaffolds, has also enabled disease modeling and potentially therapeutic intervention with greater precision.

6.9.2 Contributions to Neuroscience Research and Clinical Applications

Convergence of modern neurohistological technologies has pushed neuroscience research to the front with the capability of providing nanometric observations of neural networks, neuroplasticity, and etiology of neurodegenerative diseases, such as Alzheimer's and Parkinson's. New avenues of personalized medicine have been opened through patient-derived brain organoids with the capability of modeling disease-specific uses. In medicine, neuroimaging and neuroprosthetics are now being developed with real-time feedback systems for neurological disease treatment.

6.9.3 Vision for the Future of Neurohistology

In the future, multidisciplinary convergence of AI, bioengineering, and advanced imaging will be ready to achieve unprecedented accuracy in neuroscientific research and medical therapy. The future of neurohistology will likely see widespread use of patient-specific neurotherapy using organoids and genomic profiles for diagnosis and treatment. Furthermore, advanced imaging and modeling techniques will enable detailed studies of neuroplasticity, brain disease, and neural recovery, eventually leading to revolutionary medical breakthroughs.

References

Arizono M, Idziak A, Quici F, Nägerl UV (2023) Getting sharper: the brain under the spotlight of super-resolution microscopy. Trends Cell Biol 33(2): 148–161

Baker M (2016) The truth about reproducibility. Nature 533:452–454.https://doi.org/10.1038/533452a.4

Barkovich AJ, Guerrini R, Kuzniecky RI, Jackson GD, Dobyns WB (2005) Malformations of cortical development. Brain Dev 27(7):505–524. https://doi.org/10.1016/j.braindev.2004.12.003

Bennett ML, Zhang B, Deardorff AS, Finney A, Vainchtein ID, Zamanian JL, Barres BA (2016) New glial markers for CNS research. Neuron 91(6): 1151–1161. https://doi.org/10.1016/j.neuron.2016.07.027

Bera K, Schalper KA, Rimm DL, Velcheti V, Madabhushi A (2019) Artificial intelligence in digital pathology: new tools for diagnosis and precision oncology. Nat Rev Clin Oncol 16(11):703–715. https://doi.org/10.1038/s41571-019-0252-y

Betzig E (2015) Single-molecule microscopy in cells. Angew Chem 54(28):8034–8053. https://doi.org/10.1002/anie.201410308

Bonnicksen A, Akrami S, Charo RA, Greely HT, Hyun I, Lovell-Badge R (2020) Ethical issues in human brain organoid research. J Neuroethic 13(2):263–275. https://doi.org/10.1007/s12152-020-09428-3

Boyden ES, Zhang F, Bamberg E, Nagel G, Deisseroth K (2005) Millisecond-timescale, genetically targeted optical control of neural activity. Nat Neurosci 8(9): 1263–1268. https://doi.org/10.1038/nn1525

Braak H, Braak E (1991) Neuropathological staging of Alzheimer-related changes. Acta Neuropathol 82(4): 239–259. https://doi.org/10.1007/BF00308809

Briggman KL, Bock DD (2012) EM-based neural circuit mapping. Curr Opin Neurobiol 22(1):154–161. https://doi.org/10.1016/j.conb.2011.10.022

Broggi G, Mazzucchelli M, Salzano S, Barbagallo GMV, Certo F, Zanelli M, Palicelli A, Zizzo M, Koufopoulos N, Magro G, Caltabiano R (2024) The emerging role of artificial intelligence in neuropathology: where are we and where do we want to go? Pathol Res Pract 263:155671

Cajal SR (1889) Histology of the nervous system

Chen F, Tillberg PW, Boyden ES (2015) Expansion microscopy. Science 347(6221):543–548. https://doi.org/10.1126/science.1260088

Chen S, Lee B, Lee AY, Modzelewski AJ, He L, Patel K, Kim J (2019) CRISPRa in brain research. Nat Neurosci 22(4):471–482. https://doi.org/10.1038/s41593-019-0332-1

Callaway EM, Luo L Monosynaptic Circuit Tracing with Glycoprotein-Deleted Rabies Viruses (2015) J Neurosci Jun 17;35(24):8979-85. https://doi.org/10.1523/JNEUROSCI.0409-15.2015

Chung K, Wallace J, Kim SY, Kalyanasundaram S, Andalman AS, Davidson TJ, Mirzabekov JJ, Zalocusky KA, Mattis J, Denisin AK, Pak S, Bernstein H, Ramakrishnan C, Grosenick L, Gradinaru V, Deisseroth K (2013) Structural and molecular interrogation of intact biological systems. Nature 497(7449):332–337. https://doi.org/10.1038/nature12107

Denk W, Strickler JH, Webb WW (1990) Two-photon laser scanning microscopy. Science 248(4951):73–76. https://doi.org/10.1126/science.2321027

Dirnagl U, Iadecola C, Moskowitz MA (1999) Pathophysiology of stroke. Physiol Rev 79(3):755–810. https://doi.org/10.1152/physrev.1999.79.3.755

Doyle D (1978) The fine structure of the nervous system: the neurons and supporting cells. J Neurol Neurosurg Psychiatry 41(2):191–192. https://doi.org/10.1136/jnnp.41.2.191

Epp JR, Niibori Y, Hsiang LHL, Mercaldo V, Deisseroth K, Josselyn SA, Frankland PW (2015) Optimization of CLARITY for clearing whole-brain and other intact organs. eNeuro 2(3): ENEURO.0022-15.2015. https://doi.org/10.1523/ENEURO.0022-15.2015

Esteva A, Chou K, Yeung S, Naik N, Madani A, Mottaghi A et al (2019) A guide to deep learning in radiology. Nat Rev Cancer 19(4):237–250. https://doi.org/10.1038/s41568-019-0118-6

Fields RD, Woo DH, Basser PJ (2015) Glia in synaptic function. Neuron 86(1):74–86. https://doi.org/10.1016/j.neuron.2015.01.008

Fischer C, Shrun L, Moya J, Pasca SP, Studer L (2020) Personalized approaches using patient-derived brain organoids. Nat Biotechnol 38(9):1060–1070. https://doi.org/10.1038/s41587-020-0531-2

Golgi G (1873) Sulla struttura della sostanza grigia del cervello. Gazzetta Medica Italiana: Lombardia 33:244–246

Habib N, McCabe C, Medrano-Fernandez A, Weng Z, Khan S, Vasconcelos G et al (2020) Single-cell transcriptomics in neuroplasticity. Nat Neurosci 23(7):909–919. https://doi.org/10.1038/s41593-020-0645-7

Harris J (2019) Ethical concerns in brain-machine interface development. J Med Ethics 45(8):511–516. https://doi.org/10.1136/medethics-2019-105506

Hell SW (2009) Microscopy and its focal switch. Nat Methods 6(1):24–32

Hillman EMC, Voleti V, Li W, Yu H (2019) Light-sheet microscopy in neuroscience. Annu Rev Neurosci 42: 295–313. https://doi.org/10.1146/annurev-neuro-070918-050357

Hochgeschwender U, Zhang Q, Bruegmann T (2019) Neuroscience and bioengineering collaboration for therapeutic solutions. Nat Biomed Eng 3(10): 697–709. https://doi.org/10.1038/s41551-019-0423-x

Irwin DJ, McMillan CT, Toledo JB, Arnold SE, Shaw LM, Wang LS et al (2017) Neuropathology of neurodegenerative diseases. J Neuropathol Exp Neurol 76(2): 71–87. https://doi.org/10.1093/jnen/nlw124

Jensen MP, Qiang Z, Khan DZ, Stoyanov D, Baldeweg SE, Jaunmuktane Z, Brandner S, Marcus HJ (2024) Artificial intelligence in histopathological image analysis of central nervous system tumours: a systematic review. Neuropathol Appl Neurobiol 50(3):e12981

Jessen KR (2004) Glial cell markers in neuroscience. Glia 47(3):293–294. https://doi.org/10.1002/glia.20050

Johnson VE, Stewart W, Smith DH (2013) Inflammation and axonal injury in TBI. Acta Neuropathol 125(1): 1–23. https://doi.org/10.1007/s00401-012-1057-0

Kashiwagi Y, Okabe S (2021) Imaging of spine synapses using super-resolution microscopy. Anat Sci Int 96(3): 343–358

Khan F, Kim Y, Sharifi MS, Moradian S (2019) Neural interfaces in personalized neurotherapy. Neurosci Biobehav Rev 101:225–236. https://doi.org/10.1016/j.neubiorev.2019.03.025

Khodagholy D, Gelinas JN, Zhao Z, Yeh M, Long M, Greenlee JD et al (2013) Neuroelectronic interfaces using graphene-based materials. Nat Commun 4: 1575. https://doi.org/10.1038/ncomms2573

Kim S, Chen J, Cheng C, Wang X (2017) Advances in tissue preparation for neurohistology. J Neurosci Methods 290:16–22. https://doi.org/10.1016/j.jneumeth.2017.07.002

Knoblich JA (2013) Cerebral organoids model human brain development. Nature 501(7467):373–379. https://doi.org/10.1038/nature12517

Knöpfel T (2012) Genetically encoded optical indicators for the analysis of neuronal circuits. Nat Rev Neurosci 13(10):687–700. https://doi.org/10.1038/nrn3293

Lake BB, Ai R, Kaeser GE, Salathia NS, Yung YC, Liu R et al (2016) scRNA-seq in human neocortex. Science 352(6290):1586–1590. https://doi.org/10.1126/science.aaf1204

Lein ES, Borm LE, Linnarsson S (2017) Molecular diversity of CNS cells. Nat Rev Neurosci 18(10):530–546. https://doi.org/10.1038/nrn.2017.85

Liao S, Li Z, Zhao W (2020) Ethical concerns in CRISPR-Cas9 gene editing. J Bioethical Inq 17(1):1–10. https://doi.org/10.1007/s11673-020-09980-x

Liu X, Miller AL, Park S, Waletzki BE, Shalaby SM (2018) 3D bioprinting of neural tissue and its applications. Biomater Sci 6(6):1415–1425. https://doi.org/10.1039/C7BM01274E

Mazzolai B, Guo X, Wang X (2020) Nanotechnology applications in neurohistology. Nat Nanotechnol

15(4):266–275. https://doi.org/10.1038/s41565-020-0675-1
Mesulam MM (2000) Principles of behavioral and cognitive neurology. Oxford University Press
Michalet X, Pinaud FF, Bentolila LA, Tsay JM, Doose S, Li JJ et al (2005) Quantum dots for imaging. Science 307(5709):538–544. https://doi.org/10.1126/science.1104274
Nabi IR, Cardoen B, Khater IM, Gao G, Wong TH, Hamarneh G (2024) AI analysis of super-resolution microscopy: biological discovery in the absence of ground truth. J Cell Biol 223(8):e202311073. https://doi.org/10.1083/jcb.202311073
Nissl F (1894) Über die Veränderungen der Ganglienzellen. Allgemeine Zeitschrift für. Psychiatrie
Perl DP (2010) Neuropathology of Alzheimer's disease. Mount Sinai J Med: J Trans Personalized Med: J Trans Personalized Med 77(1):32–42
Polikov VS, Tresco PA, Reichert WM (2005) Response of brain tissue to chronically implanted neural electrodes. J Neurosci Methods 148(1):1–18. https://doi.org/10.1016/j.jneumeth.2005.08.015
Rakic P (2003) Developmental and evolutionary adaptations of cortical radial glia. Cereb Cortex 13(6):541–549. https://doi.org/10.1093/cercor/13.6.541
Rampelliwar SA, Sabnis A, Manisha AT (2025) Cytoarchitecture of the hippocampal subfields CA3, CA4, and the dentate gyrus in human cerebrum: a detailed histological analysis. Nat J Clin Anatomy 14(1):35–40
Redlich JP, Feuerhake F, Weis J, Schaadt NS, Teuber-Hanselmann S, Buck C, Luttmann S, Eberle A, Nikolin S, Appenzeller A, Portmann A (2024) Applications of artificial intelligence in the analysis of histopathology images of gliomas: a review. npj Imaging 2(1):16
Resch-Genger U, Grabolle M, Cavaliere-Jaricot S, Nitschke R, Rettig W (2008) Fluorescent dyes in bioimaging. Nat Methods 5(9):763–775. https://doi.org/10.1038/nmeth.1248
Saunders A, Macosko EZ, Wysoker A, Goldman M, Krienen FM, de Rivera H et al (2018) Single-cell atlas of the mouse brain. Cell 174(4):1015–1030. https://doi.org/10.1016/j.cell.2018.07.028
Scanziani M, Häusser M (2009) Electrophysiology in neural circuits. Nature 461(7266):930–939. https://doi.org/10.1038/nature08540
Schneider T, Faber C, Langer R (2018) Standardizing neurohistology protocols. NeuroImage 170:121–130. https://doi.org/10.1016/j.neuroimage.2017.02.004
Schröder H, Moser N, Huggenberger S, Schröder H, Moser N, Huggenberger S (2020) Basic neurohistology. In: Neuroanatomy of the mouse: an introduction, pp 7–25
Shi Y, Toga AW (2017) Connectome imaging for mapping human brain pathways. Mol Psychiatry 22(9):1230–1240
Shin J, Ming G, Song H (2017) CRISPR for neural circuit studies. Neuron 94(3):550–562. https://doi.org/10.1016/j.neuron.2017.04.002
Shroff H, White H, Betzig E (2013) Photoactivated localization microscopy (PALM) of adhesion complexes. Curr Protoc Cell Biol 58(1):4–21
Silver J, Schwab ME, Popovich PG (2015) The glial scar and CNS repair. Nat Rev Neurosci 16(8):441–452. https://doi.org/10.1038/nrn4023
Spillantini MG, Schmidt ML, Lee VM, Trojanowski JQ, Jakes R, Goedert M (1997) Alpha-synuclein in Lewy body diseases. Nature 388(6645):839–840. https://doi.org/10.1038/42166
Stelzer EH, Strobl F, Chang BJ, Preusser F, Preibisch S, McDole K, Fiolka R (2021) Light sheet fluorescence microscopy. Nat Rev Methods Prim 1(1):73
Stoner R, Chow ML, Boyle MP, Sunkin SM, Mouton PR, Roy S et al (2014) Patches of disorganization in autistic brains. N Engl J Med 370(13):1209–1219. https://doi.org/10.1056/NEJMoa1307491
Studnička FK (1900) Untersuchungen über den Bau des Ependyms der nervösen Centralorgane. Anatomische Hefte 15:301–430. https://doi.org/10.1007/BF02109682
Tan HL, Min HD, Liu J (2020) Advances in multiplex IHC. Front Neurosci 14:568. https://doi.org/10.3389/fnins.2020.00568
Tasic B, Yao Z, Graybuck LT, Smith KA, Nguyen TN, Bertagnolli D et al (2018) Single-cell transcriptomics of mouse cortex. Nat Neurosci 21(6):933–942. https://doi.org/10.1038/s41593-018-0154-9
Tognini P, Napoli D, Petrosino L (2020) Multi-omics approaches in neurohistology. Front Neurosci 14:222. https://doi.org/10.3389/fnins.2020.00222
Tomer R, Ye L, Hsueh B, Deisseroth K (2014) Whole-organ imaging with LSFM. Nat Protoc 9(4):753–763. https://doi.org/10.1038/nprot.2014.040
Toth K, Girach F, Manev R (2016) High-resolution IHC for neuroscience. J Neurosci Methods 274:183–190. https://doi.org/10.1016/j.jneumeth.2016.10.005
Trachtenberg JT, Chen BE, Knott GW, Feng G, Sanes JR, Welker E, Svoboda K (2002) Long-term in vivo imaging of experience-dependent synaptic plasticity in adult cortex. Nature 420(6917):788–794. https://doi.org/10.1038/nature01273
Uhlén M, Fagerberg L, Hallström BM, Lindskog C, Oksvold P, Mardinoglu A et al (2015) The human proteome in neuroscience. Science 347(6220):1260419. https://doi.org/10.1126/science.1260419
Vickery R, Takahashi Y, Johnson B (2019) High-cost challenges in neurotechnology. J Neurosci Res 47(1):52–58. https://doi.org/10.1002/jnr.24389
Vonsattel JP, Myers RH, Stevens TJ, Ferrante RJ, Bird ED, Richardson EP (1985) Huntington's disease neuropathology. J Neuropathol Exp Neurol 44(6):559–577. https://doi.org/10.1097/00005072-198511000-00003

Westphal V, Rizzoli SO, Lauterbach MA, Kamin D, Jahn R, Hell SW (2008) Video-rate far-field optical nanoscopy dissects synaptic vesicle movement. Science 320(5873):246–249

Zeisel A, Hochgerner H, Lönnerberg P, Johnsson A, Memic F, Van Der Zwan J et al (2018) Brain cell transcriptomics. Cell 174(5):1309–1324. https://doi.org/10.1016/j.cell.2018.07.023

Zhao X, Liu Y, Wang X, Zhang H, Ma Y (2020) Biomaterials in neural tissue engineering. Front Bioeng Biotechnol 8:135. https://doi.org/10.3389/fbioe.2020.00135

Zhu Y, Wang L, Luo Q (2020) Challenges in advanced tissue imaging technologies. Trends Neurosci 43(5): 317–328. https://doi.org/10.1016/j.tins.2020.02.006

Brain Mapping and Its Applications

7

Gulgun Sengul

Abstract

Brain mapping is a multidisciplinary field that aims to elucidate the structural and functional organization of the brain through advanced imaging techniques and computational analysis. Over the past century, the field has evolved from rudimentary anatomical studies to sophisticated methodologies integrating neuroimaging, electrophysiology, and artificial intelligence. This chapter provides a historical overview of brain mapping, highlighting key milestones and the contributions of major initiatives such as the Human Connectome Project (HCP), the Allen Brain Atlases, and the Brain Research Through Advancing Innovative Neurotechnologies (BRAIN) Initiative, which have significantly advanced our understanding of brain architecture and function. The clinical and research applications of brain mapping are vast, spanning neurosurgery, neurological disorder diagnostics, cognitive neuroscience, and neurodevelopmental studies. Emerging technologies, including brain-computer interfaces (BCIs) and the integration of artificial intelligence (AI) and machine learning, are reshaping the field by enabling more precise analyses of neural networks and facilitating novel therapeutic interventions. Despite these advances, brain mapping presents significant ethical and methodological challenges, particularly in the context of neuroprivacy, data interpretation, and translational applications. Future directions in the field will likely involve the refinement of high-resolution imaging, the development of multimodal integration techniques, and the application of brain mapping in personalized medicine and neuroeducation. By bridging fundamental neuroscience with clinical and technological innovation, brain mapping continues to revolutionize our understanding of the human brain and its vast potential applications.

Keywords

Brain mapping · Human connectome · Neuroimaging · Brain-computer interfaces

Brain mapping is a dynamic and interdisciplinary field dedicated to exploring the structure and function of the brain. By integrating imaging technologies, electrophysiology, and computational methods, scientists can create detailed representations of neural circuits, brain regions, and connectivity patterns. These advances have greatly enhanced our understanding of how the brain processes information, adapts to changes, and evolves throughout life.

A fundamental aspect of brain mapping is distinguishing between **structural** and **functional mapping**. Structural brain mapping

G. Sengul (✉)
Department of Anatomy, Ege University, School of Medicine, Bornova, Izmir, Turkey

K. S. Ravi et al. (eds.), *Visualization in Neuroanatomical Sciences*, Biomedical Visualization 11,
https://doi.org/10.1007/978-3-032-07380-8_7

focuses on the anatomical features of the brain, including cortical thickness, white matter pathways, and cellular organization. Technologies such as magnetic resonance imaging (MRI), diffusion tensor imaging (DTI), and high-resolution histological techniques provide precise visualizations of brain morphology and connectivity. Functional brain mapping, in contrast, aims to capture neural activity in real time. Functional MRI (fMRI), electroencephalography (EEG), and magnetoencephalography (MEG) allow researchers to study how different brain regions interact during cognitive functions, emotional responses, and motor activities.

One of the greatest challenges in brain mapping is integrating information from multiple modalities to develop a comprehensive model of the brain. Recent advancements in neuroinformatics and artificial intelligence (AI) have greatly facilitated large-scale data analysis, enabling researchers to uncover complex brain networks and identify connectivity patterns linked to neurological and psychiatric conditions (Shekouh et al. 2024). The emergence of connectomics—the study of the brain's intricate network of connections—has driven initiatives such as the Human Connectome Project (HCP), which seeks to map neural pathways with unprecedented detail (Glasser et al. 2016).

Beyond research, brain mapping has profound clinical applications. It plays a crucial role in neurosurgery by helping surgeons identify critical functional areas before operating, thereby minimizing the risk of impairing cognitive or motor functions. Additionally, brain mapping has significantly contributed to the study of neurodegenerative diseases such as Alzheimer's and Parkinson's, providing essential insights into disease progression and aiding in the development of targeted therapies.

With continuous technological advancements, brain mapping is becoming even more precise and integrative. Innovations in molecular imaging, single-cell transcriptomics, and high-resolution microscopy are refining our ability to examine the brain at cellular and genetic levels. These breakthroughs not only deepen our understanding of neural function but also pave the way for personalized medicine and innovative treatments for neurological disorders.

As the field of brain mapping evolves, it continues to shape our understanding of the human brain. By merging structural, functional, and computational approaches, researchers are unlocking the complexities of neural connectivity, leading to groundbreaking discoveries that have transformative implications for both neuroscience and medicine.

7.1 Historical Background of Brain Mapping

The origins of brain mapping can be traced back to the early anatomical and histological studies of the brain. During the late nineteenth and early twentieth centuries, pioneering neuroscientists such as Korbinian Brodmann and Santiago Ramón y Cajal made foundational contributions that continue to shape the field. Brodmann's meticulous classification of the cerebral cortex into distinct regions based on cellular organization remains a cornerstone of neuroscience research (Brodmann 1909). His work provided early evidence that different cortical areas perform specialized functions, a theory that was later validated through electrophysiological studies and modern neuroimaging techniques.

At the same time, Ramón y Cajal's intricate neuron drawings and detailed descriptions of synaptic connections helped establish the neuron doctrine—the concept that the brain is composed of discrete, interconnected cells (Andres-Barquin 2002). His groundbreaking work laid the foundation for later studies on neural connectivity and functional mapping (Fig. 7.1).

By the mid-twentieth century, electrophysiology brought new insights into brain function. Wilder Penfield's cortical stimulation experiments in neurosurgical patients during the 1930s and 1940s provided one of the earliest functional maps of the brain, leading to the development of the cortical homunculus, a model illustrating how different body parts are represented in the sensory and motor cortices (Penfield and Boldrey 1937).

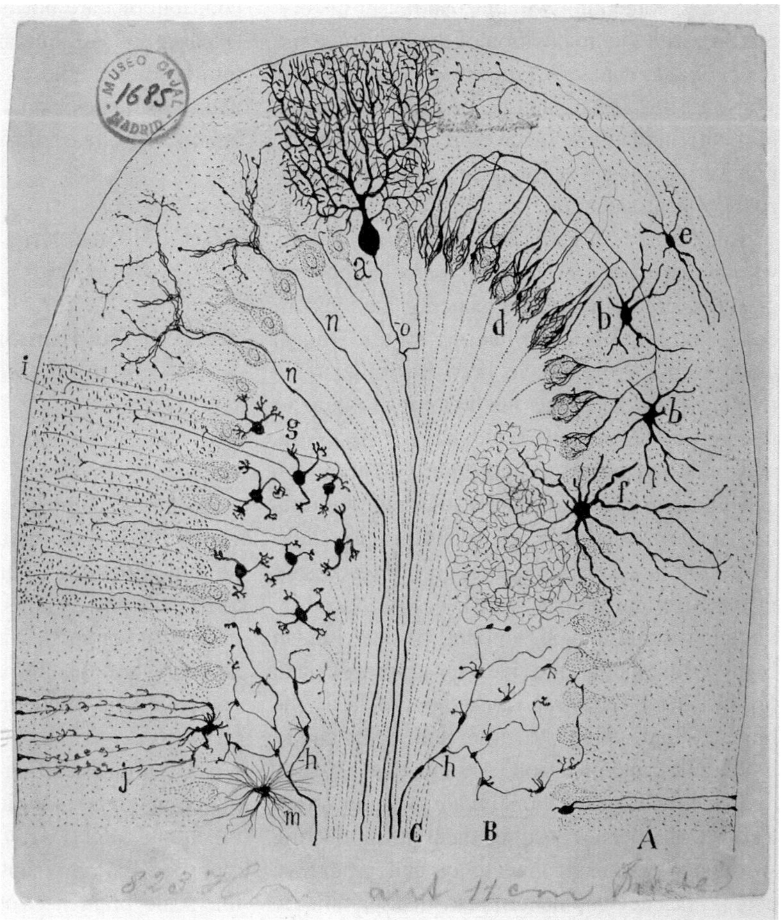

Fig. 7.1 Santiago Ramón y Cajal's drawing of a transverse section through a folium of the mammalian cerebellum (From: Les nouvelles idées sur la structure du systéme nerveux chez l'homme et chez les vertébrés. Paris: Reinwald, 1894. Cajal Legacy. Cajal Institute (CSIC). Madrid)

The latter half of the twentieth century marked a revolution in brain mapping with the advent of neuroimaging technologies. The introduction of computerized tomography (CT) and magnetic resonance imaging (MRI) provided noninvasive ways to study brain structures, while positron emission tomography (PET) and functional MRI (fMRI) allowed scientists to observe neural activity in real time. These innovations enabled researchers to investigate the brain's organization with unprecedented detail.

In recent years, large-scale projects such as the Human Connectome Project and the Brain Research Through Advancing Innovative Neurotechnologies (BRAIN) Initiative have taken brain mapping to new heights. These efforts have facilitated the mapping of neural pathways and interactions at both macro- and microlevels, enhancing our understanding of brain connectivity and function. With the continuous advancement of imaging and computational technologies, brain mapping continues to be a dynamic and evolving field, offering profound insights into the structure and workings of the human brain.

7.2 Contemporary Brain Mappers

Modern brain mapping has been significantly shaped by the contributions of leading neuroscientists who have pioneered modern staining methods and advanced imaging techniques producing influential atlases. Among the major figures in the field are **George Paxinos** and **Charles Watson**, who identified and named

94 nuclei, something without parallel in history. Starting with the first edition of the book *The Rat Brain in Stereotaxic Coordinates* (Paxinos and Watson 1982), their atlases of rodent, primate, bird, and human brains have provided a foundation for researchers studying functional and comparative neuroanatomy.

Another key figure is **David Van Essen**, whose expertise in cortical parcellation has played a crucial role in advancing the field of connectomics. His contributions to the Human Connectome Project have been instrumental in refining brain mapping methodologies, helping researchers better understand large-scale neural networks and connectivity patterns (Van Essen and Glasser 2016). Similarly, **Katrin Amunts** has made remarkable contributions to high-resolution brain mapping through her leadership in the BigBrain project, which has produced an ultrahigh-resolution 3D model of the human brain. This model provides an unprecedented level of detail, allowing researchers to study the brain's microscopic structure and connectivity with remarkable precision (Amunts et al. 2013).

In the realm of functional brain mapping, **Karl Friston** has been a leading theorist, pioneering work in functional connectivity and predictive coding. His development of statistical parametric mapping (SPM) has fundamentally transformed how researchers analyze fMRI data, offering new insights into how different brain regions interact during various cognitive tasks (Friston 2011).

Meanwhile, **Henry Markram** has made significant contributions to computational neuroscience through his work on the Blue Brain Project. This initiative aims to digitally reconstruct the mammalian brain by simulating neural circuits, bridging the gap between biological neuroscience and advanced computational models (Markram 2006). His work has been vital in enhancing our understanding of brain function and dysfunction, particularly in the context of neurological disorders.

These contemporary brain mappers continue to shape modern neuroscience by providing detailed anatomical frameworks, refining imaging methodologies, and integrating computational models into brain research. Their collective contributions are not only expanding our understanding of the human brain but also driving innovations in clinical applications, neuroprosthetics, and artificial intelligence-based neuroscience research.

7.3 Pioneering Initiatives in Brain Mapping

7.3.1 The Human Connectome Project

The Human Connectome Project (HCP) is a large-scale research initiative dedicated to mapping the complex web of neural connections within the human brain. Launched in 2010 with funding from the National Institutes of Health (NIH), the project aims to create a highly detailed map of the brain's structural and functional connectivity through the use of advanced imaging, computational modeling, and neuroinformatics techniques.

HCP utilizes cutting-edge neuroimaging tools, including high-resolution diffusion MRI (dMRI), functional MRI (fMRI), and magnetoencephalography (MEG), to capture both large-scale and fine-scale brain networks. One of the project's major advancements is its integration of multimodal MRI datasets, which allows researchers to study both the anatomical pathways and the functional interactions between different brain regions with remarkable precision (Fig. 7.2; Glasser et al. 2016; Van Essen et al. 2013).

The data generated by HCP has significantly expanded our knowledge of the connectome—the intricate network of neural connections that underlies cognition, emotions, and behavior. By analyzing these datasets, researchers have gained valuable insights into neurological and psychiatric disorders such as schizophrenia, autism, and Alzheimer's disease, identifying disruptions in network organization that may contribute to these conditions.

Beyond its core research, the HCP has inspired numerous spin-off projects, including the Lifespan Human Connectome Projects, which study brain development from infancy to old

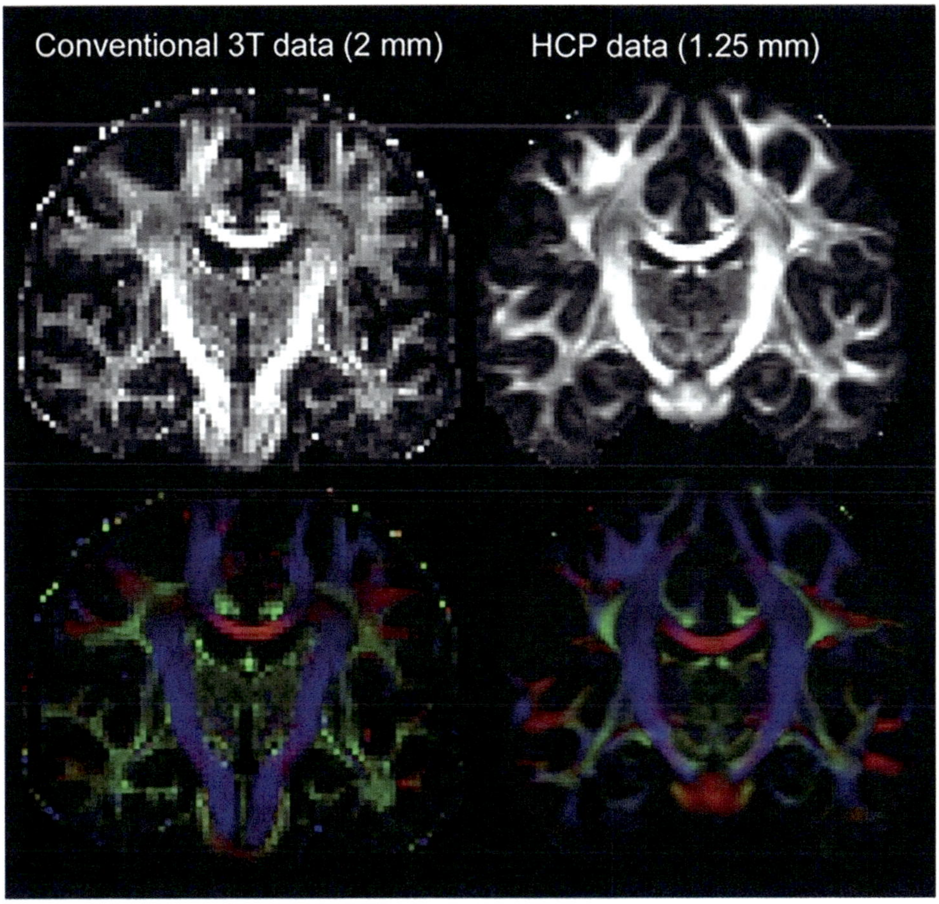

Fig. 7.2 Representative fractional anisotropy and color-encoded principal diffusion direction images from the HCP dMRI data, compared with a more conventional 2 mm dataset (from a different subject). The improvement in anatomical detail is clearly discernible. For example, many white matter tracts appear thicker (less partial voluming). The imaging protocol for the conventional data was as follows: Siemens 3 T Verio, 2 mm isotropic voxels, 64 slices, 60 directions, 2 averages with reversed phase encoding polarity, b = 1500 s/mm^2, TE/TR = 86/10,000 ms, GRAPPA = 2, scan time = 20 min (From: Van Essen et al. 2013)

age, and the Human Connectome Project for Psychiatric Disorders, which explores how abnormalities in brain connectivity relate to mental health conditions (Somerville et al. 2018).

By integrating state-of-the-art imaging technologies with computational neuroscience, the Human Connectome Project continues to drive progress in brain mapping, offering an essential framework for understanding the neural foundations of human thought and behavior.

7.3.2 The Allen Brain Atlases

The Allen Brain Atlases, developed by the Allen Institute for Brain Science, serve as an essential resource for neuroanatomical and gene expression studies. These atlases offer high-resolution, open-access maps of the brain at multiple levels, from whole-brain connectivity to single-cell gene expression. Their comprehensive datasets have accelerated research into brain structure, function, and disease, facilitating discoveries in neurodegenerative conditions and psychiatric disorders.

One of the institute's most significant achievements is the Allen Brain Atlas, a high-resolution gene expression map of the mouse and human brain (Lein et al. 2007). This invaluable resource has allowed researchers to examine how genes are expressed across different brain regions, shedding light on the molecular foundations of neurodevelopment, cognition, and disease. The accessibility of this data has made it an essential tool for neuroscientists worldwide, enabling groundbreaking discoveries in brain research.

Beyond gene expression mapping, the Allen Institute has led efforts in classifying different types of brain cells. By using single-cell transcriptomics and high-throughput molecular techniques, researchers have identified a wide variety of neuronal populations, demonstrating the remarkable diversity of cell types within the brain (Tasic et al. 2018). This research has deepened our understanding of how neural circuits function and contributed to the development of comprehensive neural cell taxonomies.

Another major contribution of the institute is the Allen Mouse Brain Connectivity Atlas, which provides detailed maps of neural circuits and their connectivity patterns (Fig. 7.3; Oh et al. 2014). By employing viral tracers and advanced imaging technologies, scientists have been able to trace the connections between different brain regions, uncovering fundamental principles of information processing and network interactions.

The Allen Institute has also made significant strides in human brain research. The Allen Human Brain Atlas has helped bridge the gap between animal models and human neuroscience by providing insights into gene expression patterns and structural organization in the human brain (Hawrylycz et al. 2012). This resource has been particularly valuable in studying neuropsychiatric and neurodegenerative disorders, aiding in the identification of genetic and anatomical markers associated with conditions like schizophrenia, Alzheimer's disease, and autism spectrum disorder.

A strong advocate for open science, the Allen Institute makes its vast datasets freely available to the global scientific community. By prioritizing transparency and collaboration, it has accelerated discoveries across multiple fields and set a standard for data-sharing in neuroscience. By ensuring that brain mapping research remains accessible and reproducible, the institute continues to push the boundaries of what is known about the brain, shaping the future of neuroscience in profound ways.

7.3.3 The Brain Initiative

The Brain Research Through Advancing Innovative Neurotechnologies (BRAIN) Initiative is a large-scale US government-funded project designed to revolutionize our understanding of the human brain. By fostering the development of next-generation neurotechnologies, including high-density neural recording systems and optogenetics, the initiative aims to map brain activity at an unprecedented resolution. The BRAIN Initiative has been instrumental in driving forward neuroengineering, neuromodulation therapies, and brain-computer interfaces, with significant implications for treating neurological and psychiatric disorders.

A central goal of the BRAIN Initiative is to refine neuroimaging and neural recording techniques, allowing scientists to study brain activity with unprecedented precision. Advances such as high-density electrode arrays, optogenetics, and next-generation functional imaging have expanded our ability to monitor neural activity in real time. The initiative also supports large-scale data integration and computational neuroscience, helping researchers analyze vast amounts of neural data and generate detailed brain maps (Jorgenson et al. 2015).

One of the most transformative aspects of the BRAIN Initiative is its commitment to interdisciplinary collaboration. By bringing together neuroscientists, engineers, physicists, and computational biologists, the initiative has driven the creation of advanced tools for studying neural

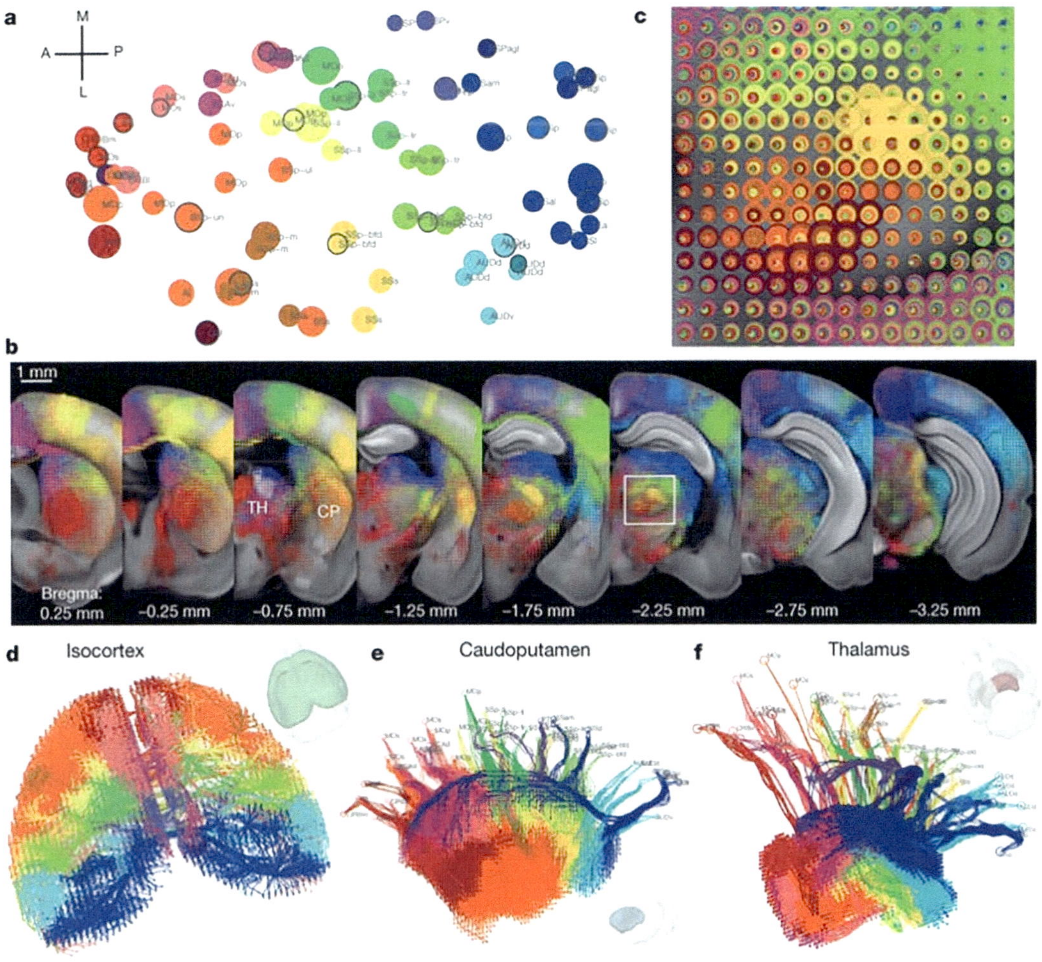

Fig. 7.3 Topography of cortico-striatal and corticothalamic projections (From: Oh et al. 2014)

connectivity. High-resolution connectomics research, for example, has provided insights into how neural circuits are organized and interact to support cognition, behavior, and neurological function (Markram et al. 2015).

The BRAIN Initiative has significantly advanced treatments for neurological and psychiatric disorders. Research funded by the initiative has led to major breakthroughs in brain-computer interfaces (BCIs), allowing individuals with paralysis to control prosthetic limbs using neural signals. Additionally, the initiative has driven the development of neuromodulation therapies for conditions like epilepsy, depression, and Parkinson's disease (Mott et al. 2018; Saha et al. 2021). These advances, made possible through refined brain mapping techniques, have enabled more personalized treatment approaches tailored to individual patient needs.

Another crucial contribution of the BRAIN Initiative is its emphasis on making brain mapping data widely accessible. The initiative has launched open-access databases and collaborative platforms, ensuring that scientists worldwide can share and utilize brain mapping data to accelerate scientific progress (Iyer et al. 2024). This open-science approach enhances the reproducibility of neuroscience research and fosters rapid innovation across disciplines.

Despite its successes, the BRAIN Initiative faces ongoing challenges. Ethical concerns surrounding neurotechnology applications, as well as the need for sustained investment in long-term research projects, remain key areas of discussion. Nonetheless, the initiative continues to shape the future of brain research, laying the groundwork for transformative discoveries in neuroscience and neurotherapeutics.

7.4 Applications of Brain Mapping

Brain mapping has a wide range of applications in both clinical and research settings, playing a critical role in neurology, psychiatry, neurosurgery, cognitive science, and brain-computer interface technologies.

7.4.1 The Role of Brain Mapping in Neurosurgery

One of the most crucial applications of brain mapping is in neurosurgical planning. Preoperative mapping techniques, such as functional MRI, magnetoencephalography (MEG), and direct cortical stimulation, allow neurosurgeons to identify essential brain regions before surgery. This approach is particularly valuable in procedures such as tumor resections and epilepsy surgeries, where preserving functional brain areas is paramount (Leone et al. 2025). Additionally, intraoperative neuroimaging and neuronavigation technologies provide real-time guidance, improving surgical precision, reducing postoperative complications, and enhancing patient outcomes.

7.4.2 Brain Mapping and Neurological Disorders

Brain mapping has also proven essential in the study and diagnosis of neurological and psychiatric disorders. Advanced imaging techniques have identified distinct structural and functional changes in conditions such as schizophrenia, depression, autism, and Alzheimer's disease (Van Essen et al. 2013). By analyzing neural connectivity patterns, researchers have been able to develop biomarkers for early diagnosis, track disease progression, and evaluate the effectiveness of therapeutic interventions. These discoveries have led to more personalized treatment approaches, including neurostimulation therapies and cognitive-behavioral interventions tailored to specific neural dysfunctions.

Neurodegenerative diseases such as Alzheimer's, Parkinson's, and multiple sclerosis have particularly benefited from brain mapping advancements. Imaging techniques have allowed researchers to visualize amyloid plaques and tau protein accumulation in Alzheimer's disease and detect functional and structural changes in Parkinson's disease, facilitating early diagnosis and monitoring disease progression. These tools provide crucial biomarkers that enable early intervention strategies, potentially slowing the progression of these conditions and improving patient outcomes.

7.4.3 Brain Mapping in Cognitive Neuroscience

Brain mapping has revolutionized cognitive neuroscience, offering profound insights into how the brain supports complex cognitive functions such as memory, attention, language, and decision-making. Functional neuroimaging techniques, particularly fMRI and MEG, have allowed researchers to investigate the intricate neural networks involved in these processes. By mapping brain activity in real time, scientists have gained a deeper understanding of how different brain regions communicate and collaborate to support cognition (Poldrack et al. 2017). These discoveries have influenced fields beyond neuroscience, contributing to advancements in artificial intelligence, human-computer interaction, and educational neuroscience.

7.4.4 Brain Mapping and Neurodevelopmental Studies

Brain mapping has also played a crucial role in understanding neurodevelopment, helping researchers track how neural circuits evolve from infancy to adulthood. By using imaging techniques such as structural MRI and DTI, scientists have been able to observe changes in brain connectivity over time, shedding light on critical developmental periods and milestones. These insights are essential for identifying atypical patterns of brain development associated with neurodevelopmental disorders such as dyslexia and ADHD. Through these studies, researchers have been able to design early interventions aimed at supporting optimal brain development and improving cognitive outcomes in children with learning and attentional difficulties (Casey et al. 2018).

7.4.5 Comparative Brain Mapping across Species

Comparative brain mapping across species has provided invaluable insights into the evolutionary development of the nervous system. By studying the brains of nonhuman primates, rodents, and other model organisms, scientists can explore fundamental neural functions, adaptations across species, and variations in brain architecture. These comparisons help researchers understand how different neural circuits have evolved to support cognition, sensory processing, and motor control.

Investigating brain similarities and differences among species has led to major discoveries about the human brain. For example, studies comparing the prefrontal cortex in primates and rodents have shed light on the evolutionary expansion of executive functions in humans (Passingham and Wise 2012). Similarly, mapping the mouse and macaque connectomes has revealed how neural networks have adapted to support increasingly complex behaviors (Van Essen et al. 2019). Beyond evolutionary insights, these studies also play a critical role in translational neuroscience, as animal models are essential for developing treatments for neuropsychiatric and neurodegenerative diseases.

Recent advances in comparative brain mapping have been driven by innovations in imaging and molecular techniques, including diffusion tensor imaging (DTI), single-cell RNA sequencing, and high-throughput histological analysis. These technologies have made it possible to identify homologous brain regions across species and analyze how connectivity patterns influence behavior and cognition (Herculano-Houzel 2016).

Additionally, comparative studies have contributed to our understanding of neurological disorders. Examining disease-related changes in different species has helped researchers develop more accurate animal models for conditions such as Alzheimer's disease, Parkinson's disease, and autism spectrum disorders (Bozzi and Fagiolini 2020). This cross-species approach continues to refine our knowledge of brain function and disorder mechanisms, ultimately leading to more effective therapeutic strategies.

7.4.6 Neuroplasticity and Brain Mapping

Neuroplasticity, or the brain's ability to reorganize itself, is a fundamental property that enables learning, memory formation, recovery from injuries, and adaptation to neurodegenerative conditions. Brain mapping technologies have been instrumental in capturing these changes, allowing scientists to study the brain's capacity to form new connections and modify existing ones over time (Pascual-Leone et al. 2005).

Research using brain mapping has highlighted two primary forms of neuroplasticity: functional and structural. Functional plasticity occurs when brain regions adapt to compensate for damage, such as in stroke recovery. Imaging techniques like functional MRI (fMRI) and magnetoencephalography (MEG) have shown how undamaged areas of the brain take over lost functions, demonstrating the brain's remarkable adaptability (Grefkes and Fink 2011). Structural plasticity, on the other hand, involves actual physical changes in neural pathways, including synaptic remodeling and dendritic growth. Diffusion tensor imaging (DTI) has provided detailed visualizations of how these changes improve connectivity, particularly in individuals engaged in cognitively demanding activities such as language learning or musical training (Zatorre et al. 2012).

Neuroplasticity plays a crucial role in learning and memory. Studies using fMRI have shown that acquiring new skills strengthens neural networks in key regions like the hippocampus, prefrontal cortex, and motor areas (Draganski et al. 2004). Structural MRI has also demonstrated increases in gray matter density associated with extensive learning (Maguire et al. 2000). In rehabilitation, neuroplasticity facilitates recovery from brain injuries, as alternative neural pathways form to restore lost functions. Techniques such as transcranial magnetic stimulation (TMS) and real-time neurofeedback have been developed to enhance neuroplastic responses, improving recovery outcomes (Cramer et al. 2011).

In neurodegenerative diseases, neuroplasticity serves as a compensatory mechanism in the early stages. For instance, in Alzheimer's disease, some brain networks exhibit increased connectivity, helping to delay cognitive decline despite neuronal loss (Voss et al. 2016). However, brain mapping has also revealed deficits in neuroplasticity in progressive disorders such as Parkinson's disease, highlighting the need for targeted therapeutic interventions. Emerging technologies, including high-resolution imaging, optogenetics, and machine learning, are refining our understanding of neuroplasticity and paving the way for personalized treatments designed to enhance brain function and recovery (Onciul et al. 2025).

7.4.7 Brain Mapping and Brain-Computer Interfaces (BCIs)

Brain mapping has also played a transformative role in the development of brain-computer interfaces (BCIs) and neuroprosthetic technologies. These innovations harness real-time neural signal decoding to enable individuals with severe motor impairments to regain some degree of independence. BCIs have provided remarkable breakthroughs in assistive technology, allowing individuals with paralysis to control prosthetic limbs or communicate solely through brain activity (Lebedev and Nicolelis 2017).

By leveraging brain mapping data, BCIs have become increasingly sophisticated, enabling more precise interactions between the brain and external devices. These advancements hold enormous promise for restoring motor function, particularly for patients recovering from stroke or spinal cord injuries. Furthermore, BCIs are expanding beyond medical rehabilitation into applications such as cognitive enhancement and human-machine symbiosis.

As brain mapping continues to refine our understanding of neural pathways and functional connectivity, the integration of BCIs into assistive technologies is becoming more seamless. Future developments are expected to enhance communication for individuals with locked-in syndrome, refine neuroprosthetic control, and even explore cognitive augmentation. With ongoing research, the synergy between brain mapping and BCIs is poised to revolutionize both clinical treatments and the broader interface between humans and technology.

7.4.8 Artificial Intelligence and Machine Learning in Brain Mapping

The integration of artificial intelligence (AI) and machine learning (ML) into brain mapping has led to remarkable advancements in the analysis and interpretation of neuroimaging data. These technologies facilitate the identification of structural and functional brain networks, accelerating research in neuroanatomy, cognitive neuroscience, and the study of neurological disorders.

AI-based models enhance neuroimaging techniques such as functional MRI (fMRI), diffusion tensor imaging (DTI), and magnetoencephalography (MEG) by enabling high-precision image processing. Deep learning algorithms, particularly convolutional neural networks (CNNs), have demonstrated superior accuracy in segmenting brain regions and detecting pathological changes associated with conditions like Alzheimer's and schizophrenia (Vieira et al. 2017). These automated systems significantly reduce human error and expedite the identification of biomarkers for various neurological diseases.

Machine learning methods have also transformed our understanding of brain connectivity. By utilizing clustering algorithms and graph theory in the analysis of fMRI and EEG data, researchers can now identify intricate neural networks associated with cognition, emotional regulation, and neuropsychiatric disorders (Finn et al. 2015). Additionally, predictive models trained on large neuroimaging datasets offer valuable insights into disease progression, particularly in conditions such as stroke and traumatic brain injury, allowing for early intervention and personalized treatment strategies (Ravi et al. 2017).

Another revolutionary application of AI in neuroscience is its role in brain-computer interfaces (BCIs). These systems, powered by machine learning algorithms trained on neural signals, enable individuals with motor impairments to control prosthetic devices, communicate, and interact with external environments. AI-enhanced BCIs are paving the way for restoring lost functions in patients suffering from paralysis and neurodegenerative conditions (Lebedev and Nicolelis 2017).

Despite its potential, AI-driven brain mapping presents challenges, including the need for standardized datasets, computational efficiency, and concerns over data privacy and algorithmic bias (Sejnowski et al. 2014). Addressing these challenges requires continued advancements in AI methodologies, cloud-based neuroinformatics, and the development of ethical frameworks for responsible AI usage in neuroscience.

As AI and machine learning continue to evolve, they promise to deepen our understanding of the human brain, improve diagnostic accuracy, and enable the development of personalized medical interventions. The integration of these technologies into neuroimaging and brain mapping represents a paradigm shift in neuroscience, offering unprecedented opportunities for research and clinical applications.

7.5 Ethical Considerations in Brain Mapping

As brain mapping technologies continue to evolve, ethical considerations must be carefully addressed to ensure their responsible application. One of the primary concerns is privacy and data security, given that brain imaging can reveal highly sensitive information about an individual's cognitive and emotional states. Researchers and clinicians must implement strict protocols to anonymize and securely store neural data, preventing unauthorized access or misuse (Ienca and Andorno 2017).

Informed consent is another key ethical issue. Since brain mapping research often involves human participants, it is essential to ensure that individuals fully understand both the potential risks and benefits of their participation. Special attention must be given when working with vulnerable populations, such as individuals with neurodegenerative diseases or psychiatric conditions, to guarantee that consent is obtained in an ethically sound manner (Illes and Sahakian 2011).

The growing use of brain mapping in neuromodulation and brain-computer interfaces (BCIs) raises important concerns about cognitive liberty and autonomy. As BCIs become more advanced, discussions surrounding their potential impact on personal identity, decision-making, and self-agency are increasing. Ethical frameworks must be established to regulate the use of BCIs in both medical and nonmedical settings, ensuring that individuals retain full control over their own neural data and cognitive processes (Goering et al. 2021).

Another pressing issue in brain mapping research is the risk of bias. Studies that rely primarily on data from specific demographic groups may lead to conclusions that do not accurately reflect the broader population. Ensuring diversity in neuroimaging studies is essential to prevent disparities in neuroscience research and its clinical applications (Poldrack et al. 2017).

As brain mapping continues to shape medical practice and neuroscience research, addressing these ethical concerns will be crucial. Establishing clear guidelines and ethical standards will help ensure that technological advancements benefit society while upholding fundamental human rights.

7.6 Technological Innovations in Brain Mapping

Technological advancements have transformed brain mapping, enabling more precise, real-time, and multimodal visualization of neural activity. Cutting-edge neuroimaging techniques, such as ultrahigh-field magnetic resonance imaging (MRI), functional MRI (fMRI), and positron emission tomography (PET), have provided researchers with unprecedented spatial and temporal resolution to study the brain in greater detail (Logothetis 2008). Diffusion tensor imaging (DTI) has further refined our ability to visualize white matter tracts, offering deeper insights into neural connectivity in both health and disease.

Beyond imaging, breakthroughs in electrophysiology, including magnetoencephalography (MEG) and high-density electroencephalography (EEG), have dramatically improved our ability to measure neural dynamics with millisecond precision. These tools have allowed scientists to capture real-time brain activity, offering critical insights into functional brain processes (Baillet 2017). At the same time, optogenetics has revolutionized experimental neuroscience by enabling precise control of neuronal activity through genetically encoded light-sensitive proteins. This technique has helped researchers study neural circuits with remarkable specificity (Deisseroth 2015).

Computational neuroscience and artificial intelligence (AI) have further accelerated brain mapping efforts. AI-driven analysis of large-scale neuroimaging datasets has enhanced the accuracy of neural activity predictions, disease biomarker identification, and treatment response assessments. The integration of deep learning in automated segmentation and classification of brain structures has streamlined data processing, making high-throughput brain mapping more efficient than ever before.

These technological advancements, alongside emerging innovations in single-cell transcriptomics and high-resolution connectomics, continue to redefine our understanding of brain structure and function. As these tools evolve, brain mapping will play an increasingly vital role in personalized medicine, early diagnosis of neurological diseases, and the development of novel therapeutic approaches.

7.6.1 Educational and Training Applications

Brain mapping has become a vital tool in education and training, transforming neuroscience curricula, clinical instruction, and public engagement with brain sciences. Advances in neuroimaging and computational modeling have allowed educators and medical professionals to integrate these technologies into teaching strategies, clinical training, and scientific outreach, ultimately improving the way we understand and study the brain.

In medical education, brain mapping has revolutionized the teaching of neuroscience. Traditional methods, such as cadaver dissections and textbook illustrations, are now being enhanced by digital brain atlases and immersive virtual reality (VR) simulations. Technologies like high-resolution MRI, diffusion tensor imaging (DTI), and functional MRI (fMRI) enable students and trainees to explore neural structures and pathways in dynamic, interactive ways (Stepan et al. 2017). This level of visualization helps future healthcare professionals develop a deeper understanding of the brain's three-dimensional organization, improving their ability to diagnose and treat neurological conditions.

Neurosurgical training has particularly benefited from brain mapping. Simulation-based platforms now integrate patient-specific brain maps, allowing surgeons to practice complex procedures in a controlled, risk-free environment. Functional mapping techniques, such as direct cortical stimulation and magnetoencephalography (MEG), help identify critical brain regions before and during surgery, reducing the risk of damage to essential functional areas (Duffau 2011). These advancements have improved surgical precision, leading to better patient outcomes and fewer complications.

Beyond medical training, brain mapping has also had a significant impact on education and cognitive sciences. Researchers use neuroimaging to study how learning and memory are processed in the brain, leading to the development of evidence-based educational strategies. fMRI studies have shown that different teaching methods activate distinct neural circuits, allowing educators to design personalized learning experiences that improve knowledge retention and problem-solving skills (Ansari et al. 2011). These findings are particularly valuable for addressing learning disabilities such as dyslexia and ADHD, helping to create targeted interventions that support students with special educational needs.

Public engagement and science communication have also benefited from brain mapping technologies. Interactive museum exhibits, VR brain tours, and augmented reality (AR) applications provide the public with immersive experiences that deepen their understanding of neuroscience. Large-scale citizen science projects, like Eyewire and the Human Brain Project, invite nonexperts to participate in mapping neural connections, fostering greater appreciation and awareness of brain research (Amunts et al. 2016).

As brain mapping technology continues to evolve, its applications in education and training will only expand. The integration of artificial intelligence, machine learning, and brain-computer interfaces into educational settings holds exciting potential for enhancing learning methodologies and making neuroscience more accessible to students, professionals, and the general public alike.

7.7 Future Directions

The future of brain mapping is poised to make remarkable strides with advancements in high-resolution imaging, artificial intelligence, and multimodal data integration. As technologies continue to evolve, brain mapping is expected to become even more precise, enabling highly individualized diagnostic tools for neurological and psychiatric conditions. Initiatives such as the **Human Connectome Project**, the **BRAIN Initiative**, and ongoing efforts at the **Allen Institute for Brain Science** are pushing the boundaries of what is possible in neuroscience. These projects, particularly those focused on single-cell resolution mapping, are shedding new light on the intricacies of neural networks, paving the way for transformative applications in brain research.

The convergence of high-resolution imaging, computational modeling, and connectomics is opening new avenues in neuroscience. These innovations are expected to enhance our ability to study neural function, improve diagnostic accuracy, and accelerate the development of novel therapeutic strategies. Additionally, emerging neurotechnologies, including brain-computer interfaces and optogenetics, will further refine our capacity to manipulate and monitor brain activity with unprecedented accuracy.

As brain mapping continues to advance, its role in medical and cognitive research will expand, offering deeper insights into how the brain functions in health and disease. The next generation of brain mapping techniques will likely integrate artificial intelligence-driven data analysis, making it possible to decode neural activity patterns in real time and tailor treatments to individual patients. These advancements will not only improve outcomes in clinical neurology and psychiatry but also contribute to fields such as cognitive enhancement and neurorehabilitation.

7.8 Conclusion

Brain mapping has fundamentally transformed our understanding of neuroanatomy and neurohistology, bridging the gap between structure and function. The ability to visualize brain activity and connectivity in ever-greater detail has reshaped the way researchers and clinicians approach the study of the brain. As visualization technologies continue to advance, their integration into clinical and research settings will further refine diagnostic accuracy, enhance therapeutic interventions, and drive new discoveries in neuroscience.

The synergy between neuroanatomy, histology, and brain mapping methodologies holds immense potential for unraveling the complexities of the human brain. As new technologies emerge and interdisciplinary collaborations grow, brain mapping will remain a cornerstone of neuroscience, continually expanding our knowledge of how the brain works and how it can be treated when impaired. The future of brain mapping is bright, with ongoing research promising breakthroughs that will deepen our understanding of the most complex organ in the human body.

References

Amunts K, Ebell C, Muller J, Telefont M, Knoll A, Lippert T (2016) The human brain project: creating a European research infrastructure to decode the human brain. Neuron 92(3):574581. https://doi.org/10.1016/j.neuron.2016.10.046. PMID: 27809997.

Amunts K, Lepage C, Borgeat L, Mohlberg H, Dickscheid T, Rousseau MÉ, Bludau S, Bazin PL, Lewis LB, Oros-Peusquens AM, Shah NJ, Lippert T, Zilles K, Evans AC (2013) BigBrain: an ultrahigh-resolution 3D human brain model. Science 340(6139):1472–1475. https://doi.org/10.1126/science.1235381. PMID: 23788795.

Andres-Barquin PJ (2002) Santiago Ramón y Cajal and the Spanish school of neurology. Lancet Neurol 1(7): 445–452. https://doi.org/10.1016/s1474-4422(02)00192-8. PMID: 12849367.

Ansari D, De Smedt B, Grabner RH (2011) Neuroeducation: A critical overview of an emerging field. Neuroethics 5(2):105–117

Baillet S (2017) Magnetoencephalography for brain electrophysiology and imaging. Nat Neurosci 20(3): 327–339. https://doi.org/10.1038/nn.4504. PMID: 28230841.

Bozzi Y, Fagiolini M (2020) Animal models of neurodevelopmental disorders. Neuroscience 445:1–2. https://doi.org/10.1016/j.neuroscience.2020.09.007. PMID: 32967770.

Brodmann K (1909) Vergleichende Lokalisationslehre der Großhirnrinde. Barth, Leipzig

Ramón y Cajal, S. (1894) Les nouvelles idées sur la structure du systéme nerveux chez l'homme et chez les vertébrés. Reinwald, Paris

Casey BJ, Cannonier T, Conley MI, Cohen AO, Barch DM, Heitzeg MM, Soules ME, Teslovich T, Dellarco DV, Garavan H, Orr CA, Wager TD, Banich MT, Speer NK, Sutherland MT, Riedel MC, Dick AS, Bjork JM, Thomas KM, Chaarani B, Mejia MH, Hagler DJ Jr, Daniela Cornejo M, Sicat CS, Harms MP, NUF D, Rosenberg M, Earl E, Bartsch H, Watts R, Polimeni JR, Kuperman JM, Fair DA, Dale AM, ABCD Imaging Acquisition Workgroup (2018) The Adolescent Brain Cognitive Development (ABCD) study: Imaging acquisition across 21 sites. Dev Cogn Neurosci 32:43–54. https://doi.org/10.1016/j.dcn.2018.03.001. Epub 2018 Mar 14. PMID: 29567376; PMCID: PMC5999559

Cramer SC, Sur M, Dobkin BH, O'Brien C, Sanger TD, Trojanowski JQ, Rumsey JM, Hicks R, Cameron J, Chen D, Chen WG, Cohen LG, deCharms C, Duffy CJ, Eden GF, Fetz EE, Filart R, Freund M, Grant SJ, Haber S, Kalivas PW, Kolb B, Kramer AF, Lynch M, Mayberg HS, McQuillen PS, Nitkin R, Pascual-Leone-A, Reuter-Lorenz P, Schiff N, Sharma A, Shekim L, Stryker M, Sullivan EV, Vinogradov S (2011) Harnessing neuroplasticity for clinical applications. Brain 134(Pt 6):1591–1609. https://doi.org/10.1093/brain/awr039. Epub 2011 Apr 10. PMID: 21482550; PMCID: PMC3102236.

Deisseroth K (2015) Optogenetics: 10 years of microbial opsins in neuroscience. Nat Neurosci 18(9): 1213–1225. https://doi.org/10.1038/nn.4091. PMID: 26308982; PMCID: PMC4790845.

Draganski B, Gaser C, Busch V, Schuierer G, Bogdahn U, May A (2004) Neuroplasticity: changes in grey matter induced by training. Nature 427(6972):311–312. https://doi.org/10.1038/427311a. PMID: 14737157.

Duffau H (2011) From neural basis of cognition to surgical applications. Springer, Vienna

Finn ES, Shen X, Scheinost D, Rosenberg MD, Huang J, Chun MM, Papademetris X, Constable RT (2015) Functional connectome fingerprinting: identifying individuals using patterns of brain connectivity. Nat Neurosci 18(11):1664–1671. https://doi.org/10.1038/nn.4135. Epub 2015 Oct 12.PMID: 26457551; PMCID: PMC5008686

Friston KJ (2011) Functional and effective connectivity: a review. Brain Connect 1(1):13–36. https://doi.org/10.1089/brain.2011.0008. PMID: 22432952.

Glasser MF, Smith SM, Marcus DS, Andersson JL, Auerbach EJ, Behrens TE, Coalson TS, Harms MP,

Jenkinson M, Moeller S, Robinson EC, Sotiropoulos SN, Xu J, Yacoub E, Ugurbil K, Van Essen DC (2016) The human connectome project's neuroimaging approach. Nat Neurosci 19(9):1175–1187. https://doi.org/10.1038/nn.4361. PMID: 27571196; PMCID: PMC6172654.

Goering S, Klein E, Specker Sullivan L, Wexler A, Arcas BAY, Bi G, Carmena JM, Fins JJ, Friesen P, Gallant J, Huggins JE, Kellmeyer P, Marblestone A, Mitchell C, Parens E, Pham M, Rubel A, Sadato N, Teicher M, Wasserman D, Whittaker M, Wolpaw J, Yuste R (2021) Recommendations for responsible development and application of neurotechnologies. Neuroethics 14(3):365–386. https://doi.org/10.1007/s12152-021-09468-6. Epub 2021 Apr 29. PMID: 33942016; PMCID: PMC8081770

Grefkes C, Fink GR (2011) Reorganization of cerebral networks after stroke: new insights from neuroimaging with connectivity approaches. Brain 134(Pt 5): 1264–1276. https://doi.org/10.1093/brain/awr033. Epub 2011 Mar 16. PMID: 21414995; PMCID: PMC3097886

Hawrylycz MJ, Lein ES, Guillozet-Bongaarts AL, Shen EH, Ng L, Miller JA, van de Lagemaat LN, Smith KA, Ebbert A, Riley ZL, Abajian C, Beckmann CF, Bernard A, Bertagnolli D, Boe AF, Cartagena PM, Chakravarty MM, Chapin M, Chong J, Dalley RA, David Daly B, Dang C, Datta S, Dee N, Dolbeare TA, Faber V, Feng D, Fowler DR, Goldy J, Gregor BW, Haradon Z, Haynor DR, Hohmann JG, Horvath S, Howard RE, Jeromin A, Jochim JM, Kinnunen M, Lau C, Lazarz ET, Lee C, Lemon TA, Li L, Li Y, Morris JA, Overly CC, Parker PD, Parry SE, Reding M, Royall JJ, Schulkin J, Sequeira PA, Slaughterbeck CR, Smith SC, Sodt AJ, Sunkin SM, Swanson BE, Vawter MP, Williams D, Wohnoutka P, Zielke HR, Geschwind DH, Hof PR, Smith SM, Koch C, Grant SGN, Jones AR (2012) An anatomically comprehensive atlas of the adult human brain transcriptome. Nature 489(7416):391–399. https://doi.org/10.1038/nature11405. PMID: 22996553; PMCID: PMC4243026.

Herculano-Houzel S (2016) The human advantage: A new understanding of how our brain became remarkable. The MIT Press, Cambridge (MA)

Ienca M, Andorno R (2017) Towards new human rights in the age of neuroscience and neurotechnology. Life Sci. Soc Policy 13(1):5. https://doi.org/10.1186/s40504-017-0050-1. Epub 2017 Apr 26. PMID: 28444626; PMCID: PMC5447561.

Illes J, Sahakian BJ (2011) The Oxford handbook of Neuroethics. Oxford University Press, Oxford

Iyer S, Maxson Jones K, Robinson JO, Provenza NR, Duncan D, Lázaro-Muñoz G, McGuire AL, Sheth SA, Majumder MA (2024) The BRAIN initiative data-sharing ecosystem: characteristics, challenges, benefits, and opportunities. elife 13:e94000. https://doi.org/10.7554/eLife.94000. PMID: 39602224; PMCID: PMC11602185

Jorgenson LA, Newsome WT, Anderson DJ, Bargmann CI, Brown EN, Deisseroth K, Donoghue JP, Hudson KL, Ling GS, MacLeish PR, Marder E, Normann RA, Sanes JR, Schnitzer MJ, Sejnowski TJ, Tank DW, Tsien RY, Ugurbil K, Wingfield JC (2015) The BRAIN initiative: developing technology to catalyse neuroscience discovery. Philos Trans R Soc Lond Ser B Biol Sci 370(1668):20140164. https://doi.org/10.1098/rstb.2014.0164. PMID: 25823863; PMCID: PMC4387507.

Lebedev MA, Nicolelis MA (2017) Brain-machine interfaces: from basic science to Neuroprostheses and Neurorehabilitation. Physiol Rev 97(2):767–837. https://doi.org/10.1152/physrev.00027.2016. PMID: 28275048

Lein ES, Hawrylycz MJ, Ao N, Ayres M, Bensinger A, Bernard A, Boe AF, Boguski MS, Brockway KS, Byrnes EJ, Chen L, Chen L, Chen TM, Chin MC, Chong J, Crook BE, Czaplinska A, Dang CN, Datta S, Dee NR, Desaki AL, Desta T, Diep E, Dolbeare TA, Donelan MJ, Dong HW, Dougherty JG, Duncan BJ, Ebbert AJ, Eichele G, Estin LK, Faber C, Facer BA, Fields R, Fischer SR, Fliss TP, Frensley C, Gates SN, Glattfelder KJ, Halverson KR, Hart MR, Hohmann JG, Howell MP, Jeung DP, Johnson RA, Karr PT, Kawal R, Kidney JM, Knapik RH, Kuan CL, Lake JH, Laramee AR, Larsen KD, Lau C, Lemon TA, Liang AJ, Liu Y, Luong LT, Michaels J, Morgan JJ, Morgan RJ, Mortrud MT, Mosqueda NF, Ng LL, Ng R, Orta GJ, Overly CC, Pak TH, Parry SE, Pathak SD, Pearson OC, Puchalski RB, Riley ZL, Rockett HR, Rowland SA, Royall JJ, Ruiz MJ, Sarno NR, Schaffnit K, Shapovalova NV, Sivisay T, Slaughterbeck CR, Smith SC, Smith KA, Smith BI, Sodt AJ, Stewart NN, Stumpf KR, Sunkin SM, Sutram M, Tam A, Teemer CD, Thaller C, Thompson CL, Varnam LR, Visel A, Whitlock RM, Wohnoutka PE, Wolkey CK, Wong VY, Wood M, Yaylaoglu MB, Young RC, Youngstrom BL, Yuan XF, Zhang B, Zwingman TA, Jones AR (2007) Genome-wide atlas of gene expression in the adult mouse brain. Nature 445(7124):168–176. https://doi.org/10.1038/nature05453. Epub 2006 Dec 6 PMID: 17151600.

Leone A, Carbone F, Spetzger U, Vajkoczy P, Raffa G, Angileri F, Germanó A, Engelhardt M, Picht T, Colamaria A, Rosenstock T (2025) Preoperative mapping techniques for brain tumor surgery: a systematic review. Front Oncol 14:1481430. https://doi.org/10.3389/fonc.2024.1481430. PMID: 39839770; PMCID: PMC11747149.

Logothetis NK (2008) What we can do and what we cannot do with fMRI. Nature 453(7197):869–878. https://doi.org/10.1038/nature06976. PMID: 18548064.

Markram H (2006) The blue brain project. Nat Rev Neurosci 7(2):153–160. https://doi.org/10.1038/nrn1848. PMID: 16429124.

Markram H, Muller E, Ramaswamy S, Reimann MW, Abdellah M, Sanchez CA, Ailamaki A, Alonso-

Nanclares L, Antille N, Arsever S, Kahou GA, Berger TK, Bilgili A, Buncic N, Chalimourda A, Chindemi G, Courcol JD, Delalondre F, Delattre V, Druckmann S, Dumusc R, Dynes J, Eilemann S, Gal E, Gevaert ME, Ghobril JP, Gidon A, Graham JW, Gupta A, Haenel V, Hay E, Heinis T, Hernando JB, Hines M, Kanari L, Keller D, Kenyon J, Khazen G, Kim Y, King JG, Kisvarday Z, Kumbhar P, Lasserre S, Le Bé JV, Magalhães BR, Merchán-Pérez A, Meystre J, Morrice BR, Muller J, Muñoz-Céspedes A, Muralidhar S, Muthurasa K, Nachbaur D, Newton TH, Nolte M, Ovcharenko A, Palacios J, Pastor L, Perin R, Ranjan R, Riachi I, Rodríguez JR, Riquelme JL, Rössert C, Sfyrakis K, Shi Y, Shillcock JC, Silberberg G, Silva R, Tauheed F, Telefont M, Toledo-Rodriguez M, Tränkler T, Van Geit W, Díaz JV, Walker R, Wang Y, Zaninetta SM, DeFelipe J, Hill SL, Segev I, Schürmann F (2015) Reconstruction and simulation of neocortical microcircuitry. Cell 163(2): 456–492. https://doi.org/10.1016/j.cell.2015.09.029. PMID: 26451489.

Maguire EA, Gadian DG, Johnsrude IS, Good CD, Ashburner J, Frackowiak RS, Frith CD (2000) Navigation-related structural change in the hippocampi of taxi drivers. Proc Natl Acad Sci U S A. Apr11;97(8): 4398-403. https://doi.org/10.1073/pnas.070039597. PMID: 10716738; PMCID: PMC18253.

Mott MC, Gordon JA, Koroshetz WJ (2018) The NIH BRAIN initiative: advancing neurotechnologies, integrating disciplines. PLoS Biol 16(11):e3000066. https://doi.org/10.1371/journal.pbio.3000066. PMID: 30475794; PMCID: PMC6283590.

Oh SW, Harris JA, Ng L, Winslow B, Cain N, Mihalas S, Wang Q, Lau C, Kuan L, Henry AM, Mortrud MT, Ouellette B, Nguyen TN, Sorensen SA, Slaughterbeck CR, Wakeman W, Li Y, Feng D, Ho A, Nicholas E, Hirokawa KE, Bohn P, Joines KM, Peng H, Hawrylycz MJ, Phillips JW, Hohmann JG, Wohnoutka P, Gerfen CR, Koch C, Bernard A, Dang C, Jones AR, Zeng H (2014) A mesoscale connectome of the mouse brain. Nature 508(7495): 207–214. https://doi.org/10.1038/nature13186. Epub 2014 Apr 2. PMID: 24695228; PMCID: PMC5102064

Onciul R, Tataru CI, Dumitru AV, Crivoi C, Serban M, Covache-Busuioc RA, Radoi MP, Toader C (2025) Artificial intelligence and neuroscience: transformative synergies in brain research and clinical applications. J Clin Med 14(2):550. https://doi.org/10.3390/jcm14020550. PMID: 39860555; PMCID: PMC11766073.

Pascual-Leone A, Amedi A, Fregni F, Merabet LB (2005) The plastic human brain cortex. Annu Rev Neurosci 28:377–401. https://doi.org/10.1146/annurev.neuro.27.070203.144216. PMID: 16022601.

Passingham RE, Wise SP (2012) The neurobiology of the prefrontal cortex. Oxford University Press, Oxford

Paxinos G, Watson C (1982) The rat brain in stereotaxic coordinates. Academic Press, Waltham

Penfield W, Boldrey E (1937) Somatic motor and sensory representation in the cerebral cortex of man. Brain 60(4):389–443

Poldrack RA, Baker CI, Durnez J, Gorgolewski KJ, Matthews PM, Munafò MR, Nichols TE, Poline JB, Vul E, Yarkoni T (2017) Scanning the horizon: towards transparent and reproducible neuroimaging research. Nat Rev Neurosci 18(2):115–126. https://doi.org/10.1038/nrn.2016.167. Epub 2017 Jan 5. PMID: 28053326; PMCID: PMC6910649.

Ramón y Cajal S (1899) Textura del Sistema Nervioso del Hombre y de los Vertebrados. Moya, Madrid

Ravi D, Wong C, Deligianni F, Berthelot M, Andreu-Perez J, Lo B, Yang GZ (2017) Deep learning for health informatics. IEEE J Biomed Health Inform 21(1):4–21. https://doi.org/10.1109/JBHI.2016.2636665. Epub 2016 Dec 29 PMID: 28055930.

Saha S, Mamun KA, Ahmed K, Mostafa R, Naik GR, Darvishi S, Khandoker AH, Baumert M (2021) Progress in brain computer interface: challenges and opportunities. Front Syst Neurosci 15:578875. https://doi.org/10.3389/fnsys.2021.578875. PMID: 33716680; PMCID: PMC7947348.

Sejnowski TJ, Churchland PS, Movshon JA (2014) Putting big data to good use in neuroscience. Nat Neurosci 17(11):1440–1441. https://doi.org/10.1038/nn.3839. PMID: 25349909; PMCID: PMC4224030

Shekouh D, Sadat Kaboli H, Ghaffarzadeh-Esfahani M, Khayamdar M, Hamedani Z, Oraee-Yazdani S, Zali A, Amanzadeh E (2024) Artificial intelligence role in advancement of human brain connectome studies. Front Neuroinform 18:1399931. https://doi.org/10.3389/fninf.2024.1399931. PMID: 39371468; PMCID: PMC11450642.

Somerville LH, Bookheimer SY, Buckner RL, Burgess GC, Curtiss SW, Dapretto M, Elam JS, Gaffrey MS, Harms MP, Hodge C, Kandala S, Kastman EK, Nichols TE, Schlaggar BL, Smith SM, Thomas KM, Yacoub E, Van Essen DC, Barch DM (2018) The lifespan Human Connectome Project in development: a large-scale study of brain connectivity development in 5–21 year olds. Neuroimage 183:456–468. https://doi.org/10.1016/j.neuroimage.2018.08.050. Epub 2018 Aug 22. PMID: 30142446; PMCID: PMC6416053.

Stepan K, Zeiger J, Hanchuk S, Del Signore A, Shrivastava R, Govindaraj S, Iloreta A (2017) Immersive virtual reality as a teaching tool for neuroanatomy. Int Forum Allergy Rhinol 7(10):1006–1013. https://doi.org/10.1002/alr.21986. Epub 2017 Jul 18 PMID: 28719062.

Tasic B, Yao Z, Graybuck LT, Smith KA, Nguyen TN, Bertagnolli D, Goldy J, Garren E, Economo MN, Viswanathan S, Penn O, Bakken T, Menon V, Miller J, Fong O, Hirokawa KE, Lathia K, Rimorin C, Tieu M, Larsen R, Casper T, Barkan E, Kroll M, Parry S, Shapovalova NV, Hirschstein D, Pendergraft J, Sullivan HA, Kim TK, Szafer A, Dee N, Groblewski P, Wickersham I, Cetin A, Harris

JA, Levi BP, Sunkin SM, Madisen L, Daigle TL, Looger L, Bernard A, Phillips J, Lein E, Hawrylycz M, Svoboda K, Jones AR, Koch C, Zeng H. Shared and distinct transcriptomic cell types across neocortical areas. Nature. 2018 ;563(7729):72–78. doi: https://doi.org/10.1038/s41586-018-06545. Epub 2018 Oct 31. PMID: 30382198; PMCID: PMC6456269.

Van Essen DC, Donahue CJ, Coalson TS, Kennedy H, Hayashi T, Glasser MF (2019) Cerebral cortical folding, parcellation, and connectivity in humans, nonhuman primates, and mice. Proc Natl Acad Sci U S A 116(52):26173–26180. https://doi.org/10.1073/pnas.1902299116. Epub 2019 Dec 23. PMID: 31871175; PMCID: PMC6936571.

Van Essen DC, Glasser MF (2016) The human connectome project: Progress and prospects. Cerebrum 2016:cer-10-16. PMCID: PMC5198757.

Van Essen DC, Smith SM, Barch DM, Behrens TE, Yacoub E, Ugurbil K, WU-Minn HCP Consortium (2013) The WU-Minn Human Connectome Project: an overview. Neuroimage 80:62–79. https://doi.org/10.1016/j.neuroimage.2013.05.041. Epub 2013 May 16. PMID: 23684880; PMCID: PMC3724347

Vieira S, Pinaya WH, Mechelli A (2017) Using deep learning to investigate the neuroimaging correlates of psychiatric and neurological disorders: methods and applications. Neurosci Biobehav Rev 74(Pt A):58–75. https://doi.org/10.1016/j.neubiorev.2017.01.002. Epub 2017 Jan 10. PMID: 28087243.

Voss MW, Weng TB, Burzynska AZ, Wong CN, Cooke GE, Clark R, Fanning J, Awick E, Gothe NP, Olson EA, McAuley E, Kramer AF (2016) Fitness, but not physical activity, is related to functional integrity of brain networks associated with aging. Neuroimage 131:113–125. https://doi.org/10.1016/j.neuroimage.2015.10.044. Epub 2015 Oct 19. PMID: 26493108; PMCID: PMC4837101.

Zatorre RJ, Fields RD, Johansen-Berg H (2012) Plasticity in gray and white: neuroimaging changes in brain structure during learning. Nat Neurosci 15(4):528–536. https://doi.org/10.1038/nn.3045. PMID: 22426254; PMCID: PMC3660656.

8 Quantitative Neurohistology and Its Applications

Hosam Eldeen Elsadig Gasmalla, Juman Almasaad, and Mohammed Alsharif

Abstract

Quantitative neurohistology is an evolving field that applies quantitative methods to the study of the nervous system's structure and function, offering objective and reproducible data that surpass traditional qualitative assessments. This review explores the key techniques in quantitative neurohistology, including tissue preparation, sectioning, staining, image capture, and image analysis. It highlights the benefits of quantitative approaches, such as enhanced objectivity, precision, and statistical rigour. The chapter next explores the applications of quantitative neurohistology that are wide-ranging, including studying neurodegenerative diseases, assessing neuroprotective compounds, investigating neural regeneration and plasticity, and studying developmental neurobiology. Furthermore, the review discusses future directions, such as integrating multi-omics approaches, leveraging advancements in imaging techniques, automation through machine learning, 3D reconstruction and modelling, and the use of AI. These advancements promise a deeper understanding of neurological disorders and the development of targeted interventions, thus underscoring the potential of quantitative neurohistology in advancing neuroscience research and clinical applications.

Keywords

Histology · Quantitative neurohistology · Neurohistology · Image analysis · Tissue processing · Neurodegenerative diseases · Neuroprotection

H. E. E. Gasmalla (✉)
Brunel Medical School, Brunel University London, London, UK
e-mail: Hosameldeen.mohammed@brunel.ac.uk

J. Almasaad
Department of Basic Medical Sciences, College of Medicine, King Saud bin Abdulaziz University for Health Sciences, Jeddah, Saudi Arabia

King Abdullah International Medical Research Centre (KAIMRC), King Abdulaziz Medical City, National Guard Health Affairs, Jeddah, Saudi Arabia

M. Alsharif
Department of Professional and Medical Education, Meharry Medical College, Nashville, TN, USA

8.1 Overview of Neurohistology

8.1.1 Definition, Key Terminologies, and Concepts

Histology is the study of cell morphology within its natural tissue environment. It has evolved with advancements in microscopy techniques, allowing for a deeper understanding of shape-function relationships at the molecular level. Modern approaches have transformed histology into an advanced experimental science capable of

K. S. Ravi et al. (eds.), *Visualization in Neuroanatomical Sciences*, Biomedical Visualization 11,
https://doi.org/10.1007/978-3-032-07380-8_8

driving knowledge progress in biology and medicine.

Neurohistology is a branch of histology that examines the structure of the nervous system, including the brain, spinal cord, and peripheral nerves, using various staining techniques and microscopic methods to understand their organisation, function, and pathology of neural cells and structures.

8.1.2 Historical Context and Evolution

During the 1500s–1600s, advancements in glassmaking inspired scientists to develop optical instruments, such as Galileo Galilei's telescope and Antonie van Leeuwenhoek's microscope. In a letter to the Academy of Sciences established by King Charles II of England, van Leeuwenhoek detailed his first microscopic discoveries, including bacteria in dental plaque, blood cells in blood, and moving spermatozoa in a drop of sperm. These observations are considered the beginning of microbiology and histology.

The term "histology" was first introduced in the early 1800s by Marie Francois Xavier Bichat and later reiterated by Karl Mayer in 1819. In 1906, histology was officially recognised as a medical discipline when the Nobel Prize for Medicine was awarded to Camillo Golgi and Santiago Ramon y Cajal for their groundbreaking histological research on nerve cells (Mazzarini et al. 2021).

In the late nineteenth and early twentieth centuries, advancements in staining techniques led to the development of neurohistology. The early development of neurohistology was spearheaded by Santiago Ramón y Cajal, whose work laid the foundation for modern neuroscience. Cajal's use of the Golgi staining method allowed for the detailed visualisation of individual neurons, leading to the formulation of the neuron doctrine, which posits that neurons are the fundamental units of the nervous system.

The Madrid School of Neurology, active between 1885 and 1939, played a crucial role in advancing neurohistology. Pioneers such as Luis Simarro, Nicolás Achúcarro, and Pío del Río-Hortega made significant contributions to the field. Achúcarro's work on glial cells and Río-Hortega's discovery of oligodendrocytes were particularly influential. These researchers combined clinical practice with cutting-edge neurohistological research, significantly advancing the understanding of neural structures and their functions (Gimenez-Roldan 2015).

Early neuropathology and neurohistology were heavily influenced by Georges Marinesco, a Romanian-born neurologist. At the Salpêtrière Hospital in Paris, Marinesco conducted pioneering research on neurodegenerative diseases under the guidance of Jean-Martin Charcot. Among his key histopathological terms were neuronophagia and chromatolysis, which he coined as a hallmark of Alzheimer's disease. A cornerstone of neurohistological research has been Marinesco's integration of clinical observations and histological findings (Buda et al. 2009).

A variety of staining techniques have played an important role in the development of neurohistology. Silver staining, using the Golgi method, was crucial to discovering the neuron theory approximately 150 years ago. A number of techniques for studying neural tissues were introduced in the 1950s and 1960s, including cholinesterase staining and fluorescence. By the mid-twentieth century, electron microscopy had become widely used, allowing detailed examination of myelinated and unmyelinated nerve fibres and providing insight into their ultrastructure (Bourlond 1976).

A significant advancement in neurohistology was made in the 1960s when modern brain banks were established. In these institutions, diseased and healthy brain tissues are collected and preserved, enabling scientists to conduct extensive research on the brain's structure and function. Postmortem neuropathological findings can now be compared with antemortem clinical data in brain banks to study neurological diseases. Integrating data is crucial to better understanding disease mechanisms and developing new treatments (Carlos et al. 2019).

8.1.3 Quantitative Approaches in Neurohistology

Measurements, such as morphometry, geometry, stereology, and anthropometry, are commonly used to assess anatomical structures. Quantitative neurohistology uses various techniques and methods to quantify morphological features such as density, frequency, coverage, and spatial patterns of histological elements.

This typically involves tissue processing to obtain histological slides and then, using image analysis tools, constructing 3D structural parameters from a series of the obtained two-dimensional histological sections.

8.1.4 Benefits over Traditional Qualitative Methods

Objectivity and Reproducibility: Quantitative methods offer data that can be reliably replicated in various studies and research facilities, minimising observer bias and enhancing the reliability of the results. **Precision and Sensitivity**: Methods such as stereology and magnetic resonance microscopy (MRM) enable accurate measurement of cellular structures and volumes, enhancing the ability to detect pathological changes. **Statistical Rigour**: Quantitative histological data can undergo thorough statistical analysis to identify important variations between experimental groups.

8.2 Methods

8.2.1 Tissue Preparation Techniques

Quantitative neurohistology involves carefully measuring and analysing neural structures. This requires precise tissue preparation to maintain the delicate structure of neural tissue and ensure accurate results. Important steps include tissue preservation, sectioning, and staining, all of which help preserve tissue integrity and allow for detailed analysis.

8.2.1.1 Preservation Process

The preservation process maintains the tissue's integrity, prevents degradation, and allows for various analytical methods. This includes routines used to preserve neural tissue, focusing on fixation, cryopreservation, and storage.

8.2.1.1.1 Fixation

Fixation is the initial and most critical step in tissue preparation, aimed at preserving the morphology and molecular composition of tissues. This process stabilises the tissue structure and prevents it from breaking down due to enzymatic activity and bacterial growth. Various types of fixatives, such as paraformaldehyde (PFA) and glutaraldehyde, offer unique benefits for preserving different tissue components. PFA is well known for its protein cross-linking properties and for preserving cellular and tissue architecture. The typical concentration of PFA used for neural tissue is at 4% in phosphate-buffered saline (PBS). Glutaraldehyde is preferred for its rapid fixation capabilities, which are important for maintaining ultrastructural details, and is often combined with PFA for electron microscopy due to its superior preservation of fine structures (Eltoum et al. 2001).

There are different fixation techniques, including immersion and perfusion fixation. Immersion fixation involves submerging tissue samples in a fixative, which is effective for smaller specimens. In contrast, perfusion fixation is used for whole-organ or whole-animal studies, ensuring uniform preservation by circulating the fixative through the vascular system. This method is particularly beneficial for preserving the integrity of larger tissues and organs. Improper fixation can lead to artefacts, altered antigenicity, and compromised results in molecular analyses. Therefore, the choice of fixative and fixation methods can affect antigen retrieval and the preservation of nucleic acids and proteins, making it crucial to select the appropriate approach for the intended analytical techniques. Furthermore, optimising fixation protocols to meet the specific requirements of quantitative neurohistology is highly recommended. This includes pursuing factors such as the concentration, pH, temperature of the fixative, and duration of the fixation process, which can significantly influence the outcome (Mufson et al. 2020).

8.2.1.1.2 Cryopreservation

Cryopreservation is a method used to freeze tissue for future study, especially when immediate fixation is not possible. This technique is essential to keep the tissue's structure and molecular details intact (Robert et al. 2016). **Fresh Freezing:** Fresh freezing involves quickly freezing brain tissue to preserve its natural state. Typically, the tissue is immersed in isopentane cooled with dry ice or liquid nitrogen. Rapid cooling helps prevent the formation of ice crystals that can damage cells. The process starts by cooling the isopentane with dry ice or liquid nitrogen. Then, the tissue is quickly placed in the cold isopentane to ensure even freezing. Once frozen, the tissue can be stored at −80 °C or in liquid nitrogen for a long time. This method is very effective as it preserves the tissue for various analyses, such as molecular, biochemical, and histological studies. However, careful handling is required to avoid partial thawing and refreezing, which can damage the tissue (Zhang and Xiong 2014). **Cryoprotectants:** Cryoprotectants like sucrose and glycerol protect neural tissues from ice damage during freezing. They reduce ice crystal formation, preserving the tissue's structure. The tissue is gradually immersed in solutions with increasing concentrations of sucrose (up to 30%) before freezing. This ensures the cryoprotectant fully permeates and protects the tissue. After cryoprotection, the tissue is rapidly frozen (Bhattacharya 2018).

8.2.1.1.3 Storage of Neural Tissue

Storage of neural tissue techniques is essential for preserving sample integrity and functionality for research and clinical purposes. Recent advancements in this area include short-term storage methods such as **refrigeration** at 4 °C, where fixed neural tissues can be kept in buffer solutions like 1× PBS with 0.1% sodium azide for several weeks to months, although the long-term effects on antigenicity need further evaluation. Additionally, **cryoprotectants** such as glycerol-DMSO are used to store tissues at −20 °C or − 80 °C, effectively maintaining tissue morphology and antigenicity. For long-term storage, cryopreservation with liquid nitrogen is widely used, with optimal formulations for engineered neural tissue including 60% Dulbecco's Modified Eagle Medium, 30% foetal bovine serum, and 10% DMSO, which ensures cell viability and structural integrity over extended periods (Estrada et al. 2017). **Hypothermic preservation** at around 4 °C using specialised media allows for up to 28 days of storage without toxic cryoprotective agents and can be performed with standard refrigeration equipment (Day et al. 2017). Emerging techniques include formaldehyde-free minimalistic preservation methods for educational purposes and advanced biopreservation strategies for living tissue-engineered nerve grafts, focusing on maintaining cellular viability and tissue architecture through both hypothermic and cryogenic methods (Frigon et al. 2024).

Embedding tissues in a supportive medium facilitates sectioning and long-term storage. Preserving neural tissue integrity for microscopic analysis is essential, and embedding techniques have evolved to meet a variety of research needs. Each embedding technique provides distinct advantages, ensuring that neural tissues are optimally preserved and prepared for detailed histopathological examination. By choosing the appropriate method, researchers can achieve high-quality neurohistological results. **Paraffin Embedding:** This method involves formalin fixation, ethanol dehydration, xylene clearing, and paraffin infiltration. While it supports thin sectioning well, it may alter some molecular markers. **Resin Embedding:** Ideal for high-resolution imaging like electron microscopy, this process includes glutaraldehyde and paraformaldehyde fixation, osmium tetroxide post-fixation, dehydration, resin infiltration, and polymerisation (Ghnenis et al. 2018). **Gelatin Embedding:** Used for specific staining, this technique involves immersing tissues in gelatin and solidifying at 4 °C. **Agar Embedding:** Suitable for whole brain slicing or multiple samples, this method immerses tissues in warm agar, which solidifies at room temperature or 4 °C. Recent techniques include **cryoembedding:** Supports frozen tissues for sectioning while preserving enzyme activity and antigenicity by embedding tissues in an OCT compound and snap-freezing them. **Polyethylene glycol (PEG) embedding** is an eco-friendly alternative involving fixation, dehydration, and PEG infiltration, which solidifies at room temperature (Zhang and Xiong 2014). Microwave embedding is an advanced technique that significantly enhances the efficiency and quality of neural tissue preparation for neurohistological analysis. Microwave embedding accelerates the embedding process through rapid microwave fixation, expedited ethanol dehydration, quick resin infiltration, and microwave-assisted polymerisation, maintaining tissue integrity while drastically reducing processing time (Steyer et al. 2019).

8.2.1.2 Sectioning

Different sectioning methods are used depending on the specific needs of the study, each with its own benefits and challenges. Traditional techniques such as microtome sectioning, cryosectioning, vibratome sectioning, and ultramicrotomy have been essential in histological analysis for many years. Choosing the right sectioning technique in quantitative neurohistology depends on the specific goals of the study, the required resolution, and the availability of resources.

Microtome sectioning is one of the most widely used methods. It involves embedding neural tissue in paraffin wax or resin to support the tissue during cutting. The process includes trimming the tissue blocks, slicing the tissue with a microtome, floating the sections on a water bath, and drying them. This method produces thin, consistent sections (5–20 μm thick) suitable for various staining techniques and microscopic analysis. However, it is time-consuming and requires careful preparation to avoid artefacts and ensure high-quality sections (Slaoui and Fiette 2011). **Cryosectioning** is a rapid technique that preserves enzyme activity and antigenicity, important for immunohistochemical analyses. Tissues are rapidly frozen in a cryostat and then sectioned while still frozen. This technique minimises heat-induced changes in cellular components and is less labour-intensive than traditional methods. However, it can cause tissue cracking and artefacts due to ice crystal formation, requiring specialised handling and equipment (Pinskiy et al. 2013). **Vibratome sectioning** is an ideal technique for creating thick tissue sections with minimal tissue distortion. This method uses a vibrating blade to cut through fresh or fixed tissue without embedding. It is particularly useful for visualising detailed histological and 3D structures in tissues and embryos that have undergone various staining procedures, such as β-galactosidase, alkaline phosphatase, fluorescent, and DAB (diaminobenzidine) immunostaining. Since it does not use harsh solvents, vibratome sectioning is well suited for specimens stained with fluorescent antibodies or dyes. Additionally, it can be used to section live tissue for further culture, manipulation, and live imaging. Although it requires careful handling to avoid damage and typically produces thicker, less uniform sections than microtome sectioning, this technique is straightforward, requires minimal materials, and can be quickly mastered (Iulianella 2017). **Ultramicrotomy** is essential for high-resolution imaging of cellular structures, especially in electron microscopy. It involves cutting resin-embedded tissue with an ultramicrotome equipped with a diamond or glass knife to produce ultrathin sections, often less than 100 nm thick. This technique provides detailed images of cellular and subcellular structures but is delicate, is time-consuming, and requires expensive, specialised equipment (Baena et al. 2019).

In addition to these traditional methods, advanced sectioning techniques have been developed to improve the precision and quality of tissue analysis. **Laser capture microdissection (LCM)** allows the isolation of morphologically distinct subpopulations of cells or regions from heterogeneous tissue sections with high precision, making it ideal for molecular analyses like gene expression studies. Though LCM requires specialised equipment and expertise, it minimises contamination from surrounding tissues. It provides a powerful tool for retrospective and clinically relevant studies of tissue proteins when used with formalin-fixed and paraffin-embedded (FFPE) tissue in healthy and diseased conditions (Griesser et al. 2020). **Automated serial sectioning and imaging (ASI)** combines robotics and imaging systems to automatically cut and image serial sections of neural tissue. This method is essential for creating detailed three-dimensional reconstructions of brain architecture, allowing high-throughput sectioning and imaging while reducing human error and variability. Despite the complex and expensive setup, ASI significantly improves precision and throughput (Horstmann et al. 2012).

Optical coherence tomography (OCT) offers a nondestructive, real-time imaging solution by directing a beam of light into the tissue and using the reflected light to create detailed images. OCT provides high-resolution cross-sectional views of tissue structures, making it useful for dynamic studies and monitoring tissue changes over time. However, its penetration depth is limited compared to traditional sectioning methods (Lains et al. 2021). **Knife-edge scanning microscopy (KESM)** combines microtome sectioning with high-resolution imaging to produce detailed, large-scale datasets of neural tissues. This method is invaluable for mapping neural circuits and brain connectivity, providing high-resolution images efficiently. However, the specialised equipment required is costly, and data processing can be computationally intensive (Choe et al. 2011).

8.2.1.3 Staining

In the field of neurohistology, staining techniques have evolved significantly, enhancing our ability to visualise and analyse neural tissues. These advancements have led to the development of specialised dyes and protocols, which are essential for detailed tissue examination.

Common Staining Techniques in Neurohistology

Haematoxylin and eosin (H&E) is one of the oldest and most widely used techniques in neurohistology, fundamental in basic histology and clinical pathology. Haematoxylin stains acidic structures like DNA and RNA a dark blue or purple, while eosin stains proteins and cytoplasmic components pink. Despite its effectiveness for general tissue visualisation, H&E staining has limitations, such as not providing specific information about cellular functions and struggling with staining lipids and carbohydrates.

Mallory trichrome stain uses a combination of dyes to differentiate tissue components. This stain is particularly useful for identifying collagen deposition in tissues, aiding in the diagnosis and monitoring of fibrosis. However, its inability to distinguish between different types of collagens can limit diagnostic accuracy.

Periodic acid-Schiff (PAS) reaction represents another significant advancement. This method chemically transforms hydroxyl groups in sugars into aldehydes, which then react with Schiff reagent to produce a red colour. PAS staining effectively identifies polysaccharides such as glycogen and glycosaminoglycans. The **Feulgen reaction**, a variation of PAS, can identify DNA, and enzymatic digestion can further differentiate PAS-positive components, distinguishing glycogen from glycoproteins (Suvarna et al. 2018).

Metal impregnation methods have also been pivotal in neurohistology. Using metal salts like silver, gold, and mercury, these techniques visualise specific cell elements and extracellular matrix fibres, highlighting intricate neural structures. Depending on factors like tissue size and staining technique, the process can take from 12 h to 2 days (Frigon et al. 2024).

Fluorescent staining has become increasingly popular due to its ability to provide high-resolution images. This technique uses fluorescent dyes or tagged antibodies to visualise specific cellular components. It allows for multiplexing, enabling multiple targets to be visualised simultaneously, which is particularly useful for complex tissue analysis (Ueda et al. 2020).

Immunocytochemistry and immunohistochemistry (IHC) are highly specific methods for detecting enzymes, macromolecules, or antigens within cells or tissue sections using antigen-antibody reactions. Antibodies conjugated to fluorescent dyes such as fluorescein isothiocyanate (FITC) or rhodamine B are excited by specific wavelengths of light and emit fluorescence, enabling precise localization of target molecules. These techniques are invaluable for studying cellular interactions and disease mechanisms in neurobiology. For example, IHC is used to identify beta-amyloid plaques in brain tissue samples from Alzheimer's disease patients. By using antibodies that specifically bind to beta-amyloid, researchers can visualise and quantify plaque deposition, providing detailed insights into disease mechanisms and progression (Ramos-Vara 2011).

Advanced Staining Techniques in Neurohistology

Recent advancements in neurohistology have introduced various sophisticated staining methods, significantly enhancing the ability to visualise and analyse neural tissues. These methods provide detailed insights into the structure and function of neural tissues, enabling researchers to conduct precise quantitative assessments and furthering our understanding of neurological diseases and conditions.

Silver-based stains have advanced into specialised techniques that highlight specific neural structures, such as axons and dendrites. Key techniques include the **modified Bielschowsky stain**, which enhances the visibility of neurofibrillary tangles and plaques in Alzheimer's disease using silver nitrate. The **Gallyas stain** is another advanced method that effectively highlights pathological deposits in neurodegenerative diseases with high sensitivity. **Immunogold-silver staining** combines **immunogold labelling** with silver enhancement to visualise proteins at the subcellular level, improving the resolution of light microscopy. Lastly, combining **silver impregnation with Nissl staining** allows for better identification of neuronal structures and enhances the morphological assessment of nervous tissue (Ikenari et al. 2020).

Fluoro-jade staining is another recent advancement designed specifically to label degenerating neurons. This technique allows researchers to assess neuronal death and degeneration in various conditions, providing valuable information about neural damage and recovery processes (Ikenari et al. 2020).

Virtual stain transfer is one of the most innovative recent advancements. Virtual stain transfer utilises deep learning algorithms to generate virtual images of stained tissues from unstained samples. This approach facilitates quantitative analysis without the need for physical staining, revolutionising the way tissue samples are analysed. This innovative approach allows for the quantitative analysis of tissue architecture without the need for traditional staining procedures, streamlining the workflow and reducing the use of reagents (Bai et al. 2023).

8.2.2 Image Capture Techniques

8.2.2.1 Basic-Level Techniques

Light Microscopy: This includes various methods that use visible light to magnify and view specimens. The three main types are brightfield microscopy, phase contrast microscopy, and differential interference contrast (DIC) microscopy. **Brightfield Microscopy:** This common technique involves shining light through a specimen to produce an image based on light absorption. It is especially good for examining stained samples like tissue sections, allowing for detailed observation of cellular structures. **Phase Contrast Microscopy:** This method enhances the contrast of clear and colourless specimens without staining. It is useful for watching live cells and understanding their behaviours and interactions, commonly used in studies like neurological research. **Differential interference contrast (DIC) microscopy** creates high-contrast images that look three-dimensional, helping to reveal fine details in specimens. It is great for examining living cells and tissues, such as nerve cell structures, showing the morphology and organisation of cellular components more naturally than staining methods (Alturkistani et al. 2016).

8.2.2.2 Intermediate-Level Techniques

Fluorescence Microscopy: It is important in cell and molecular biology, especially in neurobiology studies. It enables subcellular resolution visualisation of living cells, allowing scientists to observe physiological processes within cells. This technique is needed for exploring phenomena like protein localisation, cellular motility, ion transport, and metabolism, significantly enhancing our understanding of neuron structure, function, and dynamics in living tissues. The method involves various specialised techniques: **Widefield fluorescence microscopy** marks specific cellular components with fluorescent dyes or proteins, useful for visualising protein or neurotransmitter receptor distribution within cells and helping in mapping neuronal pathways. **Confocal microscopy** uses a focused laser to generate high-resolution images by removing out-of-focus light, ideal for 3D imaging of neuronal structures. **Two-photon microscopy** leverages near-infrared lasers to excite fluorophores, optimising imaging deep within tissues, and is suitable for real-time observation of neuronal activity (Sanderson et al. 2014).

Electron Microscopy: In neurohistology, researchers use two main electron microscopy techniques to examine the structures of the nervous system: transmission electron microscopy (TEM) and scanning electron microscopy (SEM). **Transmission Electron Microscopy (TEM)**: This technique involves directing a beam of electrons through thin brain tissue slices to generate detailed nanometre-scale images. It allows for a close examination of synaptic structures, including neurotransmitter vesicles and transmission areas, providing critical insights into synaptic function and plasticity. **Scanning Electron Microscopy (SEM)**: It focuses on specimen surfaces, scanning them with a fine electron beam to create detailed 3D images. It is particularly effective for studying neuronal dendrites, the structures that receive synaptic inputs, by analysing their texture and shape to understand their impact on synaptic connections and neuronal communication (Pease 2013).

Optical Coherence Tomography (OCT): It offers a nondestructive, real-time imaging solution by directing a beam of light into the tissue and using the reflected light to create detailed images. OCT provides high-resolution cross-sectional views of tissue structures, making it useful for dynamic studies and monitoring tissue changes over time. However, its penetration depth is limited compared to traditional sectioning methods (Lains et al. 2021).

Magnetic Resonance Imaging (MRI): Nowadays, magnetic resonance imaging (MRI) has become an essential tool in neurohistology, providing insights into brain structure and function in both healthy and diseased conditions. MRI offers quantitative measurements of brain microstructure, which can supplement or even replace traditional invasive histology techniques. Here are some of the applications of MRI in neurohistology. **Quantitative MRI (qMRI):** This includes relaxometry, which measures the time protons take to return to their normal state after being disturbed by a magnetic pulse. This helps identify different brain tissues based on their unique relaxation properties. **Quantitative susceptibility mapping (QSM)** is used to evaluate the magnetic susceptibility of tissues, which is important for studying iron content in the brain, particularly in neurodegenerative diseases. **Diffusion Tensor Imaging (DTI):** This examines the movement of water molecules in the brain to map and assess white matter tracts, essential for diagnosing conditions like brain injuries and diseases affecting white matter. **Structural MRI:** It provides detailed anatomical images of the brain, important for detecting structural changes linked to neurological disorders. It is particularly useful in observing brain atrophy and tracking pathological changes in diseases such as Alzheimer's and Parkinson's over time. **Functional MRI (fMRI):** It monitors changes in blood flow associated with neural activity and offers insights into brain function. It identifies active brain regions during various tasks, helping to understand where specific cognitive functions occur. It also studies the impact of neurological disorders on brain function and evaluates treatment outcomes (Alkemade et al. 2023).

Neurosonology: This refers to the use of various noninvasive ultrasound imaging and non-imaging techniques to evaluate patients with neurological disorders. This includes assessing blood flow, detecting emboli, evaluating vascular stenosis or occlusion, and providing hemodynamic and structural information about intracranial and extracranial vessels.

Cervical Duplex Ultrasonography: It combines brightness-mode (B-mode) and Doppler waveform imaging to assess extracranial arteries, including the internal carotid, common carotid, external carotid, and vertebral arteries. **Transcranial Doppler (TCD)**: It uses the Doppler effect to measure blood flow velocity in the intracranial arteries, such as the middle cerebral, anterior cerebral, posterior cerebral, and basilar arteries. **Transcranial Colour-Coded Duplex Sonography**: It provides colour-coded images of blood flow in the intracranial arteries, helping to identify stenoses, emboli, and collateral circulation. **Transcranial Parenchymal Sonography**: It is used for brain parenchyma imaging, particularly in the diagnosis of movement disorders. **Peripheral Nerve Ultrasound**: High-frequency probes visualise the morphology of peripheral nerves, aiding in the diagnosis of entrapment neuropathies and other peripheral nervous system disorders (Tsivgoulis and Alexandrov 2016).

8.2.2.3 Advanced-Level Techniques

Photoacoustic Imaging: Photoacoustic imaging (PAI) is a new biomedical technique that combines light and sound to create high-resolution images of tissues deep inside the body. It has several potential medical uses, such as breast imaging to detect tumour growth and differences between benign and cancerous masses; skin imaging to visualise layers, blood vessels, and pigments for diagnosing skin cancers and other conditions; and vascular imaging to study blood vessels and measure oxygen levels without needing contrast agents (Attia et al. 2019). This noninvasive technique combines light and sound to create detailed images of tissues, which is especially useful for seeing blood oxygen levels, detecting fat, and measuring water content. It has the potential to track blood flow and inflammation, making it valuable for studying vascular health and heart disease (Karlas et al. 2019). PAI provides clear, detailed images and important functional information, but more research is needed before it can be widely used in clinics.

Expansion Microscopy (ExM)**:** It is a new technique that enhances the resolution of microscopes by physically expanding biological samples. By embedding the sample in a swellable polymer gel, researchers can make the specimen larger, allowing standard microscopes to capture high-resolution images. This method provides nanoscale details, around ~70 nm, without the need for specialised, expensive equipment. The technique works on both cells and tissues, such as brain samples, and is useful for studying detailed structures like proteins, neurons, and synapses. It offers a simpler and more affordable way for scientists to perform super-resolution imaging, making it accessible for a wide range of biological research (Chen et al. 2015). ExM has emerged as a powerful tool in neurobiology, offering nanoscale-resolution imaging of neural structures using conventional microscopes. ExM has been successfully applied to study various aspects of neurobiology, including synaptic organisation, neuronal morphology, and neurotransmitter receptor distribution in intact brains. Several protocols have been developed for neuroscience applications, such as protein retention ExM (proExM) and ExFISH for RNA visualisation. Recent advances include combining ExM with other super-resolution techniques and applying it to study neurodevelopmental and neurodegenerative disorders. While ExM offers numerous advantages, it has limitations, including incompatibility with live imaging, increased imaging time, and data storage requirements (Derevtsova et al. 2021).

8.2.3 Image Analysis Techniques

Image analysis techniques in neurohistology have become essential for studying the nervous system at the microscopic level, enabling researchers to extract quantitative information from histological images. These techniques include a wide range of approaches, from basic preprocessing to advanced machine learning methods. The choice of method depends on the specific research question, tissue type, and available imaging modalities. As technology progresses, these techniques offer increasingly sophisticated ways to extract meaningful information from complex neural tissues, advancing our understanding of neuroanatomy, neuropathology, and neurodevelopment. Here's a comprehensive overview of key image analysis techniques used in neurohistology:

Image Preprocessing

- **Colour normalisation** is an important preprocessing step that aims to reduce variability in histological images caused by differences in staining techniques and imaging conditions. This process involves transforming the colour attributes of an image to match a standardised colour distribution, thereby improving consistency across images and enhancing the reliability of the subsequent analysis. Techniques such as intensity thresholding, histogram normalisation, stain separation, and colour deconvolution are commonly used to achieve this (Hoque et al. 2024).
- **Noise reduction and artefact removal** are used to enhance the accuracy and reliability of images by eliminating unwanted distortions and interferences that can blur details in brain scans or microscopy. Common methods include filtering to smooth out random noise, motion correction to align images affected by subject movement, and specialised algorithms to remove specific artefacts. Advanced techniques, such as machine learning approaches, are increasingly being used to tackle more difficult noise and artefact patterns (Das and Chandra 2022).
- **Image registration**
  Image registration in neurohistology is the process of accurately aligning tissue sections for analysis, which is necessary for understanding brain structure and function. Recent advancements have significantly enhanced the accuracy and reproducibility of neurohistological studies, enabling comprehensive data integration across various scales and modalities. Notable innovations include the SHARCQ workflow, which semi-automates slice alignment, atlas registration, and cell quantification, streamlining the analysis process (Lauridsen et al. 2022). Additionally, the use of diffeomorphic mappings for multimodal cross-registration in marmoset brain histology allows for precise alignment of histological sections with MRI and blockface photographs, effectively addressing metric distortions and facilitating accurate mapping across different imaging modalities (Lauridsen et al. 2022).

Morphometry involves the measurement of the shape, size, and structure of neurons and glial cells. Morphometry in quantitative neurohistology involves techniques like voxel-based, deformation-based, and tensor-based morphometry to analyse brain structures. Voxel-based morphometry (VBM) compares local concentrations of grey matter between groups by normalising and segmenting MR images, followed by statistical analysis. Deformation-based morphometry (DBM) identifies differences in the positions of brain structures using deformation fields, while tensor-based morphometry (TBM) examines local shape differences through tensor fields. These methods provide detailed insights into brain morphology, enabling the study of structural differences and aiding in the classification and characterisation of neurological conditions (Ashburner and Friston 2011).

Stereology is a powerful image analysis technique used in quantitative neurohistology to analyse 3D tissue structures obtained from 2D sections accurately. Stereology uses rigorous sampling and statistical principles to provide unbiased estimates of various structural parameters such as number, volume, surface area, and length (Kipp et al. 2017). Reliable stereological methods provide essential information for observing cell degeneration and assessing the impact of therapeutic agents. With technological advancements making stereological techniques more accessible, these methods offer a more efficient approach to quantitative analysis, supporting the generation of essential numerical data for future research (Delibas et al. 2024).

Segmentation is a process that allows for the identification and analysis of different neural components, utilising advanced algorithms like thresholding, region-growing, and machine learning-based methods for accurate delineation. Recent advancements in brain image segmentation have significantly improved diagnostic processes in clinical environments, driven by the increasing incidence of brain disorders, which has led to a heightened demand for automated segmentation methods to assist medical professionals in early diagnosis and treatment planning. Segmentation techniques range from basic intensity-based methods to more sophisticated approaches involving machine learning, metaheuristics, deep learning, and hybrid models. Among these, deep learning and hybrid metaheuristic methods have shown superior efficiency in reliably segmenting brain tumours, although they are challenged by computational and memory demands (Fawzi et al. 2021). Intensity-based segmentation is one of the simplest approaches, using pixel intensity distributions to differentiate between brain and non-brain regions in neuroimaging datasets. Advanced image segmentation techniques, such as those utilising convolutional neural networks (CNNs) and Unet models, have been applied to analyse neural activity markers like C-fos gene expression, demonstrating improved accuracy and reduced manual intervention (Cai 2023). Additionally, innovative methods combining attention mechanisms with deep residual networks have enhanced the precision of brain tumour segmentation in MRI images, thereby improving diagnostic efficiency and reducing medical manpower requirements (Yuan 2024).

Texture analysis techniques assess the spatial distribution of intensity patterns in images to characterise tissue organisation and detect abnormalities in neural structures. Features like entropy, contrast, and correlation provide valuable insights into tissue heterogeneity and microstructural changes. Co-localisation analysis, particularly useful in immunohistochemistry and fluorescence microscopy studies, identifies the spatial overlap of different markers, such as proteins or genes, within the same tissue sample. This method helps reveal functional relationships between molecular targets in neural tissues.

Tractography utilising diffusion tensor imaging (DTI) maps out the pathways of white matter tracts in the brain, allowing for visualisation and quantification of neural connections. This technique is particularly important in studying brain connectivity and alterations in neurological disorders.

Recent advancements in the field include the use of machine learning algorithms and neural networks for automated detection, classification, and prediction tasks in neurohistology. These cutting-edge techniques offer increased accuracy and efficiency in analysing large datasets and complex tissue structures, pushing the boundaries of what's possible in quantitative neurohistological analysis.

8.3 Applications of Quantitative Neurohistology

8.3.1 Studying Neurodegenerative Diseases

Quantifying pathological lesions can aid in the diagnosis and monitoring of neurodegenerative diseases such as Alzheimer's disease, Parkinson's disease, and other neurodegenerative disorders. A thin section of brain tissue can reveal many quantifiable histological objects, including neurons' cell bodies, glial cells, and discrete lesions caused by pathological processes. For example, aggregated or misfolded proteins are deposited in the brain as neurofibrillary tangles (NFT), Lewy bodies (LB), or Pick bodies (PB) in neurodegenerative diseases. Furthermore, proteins can form extracellular deposits such as β-amyloid (Aβ) in Alzheimer's disease (AD) or the disease form of prion protein (PrPSc) in Creutzfeldt-Jakob disease (CJD) (Armstrong 2010; Bolender et al. 1991).

8.3.2 Assessing Neuroprotective Compounds

Quantitative neurohistology can greatly benefit research on neurotoxicity, including the effects of oxidative stress on brain tissue, as well as investigating the effects of different neuroprotective substances. This approach involves techniques like immunofluorescence to measure and pinpoint specific oxidative stress indicators in brain tissue samples (Frijhoff et al. 2015; Gasmalla and Kaplan 2024; Scallet 1995; Yurt et al. 2018).

8.3.3 Regeneration and Neuronal Plasticity Studies

Quantitative neurohistology plays a crucial role in studying nerve regeneration and neuronal plasticity by providing detailed, measurable insights into structural changes at the cellular and subcellular levels. It allows assessing synaptic plasticity, axon regrowth, myelin changes, and outcomes of peripheral nerve regeneration therapies (Delibas et al. 2024; Sampaio-Baptista and Johansen-Berg 2017).

8.3.4 Studying Developmental Neurobiology

Quantitative neurohistology provides a robust framework for studying the complex processes underlying brain development. By integrating advanced imaging techniques, stereological methods, and comprehensive brain mapping, researchers can gain a deeper understanding of the cellular and structural changes that occur throughout the lifespan. These insights are vital for identifying the neurobiological basis of developmental disorders and for developing targeted interventions (Deniz et al. 2018; Garman et al. 2016; Yurt et al. 2018).

8.3.5 Validating the Effectiveness of Other Techniques

Validating the effectiveness of advanced MRI techniques used to study myelin in multiple sclerosis (MS) is an example; histological staining methods are utilised to validate MRI metrics by demonstrating correlation with myelin content. These methods play a crucial role in confirming the accuracy of MRI-derived metrics in reflecting the myelin content in brain tissues, including distinguishing between demyelinated, remyelinated, and normal-appearing white matter. This is crucial for monitoring the disease progress and the effectiveness of therapeutic interventions aimed at promoting remyelination in MS patients (Khormi et al. 2023).

8.4 Future Directions

Quantitative neurohistology utilises cutting-edge techniques to obtain highly precise measurements of neural structures and connections. By combining traditional histological staining methods with high-powered microscopy and sophisticated image analysis software, neurophysiologists can quantify characteristics like neuron size, density, and number with unprecedented accuracy. This level of quantification provides insights that simple visual observation alone cannot (Gazula et al. 2024).

One area that is actively developing is automated image processing tools for digital histology slides. Advanced machine learning algorithms show promise in streamlining previously manual tasks like cell segmentation and counting. This could allow neurophysiologists to obtain quantitative data from vast fields of whole-brain or whole-spinal cord sections much more quickly than present methods permit. Larger datasets would facilitate the discovery of novel structure-function relationships (Gazula et al. 2024).

Obtaining serial measurements over time is also an important goal. Through techniques like repeated in vivo brain imaging and postmortem analysis, quantitative neurohistology may help chart the precise time course of structural changes in conditions like Alzheimer's disease, multiple sclerosis, or normal ageing. This could yield new targets for therapeutic intervention or a deeper understanding of disease mechanisms (Tian et al. 2023).

8.4.1 Integration with Multi-omics Approaches

Holistic Analysis Combining quantitative neurophysiology with genomics, proteomics, and metabolomics will provide a more comprehensive analysis of neural tissues. This integration can reveal complex interactions between genes, proteins, and metabolites, which is essential for understanding the multifaceted nature of neurological diseases (Armstrong 2010; Bhattacharya 2018).

Understanding Disease Mechanisms Quantitative neurohistology is a powerful tool that can improve our understanding of neurological diseases by precisely tracking structural changes in the nervous system. It allows researchers to map how diseases like Alzheimer's and multiple sclerosis spread in the brain, link structural changes to symptoms, analyse how different cell types respond to diseases, and measure treatment effects accurately. By combining histology with advanced computational methods, this approach can reveal the underlying mechanisms of neurological disorders and help develop better diagnostics and treatments (Beach et al. 2015; Pakkenberg and Gundersen 1997).

8.4.2 Advancements in Imaging Techniques

Super-resolution Microscopy The development of super-resolution microscopy techniques has been transformative for quantitative neurohistology and understanding disease mechanisms. New microscopy techniques, such as STED (stimulated emission depletion) and PALM (photoactivated localisation microscopy), which overcome the diffraction limit, have enabled unprecedented visualisation of neural systems. Structures like synapses, vesicles, and protein aggregates can now be seen at the nanoscale level, providing previously impossible insights. Subtle disruptions to synaptic architecture or abnormal protein clustering can be observed at 10–20 nm resolution. These advanced microscopies, paired with tissue from disease models and postmortem samples, allow a direct correlation between ultrastructural changes and clinical progression. Molecular-level defects can be seen in fine detail, uncovering previously hidden subtypes and novel targets. Longitudinal tracking of network degradation and oligomeric aggregate nucleation offers a game-changing understanding of pathogenesis. The quantification of super-resolution data feeds into computational studies that model neural circuit dysfunction.

Advanced MRI Techniques Innovative MRI techniques continue pushing the boundaries of what can be noninvasively learned about neural systems in vivo. Alongside other quantitative methods, they constitute an incredibly valuable toolset. Diffusion-based methods like DTI allow for the reconstruction of microscopic white matter structures with unprecedented clarity, providing fundamental new insights into disorders by visualising fine-scale disruptions to structural connectivity between regions. FMRI further expands the understanding of functional reorganisations that may compensate for these disruptions. Multimodal studies integrating DTI, fMRI, and quantitative histological data offer a holistic view of structure-function relationships from micro- to macroscales, crucial for tracking disease progression and observing compensatory plasticity as pathology spreads. Innovations enabling higher resolution and specialised contrasts continue refining what's visible, with real-time in vivo observation bypassing many limitations of postmortem analyses. Monitoring molecular-level changes or microstructural rewiring throughout disease progression and treatment can power predictive models to optimise interventions. Advanced MRIs enable genuinely longitudinal investigations spanning decades in living cohorts, and when paired with AI and big data tools, they promise to revolutionise our dynamic understanding of mechanisms in ageing and disease like never before. The future of using these noninvasive methods for quantitative studies appears tremendously promising (Bosma et al. 2016; Cadotte et al. 2012; Le Bihan 2003).

8.4.3 Automation and Machine Learning

Automated Image Analysis The integration of machine learning and AI to automate quantitative analysis is seen as having immense potential to scale up work in the field. Manual segmentation and counting of cells, synapses, or other features are known to be time-consuming for researchers. However, algorithms can be trained using expert annotations, allowing rapid and objective analysis of large image stacks. Throughput is maximised, and reproducibility is promoted by removing human bias from calculations. Subtle patterns too complex for human observation are discovered by deep learning, enabling networks to identify abnormal molecular distributions or early signs of synaptic loss that even experts might miss. These insights can drive new hypotheses about early disease triggers. Large-scale studies relating neural architecture to variables like age, sex, or disease severity are facilitated by objective and systematic analysis of extensive histology datasets. Targets for mechanistic follow-up that may elucidate pathogenesis are revealed by the correlations identified. User-friendly algorithm outputs enhance accessibility, empowering more scientists to easily obtain quantitative readouts guiding their research. AI promises to accelerate discovery in the field by augmenting human cognition at the scale of big data, making the future of automated digital pathology very promising (Hinton et al. 2012; Wang et al. 2022).

Predictive Modelling Integrating machine learning and AI into predictive modelling holds immense potential for revolutionising quantitative analysis in the medical field. Manual segmentation and counting of cells, synapses, and other features are traditionally time-consuming, but AI algorithms can be trained with expert annotations to analyse large image stacks quickly and objectively, enhancing throughput and promoting reproducibility by eliminating human bias. Deep learning can uncover subtle patterns that are too complex for human observation, such as abnormal molecular distributions or early signs of synaptic loss, driving new hypotheses about early disease triggers. The objective and systematic analysis of extensive histology datasets facilitates large-scale studies that correlate neural architecture with variables like age, sex, or disease severity, providing targets for mechanistic follow-up that may elucidate pathogenesis. User-friendly algorithm outputs further enhance accessibility, enabling more scientists to obtain quantitative readouts to guide their research. AI's capability to handle multimodal data, integrating biomarkers from histology, liquid biopsies, imaging, and other omics, further bolsters predictive power by discovering hidden combinatorial markers with significant prognostic value. Ultimately, developing digital replicas that mimic real disease dynamics promises to usher in an era of truly personalised medicine. Simulating how interventions like lifestyle changes, drugs, or compound treatments may alter a specific disease trajectory empowers prevention and mitigation strategies like never before. Leveraging the twenty-first-century technologies will accelerate advancements in diagnosis, treatment, and prevention, dramatically improving patient outcomes on a large scale. Exciting times lie ahead as these approaches revolutionise patient outcomes (Wang et al. 2022).

8.4.4 3D Reconstruction and Modelling

Detailed Spatial Mapping Three-dimensional reconstruction techniques are poised to revolutionise the understanding of brain architecture and pathology by capturing detailed spatial information. Techniques like serial block-face electron microscopy enable the digital reconstruction of entire neural volumes with micron-scale precision, allowing for visualising complete neuronal and glial morphologies within intact microcircuitry in 3D. This enhances comprehension of functional organisation and mapping of long-range projections and synaptic networks between cell types, revealing principles governing normal and disrupted connectivity. Integrating these methods with optogenetics, imaging, and electrophysiology provides a platform for multimodal investigations, offering a comprehensive picture of structure-function relationships in health and disease. Computational tools leverage these vast datasets for simulations, predictive models, and network analyses based on 3D connectomics, driving targeted experiments and treatments. Quantitative connectomics is anticipated to be a groundbreaking technology for neuroscience, making the future in this area extraordinarily bright (Denk and Horstmann 2004; Hughes et al. 2014; Jurrus et al. 2009).

Virtual Reality (VR) Applications VR applications of 3D neural reconstructions are set to transform multiple areas of neuroscience. Intuitively navigating fully reconstructed neural volumes with six degrees of freedom will revolutionise how researchers explore and understand neural architecture, enabling the visualisation of intricate cell morphologies and long-range projections within entire circuits. This immersive quality of VR promotes scientific accessibility by allowing students and clinicians to visually learn neuroanatomy in situ, potentially standardising knowledge and skills. Interactive brain atlases could assist in diagnosing by intuitively localising pathologies. Public outreach using VR can help nonscientists appreciate the complexity of the brain, fostering support for neuroscience and addressing issues related to brain disorders. Additionally, VR simulations incorporating AI-derived predictive models may guide the development of novel targeted therapies, providing a powerful preclinical tool by validating the predicted impacts of interventions within virtual reconstructed neural volumes. Overall, VR promises to advance our understanding of brain structure-function relationships, promote education, and enhance therapeutic development, showcasing the revolutionary integrative power of quantitative neurohistology (Alhmoudi et al. 2023; Inkarbekov et al. 2023).

8.4.5 Longitudinal Studies

In Vivo Imaging Techniques for in vivo longitudinal imaging are being developed to significantly enhance our capability to monitor dynamic changes in neural tissues over time, which holds substantial implications for disease research. While postmortem analysis offers only snapshots, the ability to repeatedly and noninvasively visualise the same individual's brain and spinal cord over years or even decades provides an unprecedented window into the progression of pathogenesis. Combining minimally invasive histological sampling with advanced in vivo imaging techniques such as MRI, PET, ultrasound, or optical methods can synergistically capture both micro- and macroscale changes. This multimodal fusion produces a rich, multidimensional view of structural-functional relationships as neural networks degrade. Longitudinal monitoring of therapeutic impacts enables the direct observation of drug effects on molecular-, cellular-, and systems-level endpoints over extended periods, facilitating more efficient iterative optimisation compared to traditional trials. Precisely mapping the remodelling that occurs during natural compensatory responses also sheds light on circuit-level resilience and plasticity mechanisms, potentially informing the development of novel rehabilitation methods (Young et al. 2020).

Time-Lapse Imaging The potential of time-lapse imaging to transform our understanding of dynamic neural processes is recognised, as real-time observation of structural changes is enabled. Growth cone dynamics, synaptic modifications, and molecular aggregation are captured, revealing mechanisms of neural plasticity and disruptions, thus allowing therapeutic targets to be identified. Differences between reversible and irreversible degeneration are distinguished, informing personalised rehabilitation strategies based on degradation timelines. Real-time histological data are paired with in vivo imaging and simulations to validate computational models for screening treatment candidates. Technological advancements like light-sheet microscopy now permit extended time-lapse imaging at cellular resolution, and the integration with microfluidic organoid and slice culture platforms facilitates automated longitudinal analysis, offering groundbreaking insights into neural processes (Gritti et al. 2022; Tabata et al. 2024).

Role of AI in Quantitative Neurohistology
Artificial intelligence (AI) is set to play a transformative role in the future of quantitative neurohistology, enhancing both the efficiency and accuracy of research. AI algorithms can automate the analysis of histological images, drastically reducing the time required for manual annotation and quantification (Chaudhary et al. 2024). Machine learning models, especially those based on deep learning, can be trained to recognise and quantify various histological features such as neuron density, cell morphology, and synaptic connections, ensuring consistency and reducing human error (Poalelungi et al. 2024). AI excels in pattern recognition, making it invaluable for identifying subtle histological changes indicative of early disease stages. AI can analyse large datasets to uncover patterns and correlations that might be overlooked by human researchers, which is particularly useful for studying neurodegenerative diseases, where early detection is crucial (Javed et al. 2024). AI can also develop predictive models that forecast disease progression based on histological data, integrating various data types to provide comprehensive risk assessments and prognoses for individual patients, which is essential for personalised medicine. Additionally, AI can integrate data from diverse sources, enabling a more comprehensive understanding of neurological conditions and supporting the development of holistic treatment approaches. AI can assist in interpreting complex datasets, highlighting key findings, and generating hypotheses for further investigation (Peng and Ren 2024). Furthermore, AI can enhance virtual and augmented reality applications in neurohistology, such as AI-driven VR simulations that provide immersive educational experiences and augmented reality overlays that facilitate real-time analysis and interpretation (Hamam 2024). AI can streamline the analysis of longitudinal studies by efficiently processing large volumes of data collected over time, tracking changes in neural tissues, identifying trends, and correlating these changes with clinical outcomes, which is vital for understanding the long-term effects of diseases and treatments (Chatterjee et al. 2024).

8.5 Conclusion

Quantitative neurohistology represents a transformative evolution in neuroscience research, fundamentally shifting the field from traditional qualitative assessments to objective, reproducible, and statistically rigorous methodologies. This comprehensive review has illuminated how the integration of advanced tissue preparation techniques, sophisticated imaging technologies, and computational analysis methods has revolutionised our understanding of neural structure and function.

The field's remarkable journey from the pioneering work of Santiago Ramón y Cajal and the Madrid School of Neurology to today's cutting-edge applications demonstrates the continuous innovation that drives scientific progress. The development of modern brain banks, the establishment of standardised protocols, and the integration of multiple analytical approaches have created a robust foundation for contemporary neurohistological research.

Methodological Advancements and Their Impact

The methodological framework of quantitative neurohistology encompasses three critical domains: tissue preparation, image capture, and image analysis. Each component has undergone significant refinement, with innovations in fixation protocols, embedding techniques, and sectioning methods ensuring optimal preservation of neural architecture. The evolution from basic light microscopy to advanced techniques such as super-resolution microscopy, expansion microscopy, and photoacoustic imaging has dramatically enhanced our ability to visualise and quantify neural structures at unprecedented resolution.

The integration of stereology, morphometry, and automated image analysis has provided researchers with powerful tools to extract meaningful quantitative data from complex neural tissues. These methodological advances have enabled the objective measurement of cellular parameters, spatial relationships, and structural changes that were previously impossible to assess with traditional qualitative approaches.

Clinical and Research Applications

The clinical applications of quantitative neurohistology have proven invaluable across multiple domains of neuroscience research. In neurodegenerative disease research, the ability to quantify pathological lesions, protein aggregations, and cellular changes has provided crucial insights into disease mechanisms and progression. The assessment of neuroprotective compounds, the study of neural regeneration and plasticity, and the investigation of developmental neurobiology have all benefited from the precision and objectivity that quantitative approaches provide.

Furthermore, the validation of other techniques, particularly advanced MRI methods, through histological correlation has strengthened the bridge between in vivo and postmortem studies, enhancing our understanding of brain structure-function relationships in both health and disease.

Future Directions and Emerging Technologies

The future of quantitative neurohistology is characterised by unprecedented technological convergence and methodological sophistication. The integration of multi-omics approaches promises to provide holistic insights into neural systems by combining structural, molecular, and functional data. The advancement of artificial intelligence and machine learning technologies is poised to revolutionise automated image analysis, pattern recognition, and predictive modelling capabilities.

The development of 3D reconstruction and modelling techniques, coupled with virtual reality applications, will transform how researchers visualise and interact with neural data. These innovations will not only enhance scientific understanding but also improve educational approaches and clinical diagnostic capabilities.

The Role of Artificial Intelligence

Artificial intelligence emerges as a transformative force in quantitative neurohistology, offering solutions to longstanding challenges in image analysis, pattern recognition, and data interpretation. The automation of previously labour-intensive tasks, the discovery of subtle patterns invisible to human observation, and the development of predictive models for disease progression represent just the beginning of AI's potential impact on the field.

The integration of AI with longitudinal studies, multimodal imaging, and personalised medicine approaches promises to usher in an era of precision neuroscience, where individual patient outcomes can be predicted and optimised based on comprehensive histological and molecular profiles.

Challenges and Opportunities

While the field of quantitative neurohistology has achieved remarkable progress, several challenges remain. The need for standardisation across laboratories, the development of more sophisticated computational tools, and the integration of increasingly complex datasets require continued innovation and collaboration. The balance between technological advancement and practical accessibility must be carefully maintained to ensure that these powerful tools remain available to the broader scientific community.

The ongoing development of minimally invasive techniques, the refinement of longitudinal study methodologies, and the expansion of automated analysis capabilities will continue to shape the field's evolution. The potential for real-time tissue analysis, the development of portable imaging systems, and the creation of comprehensive digital atlases represent exciting frontiers for future exploration.

Summary

Quantitative neurohistology has matured into an indispensable discipline that bridges traditional morphological analysis with modern computational approaches. The field's commitment to objectivity, reproducibility, and statistical rigour has elevated the quality of neuroscience research and enhanced our understanding of the nervous system's complexity. As we look toward the future, the convergence of advanced imaging technologies, artificial intelligence, and multi-omics approaches promises to unlock new dimensions of neural understanding.

The continued evolution of quantitative neurohistology will undoubtedly contribute to breakthrough discoveries in neurodegenerative disease research, therapeutic development, and our fundamental understanding of brain function. By maintaining a focus on methodological excellence while embracing technological innovation, the field is well positioned to address the most pressing challenges in neuroscience and contribute to the development of more effective treatments for neurological disorders. The journey from qualitative observation to quantitative analysis represents more than a methodological shift; it embodies a fundamental transformation in how we approach the study of the nervous system. As quantitative neurohistology continues to evolve, it will remain at the forefront of neuroscience research, driving discoveries that will ultimately benefit patients and advance our understanding of the most complex organ in the human body.

References

Alhmoudi S, Alkatheri S, Alghafri S, Idrais AR, Damseh R (2023) A virtual reality educational tool for complex brain anatomy. Paper presented at the 2023 14th international conference on information and communication systems (ICICS)

Alkemade A, Grossmann R, Bazin PL, Forstmann BU (2023) Mixed methodology in human brain research: integrating MRI and histology. Brain Struct Funct 228(6):1399–1410. https://doi.org/10.1007/s00429-023-02675-2

Alturkistani HA, Tashkandi FM, Mohammedsaleh ZM (2016) Histological stains: a literature review and case study. Global J Health Sci 8(3):72

Armstrong RA (2010) Review paper. Quantitative methods in neuropathology. Folia Neuropathol 48(4): 217–230. Retrieved from https://www.ncbi.nlm.nih.gov/pubmed/21225504

Ashburner J, Friston K (2011) Voxel-based morphometry. In: Penny W, Friston K, Ashburner J, Kiebel S, Nichols T (eds) Statistical parametric mapping: the analysis of functional brain images. Elsevier, London, pp 92–98

Attia ABE, Balasundaram G, Moothanchery M, Dinish U, Bi R, Ntziachristos V, Olivo M (2019) A review of clinical photoacoustic imaging: current and future trends. Photoacoustics 16:100144

Baena V, Schalek RL, Lichtman JW, Terasaki M (2019) Serial-section electron microscopy using automated tape-collecting ultramicrotome (ATUM). In: Methods in cell biology, vol 152. Elsevier, London, pp 41–67

Bai B, Yang X, Li Y, Zhang Y, Pillar N, Ozcan A (2023) Deep learning-enabled virtual histological staining of biological samples. Light Sci Appl 12(1):57. https://doi.org/10.1038/s41377-023-01104-7

Beach TG, Adler CH, Sue LI, Serrano G, Shill HA, Walker DG et al (2015) Arizona study of aging and neurodegenerative disorders and brain and body donation program. Neuropathology 35(4):354–389. https://doi.org/10.1111/neup.12189

Bhattacharya S (2018) Cryoprotectants and their usage in cryopreservation process. In: Cryopreservation biotechnology in biomedical and biological sciences. IntechOpen, London

Bolender RP, Charleston J, Mottet K, McCabe JT (1991) Quantitative morphology of the nervous system: expanding horizons. Anat Rec 231(4):407–415. https://doi.org/10.1002/ar.1092310403

Bosma RL, Mojarad EA, Leung L, Pukall C, Staud R, Stroman PW (2016) FMRI of spinal and supra-spinal correlates of temporal pain summation in fibromyalgia patients. Hum Brain Mapp 37(4):1349–1360

Bourlond A (1976) Cutaneous innervation. J Invest Dermatol 67(1):106–109. https://doi.org/10.1111/1523-1747.ep12512513

Buda O, Arsene D, Ceausu M, Dermengiu D, Curca GC (2009) Georges Marinesco and the early research in neuropathology. Neurology 72(1):88–91. https://doi.org/10.1212/01.wnl.0000338626.93425.74

Cadotte DW, Bosma R, Mikulis D, Nugaeva N, Smith K, Pokrupa R et al (2012) Plasticity of the injured human spinal cord: insights revealed by spinal cord functional MRI. PLoS One 7(9):e45560

Cai P (2023) Advanced image segmentation techniques for neural activity detection via C-fos immediate early gene expression. Paper presented at the 2023 4th international conference on computers and artificial intelligence technology (CAIT)

Carlos AF, Poloni TE, Medici V, Chikhladze M, Guaita A, Ceroni M (2019) From brain collections to modern brain banks: a historical perspective. Alzheimers Dement (N Y) 5:52–60. https://doi.org/10.1016/j.trci.2018.12.002

Chatterjee I, Ghosh R, Sarkar S, Kundu M, Das K (2024) Revolutionizing innovations and impact of artificial intelligence in healthcare. Int J Multidiscipl Res. https://doi.org/10.36948/ijfmr.2024.v06i03.19333

Chaudhary S, Voigts A, Bereket M, Albert ML, Schwamborn K, Zeggini E, Casale FP (2024) HistoGWAS: an AI-enabled framework for automated genetic analysis of tissue phenotypes in histology cohorts. bioRxiv. 2024.2006.2009.597752

Chen F, Tillberg PW, Boyden ES (2015) Optical imaging. Expansion microscopy. Science 347(6221):543–548. https://doi.org/10.1126/science.1260088

Choe Y, Mayerich D, Kwon J, Miller DE, Sung C, Chung JR et al (2011) Specimen preparation, imaging, and analysis protocols for knife-edge scanning microscopy. J Vis Exp 58:e3248. https://doi.org/10.3791/3248

Das KP, Chandra J (2022) A review on preprocessing techniques for noise reduction in PET-CT images for lung cancer. Paper presented at the Congress on intelligent systems: proceedings of CIS 2021, vol 2

Day AGE, Bhangra KS, Murray-Dunning C, Stevanato L, Phillips JB (2017) The effect of hypothermic and cryogenic preservation on engineered neural tissue. Tissue Eng Part C Methods 23(10):575–582. https://doi.org/10.1089/ten.TEC.2017.0244

Delibas B, Vianney JM, Kaplan S (2024) The assessment of neuronal plasticity following sciatic nerve injuries in rats using electron microscopy and stereological methods. J Chem Neuroanat 136:102396. https://doi.org/10.1016/j.jchemneu.2024.102396

Deniz OG, Altun G, Kaplan AA, Yurt KK, von Bartheld CS, Kaplan S (2018) A concise review of optical, physical and isotropic fractionator techniques in neuroscience studies, including recent developments. J Neurosci Methods 310:45–53. https://doi.org/10.1016/j.jneumeth.2018.07.012

Denk W, Horstmann H (2004) Serial block-face scanning electron microscopy to reconstruct three-dimensional tissue nanostructure. PLoS Biol 2(11):e329. https://doi.org/10.1371/journal.pbio.0020329

Derevtsova K, Pchitskaya E, Rakovskaya A, Bezprozvanny I (2021) Applying the expansion microscopy method in neurobiology. J Evol Biochem Physiol 57(3):681–693

Eltoum I, Fredenburgh J, Myers RB, Grizzle WE (2001) Introduction to the theory and practice of fixation of tissues. J Histotechnol 24(3):173–190

Estrada LI, Robinson AA, Amaral AC, Giannaris EL, Heyworth NC, Mortazavi F et al (2017) Evaluation of long-term cryostorage of brain tissue sections for quantitative histochemistry. J Histochem Cytochem 65(3):153–171. https://doi.org/10.1369/0022155416686934

Fawzi A, Achuthan A, Belaton B (2021) Brain image segmentation in recent years: a narrative review. Brain Sci 11(8):1055. https://doi.org/10.3390/brainsci11081055

Frigon EM, Gerin-Lajoie A, Dadar M, Boire D, Maranzano J (2024) Comparison of histological procedures and antigenicity of human post-mortem brains fixed with solutions used in gross anatomy laboratories. Front Neuroanat 18:1372953. https://doi.org/10.3389/fnana.2024.1372953

Frijhoff J, Winyard PG, Zarkovic N, Davies SS, Stocker R, Cheng D et al (2015) Clinical relevance of biomarkers of oxidative stress. Antioxid Redox Signal 23(14):1144–1170. https://doi.org/10.1089/ars.2015.6317

Garman RH, Li AA, Kaufmann W, Auer RN, Bolon B (2016) Recommended methods for brain processing and quantitative analysis in rodent developmental neurotoxicity studies. Toxicol Pathol 44(1):14–42. https://doi.org/10.1177/0192623315596858

Gasmalla HEE, Kaplan S (2024) Empirical experience using the physical disector counting technique for estimating pyramidal cell numbers in the rat hippocampus. In: Kaplan S (ed) A manual of the stereological techniques. Nova Science Publishers, New York

Gazula H, Tregidgo HF, Billot B, Balbastre Y, Williams-Ramirez J, Herisse R et al (2024) Machine learning of dissection photographs and surface scanning for quantitative 3D neuropathology. elife 12:RP91398

Ghnenis AB, Czaikowski RE, Zhang ZJ, Bushman JS (2018) Toluidine blue staining of resin-embedded sections for evaluation of peripheral nerve morphology. J Vis Exp 137:e58031. https://doi.org/10.3791/58031

Gimenez-Roldan S (2015) The Madrid School of Neurology (1885–1939). Rev Neurol (Paris) 171(1):5–15. https://doi.org/10.1016/j.neurol.2014.08.002

Griesser E, Wyatt H, Ten Have S, Stierstorfer B, Lenter M, Lamond AI (2020) Quantitative profiling of the human substantia nigra proteome from laser-capture microdissected FFPE tissue. Mol Cell Proteomics 19(5):839–851. https://doi.org/10.1074/mcp.RA119.001889

Gritti N, Power RM, Graves A, Huisken J (2022) Image restoration of degraded time-lapse microscopy data mediated by infrared-imaging. bioRxiv, 2022.2011.2010.515910

Hamam H (2024) Revolutionizing neurosurgery and neurology: the transformative impact of artificial intelligence in healthcare. Comput Artif Intell 2(1):416

Hinton G, Srivastava N, Swersky K (2012) Lecture 6a overview of mini–batch gradient descent. Coursera Lecture slides https://class.coursera.org/neuralnets-2012-001/lecture, [Online]

Hoque MZ, Keskinarkaus A, Nyberg P, Seppänen T (2024) Stain normalization methods for histopathology image analysis: a comprehensive review and experimental comparison. Inf Fusion 102:101997

Horstmann H, Korber C, Satzler K, Aydin D, Kuner T (2012) Serial section scanning electron microscopy (S3EM) on silicon wafers for ultra-structural volume imaging of cells and tissues. PLoS One 7(4):e35172. https://doi.org/10.1371/journal.pone.0035172

Hughes L, Hawes C, Monteith S, Vaughan S (2014) Serial block face scanning electron microscopy—the future of cell ultrastructure imaging. Protoplasma 251(2): 395–401. https://doi.org/10.1007/s00709-013-0580-1

Ikenari T, Kurata H, Satoh T, Hata Y, Mori T (2020) Evaluation of Fluoro-Jade C staining: specificity and application to damaged immature neuronal cells in the normal and injured mouse brain. Neuroscience 425: 146–156. https://doi.org/10.1016/j.neuroscience.2019.11.029

Inkarbekov M, Pearlmutter BA, Monahan R (2023). Immersive neural network exploration: a VR approach to human-centered AI understanding. Paper presented at the proceedings of the 2023 conference on human Centered artificial intelligence: education and practice

Iulianella A (2017) Cutting thick sections using a vibratome. Cold Spring Harb Protoc 2017(6):pdb.prot094011. https://doi.org/10.1101/pdb.prot094011

Javed M, Bajwa MH, Bakhshi SK (2024) Artificial intelligence- image learning and its applications in neurooncology: a review. J Pak Med Assoc 74-(4 (Supple-4)):S158–S160. https://doi.org/10.47391/JPMA.AKU-9S-24

Jurrus E, Hardy M, Tasdizen T, Fletcher PT, Koshevoy P, Chien CB et al (2009) Axon tracking in serial block-face scanning electron microscopy. Med Image Anal 13(1):180–188. https://doi.org/10.1016/j.media.2008.05.002

Karlas A, Fasoula NA, Paul-Yuan K, Reber J, Kallmayer M, Bozhko D et al (2019) Cardiovascular optoacoustics: from mice to men – a review. Photoacoustics 14:19–30. https://doi.org/10.1016/j.pacs.2019.03.001

Khormi I, Al-Iedani O, Alshehri A, Ramadan S, Lechner-Scott J (2023) MR myelin imaging in multiple sclerosis: a scoping review. J Neurol Sci 455:122807. https://doi.org/10.1016/j.jns.2023.122807

Kipp M, Kiessling MC, Hochstrasser T, Roggenkamp C, Schmitz C (2017) Design-based stereology for evaluation of histological parameters. J Mol Neurosci 61(3): 325–342. https://doi.org/10.1007/s12031-016-0858-9

Lains I, Wang JC, Cui Y, Katz R, Vingopoulos F, Staurenghi G et al (2021) Retinal applications of swept source optical coherence tomography (OCT) and optical coherence tomography angiography (OCTA). Prog Retin Eye Res 84:100951. https://doi.org/10.1016/j.preteyeres.2021.100951

Lauridsen K, Ly A, Prevost ED, McNulty C, McGovern DJ, Tay JW et al (2022) A semi-automated workflow for brain slice histology alignment, registration, and cell quantification (SHARCQ). eneuro 9(2). https://doi.org/10.1523/ENEURO.0483-21.2022

Le Bihan D (2003) Looking into the functional architecture of the brain with diffusion MRI. Nat Rev Neurosci 4(6):469–480. https://doi.org/10.1038/nrn1119

Mazzarini M, Falchi M, Bani D, Migliaccio AR (2021) Evolution and new frontiers of histology in bio-medical research. Microsc Res Tech 84(2): 217–237. https://doi.org/10.1002/jemt.23579

Mufson EJ, Perez SE, Kelley CM, Alldred MJ, Ginsberg SD (2020) Fixation protocols for neurohistology: neurons to genes. In: Neurohistology and imaging techniques. Humana, New York, pp 49–71

Pakkenberg B, Gundersen HJ (1997) Neocortical neuron number in humans: effect of sex and age. J Comp Neurol 384(2):312–320. Retrieved from https://www.ncbi.nlm.nih.gov/pubmed/9215725

Pease DC (2013). Histological techniques for electron microscopy. Elsevier

Peng Z, Ren X (2024) Application and development of artificial intelligence-based medical imaging diagnostic assistance system. Int J Biol Life Sci 6(1):39–43

Pinskiy V, Tolpygo AS, Jones J, Weber K, Franciotti N, Mitra PP (2013) A low-cost technique to cryo-protect and freeze rodent brains, precisely aligned to stereotaxic coordinates for whole-brain cryosectioning. J Neurosci Methods 218(2):206–213

Poalelungi DG, Neagu AI, Fulga A, Neagu M, Tutunaru D, Nechita A, Fulga I (2024) Revolutionizing pathology with artificial intelligence: innovations in

immunohistochemistry. J Pers Med 14(7):693. https://doi.org/10.3390/jpm14070693
Ramos-Vara JA (2011) Principles and methods of immunohistochemistry. Methods Mol Biol 691:83–96. https://doi.org/10.1007/978-1-60761-849-2_5
Robert MC, Juan de Paz L, Graf DA, Gazzin S, Tiribelli C, Bottai H, Rodriguez JV (2016) Cryopreservation by slow cooling of rat neuronal cells. Cryobiology 72(3):191–197. https://doi.org/10.1016/j.cryobiol.2016.05.003
Sampaio-Baptista C, Johansen-Berg H (2017) White matter plasticity in the adult brain. Neuron 96(6):1239–1251. https://doi.org/10.1016/j.neuron.2017.11.026
Sanderson MJ, Smith I, Parker I, Bootman MD (2014) Fluorescence microscopy. Cold Spring Harb Protoc 2014(10):pdb.top071795. https://doi.org/10.1101/pdb.top071795
Scallet AC (1995) Quantitative histological evaluation of neuroprotective compounds. Ann N Y Acad Sci 765:47–58; discussion 59–61. https://doi.org/10.1111/j.1749-6632.1995.tb16559.x
Slaoui M, Fiette L (2011) Histopathology procedures: from tissue sampling to histopathological evaluation. Methods Mol Biol 691:69–82. https://doi.org/10.1007/978-1-60761-849-2_4
Steyer AM, Ruhwedel T, Mobius W (2019) Biological sample preparation by high-pressure freezing, microwave-assisted contrast enhancement, and minimal resin embedding for volume imaging. J Vis Exp 145:e59156. https://doi.org/10.3791/59156
Suvarna KS, Layton C, Bancroft JD (2018) Bancroft's theory and practice of histological techniques. Elsevier Health Sciences, Philadelphia
Tabata H, Nagata KI, Nakajima K (2024) Time-lapse imaging of migrating neurons and glial progenitors in embryonic mouse brain slices. J Vis Exp 205. https://doi.org/10.3791/66631
Tian Y, Johnson GA, Williams RW, White LE (2023) A rapid workflow for neuron counting in combined light sheet microscopy and magnetic resonance histology. Front Neurosci 17:1223226
Tsivgoulis G, Alexandrov AV (2016) Ultrasound in neurology. Continuum (Minneap Minn) 22-(5, Neuroimaging):1655–1677. https://doi.org/10.1212/CON.0000000000000374
Ueda HR, Ertürk A, Chung K, Gradinaru V, Chédotal A, Tomancak P, Keller PJ (2020) Tissue clearing and its applications in neuroscience. Nat Rev Neurosci 21(2):61–79
Wang H, Xu G, Pan X, Liu Z, Lan R, Luo X (2022) Multi-task generative adversarial learning for nuclei segmentation with dual attention and recurrent convolution. Biomed Signal Process Control 75:103558
Young PNE, Estarellas M, Coomans E, Srikrishna M, Beaumont H, Maass A et al (2020) Imaging biomarkers in neurodegeneration: current and future practices. Alzheimer's Res Ther 12(1):49. https://doi.org/10.1186/s13195-020-00612-7
Yuan J (2024) Brain tumor image segmentation method using hybrid attention module and improved mask RCNN. Sci Rep 14(1):20615. https://doi.org/10.1038/s41598-024-71250-4
Yurt KK, Kivrak EG, Altun G, Mohamed H, Ali F, Gasmalla HE, Kaplan S (2018) A brief update on physical and optical disector applications and sectioning-staining methods in neuroscience. J Chem Neuroanat 93:16–29. https://doi.org/10.1016/j.jchemneu.2018.02.009
Zhang J, Xiong H (2014) Brain tissue preparation, sectioning, and staining. In: Current laboratory methods in neuroscience research. Springer, New York, pp 3–30

Clinical Applications and Implications of Neuroimaging

9

Vykuntaraju K. Gowda, C. Viveka Santhosh Reddy, Varunvenkat M. Srinivasan, and Amena Nayyer

Abstract

Neuroimaging has revolutionized the field of clinical neuroscience by providing essential insights into the structure and function of the human brain. It has played a pivotal role in developing targeted treatment approaches. Brain imaging techniques have seen significant progress, and functional magnetic resonance imaging (fMRI) has added a whole new dimension to the traditional morphological images obtained from structural MRI and computed tomography (CT). The use of these tools has increased our understanding of central nervous system (CNS) pathology, and prognostication. In epilepsy, based on imaging findings, we can classify epilepsy as structural versus nonstructural etiology which aids further investigation of the underlying causes and its management. MRI of the brain is the most used modality in primary diagnosis, treatment planning like surgery or radiation therapy, and surveillance for recurrence in brain tumors. The Surveillance of Cerebral Palsy in Europe (SCPE) proposed a classification system with cerebral palsy based on MRI termed as MRI classification system (MRICS) primarily based on the pathogenic mechanisms according to the timing of injury to the developing brain. MRICS is a mainly qualitative classification system for cerebral palsy describing pathogenic neuroimaging patterns related to the timing of brain compromise that helps in selecting patients for genetic evaluation. Neuroimaging techniques like CT, MRI, and further magnetic resonance spectroscopy have helped delineate the pathophysiology and arrive at a specific diagnosis in more than half of the neurometabolic cases. In this chapter, we will explore the clinical applications and implications of neuroimaging, highlighting its profound impact on the field of neuroscience and patient care.

Keywords

Magnetic resonance imaging (MRI) · Computed tomography (CT) · Magnetic resonance spectrometry (MRS) · Cerebral palsy · MRI classification system (MRICS) · Brain tumors · Neurometabolic disorders · Epilepsy

V. K. Gowda (✉) · C. Viveka Santhosh Reddy · V. M. Srinivasan
Department of Pediatric Neurology, Indira Gandhi Institute of Child Health, Bangalore, Karnataka, India

A. Nayyer
Pediatric Radiology Fellow Department of Pediatric Radiology, Indira Gandhi Institute of Child Health, Bangalore, India

9.1 Introduction

As Alexander Pope once said, "Nature and Nature's laws lay hidden in the night, God said,

K. S. Ravi et al. (eds.), *Visualization in Neuroanatomical Sciences*, Biomedical Visualization 11,
https://doi.org/10.1007/978-3-032-07380-8_9

'Let Newton be' and all was light." With modern technology, more subtle areas of nervous system development that were in darkness for our understanding are being increasingly explored. Neuroimaging has revolutionized the field of clinical neuroscience by providing essential insights into the structure and function of the human brain. It has significantly enhanced our understanding of various neurological and psychiatric disorders and has played a pivotal role in developing targeted treatment approaches. This technology enables clinicians to visualize and analyze the brain in previously unimaginable ways, leading to more accurate diagnoses and personalized treatment plans for patients (Krishnan et al. 2015).

Imaging of the head and neck structures started with imaging of the skull and spine for locating bullets using X-ray technology. In the early twentieth century, Viennese neuropsychiatrist Arthur Schuller popularly known as the "father of neuroradiology" studied intracranial disorders which cause abnormalities detectable on the X-ray of the skull. Walter Dandy in the year 1918 proposed air as a good radiolucent contrast agent which was subsequently adapted in techniques like pneumography, ventriculography, and encephalography (Alper 1999). Further, in 1927, Egas Moniz used positive contrast to visualize the carotid artery which was later adapted in the technique of cerebral angiogram (Houser 1995). Subsequently, Hounsfield and Lauterbur furthered the science of neuroradiology with their contributions to the development of computed tomography (CT) and magnetic resonance imaging (MRI) technologies, both of which have revolutionized the field. The process of discovery of these technologies has many contributors and needs to be acknowledged as well (Leeds and Kieffer 2000; Lev et al. 2022).

The unique challenges encountered in imaging the pediatric population, handling them, and the utility of specific and often repeated MRI sequences have also been highlighted and discussed. Pediatric practice often requires the use of sedation or specific imaging protocols. These stresses are experienced by the imaging team. A well-framed clinical question, strict adherence to imaging protocols, and a thorough understanding of the peculiarities of pediatric anatomy and physiology form the basis of successful imaging in the pediatric population. Formal reading of images, interpretation, support, and guidance enhance the role of neuroradiology in pediatric imaging in many ways (Barkovich et al. 2019; Turesky et al. 2021; Wisnowski et al. 2013).

In the pediatric population, neuroimaging plays a pivotal role in early detection, measurement of developmental quotient, risk stratification and prognosticating, and therapy. This review attempts to touch upon various aspects of brain imaging as applied to the pediatric population, keeping content highly practical. These techniques must be used in conjunction with the clinical examination, and the use of structured neurodevelopment assessment tools can avoid underestimation of cases. The application of imaging is not just isolated to the diagnosis; in recent times, these ethical issues have moved to risk prediction and even therapeutic interventions using functional MRI.

Brain imaging techniques have seen significant progress in the past two decades, and functional magnetic resonance imaging (fMRI) has added a whole new dimension to the traditional morphological images obtained from structural MRI and computed tomography (CT). The use of these tools has increased our understanding of central nervous system (CNS) pathology in various disease conditions, and prognostication too has been simplified. In this chapter, we will explore the clinical applications and implications of neuroimaging, highlighting its profound impact on the field of neuroscience and patient care (Chavhan 2016; Dubois et al. 2021; Jones et al. 2020).

9.2 Common Neuroimaging Techniques in Pediatric Population

In this section, we describe common neuroimaging techniques widely used in the pediatric population.

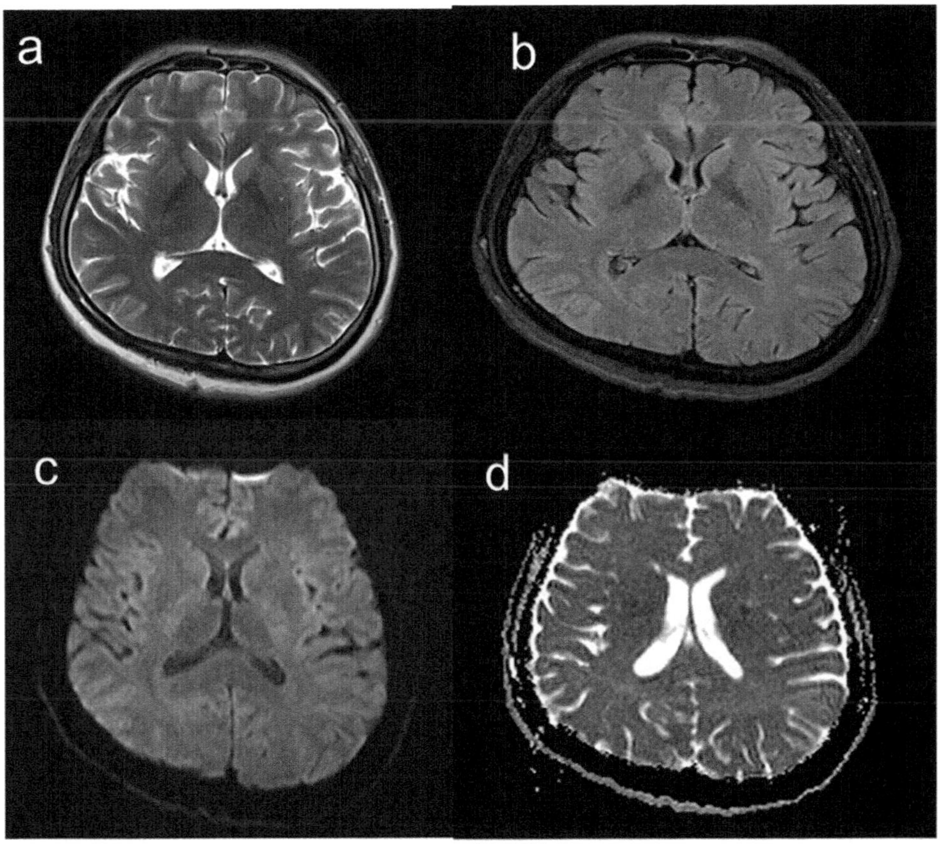

Fig. 9.1 MRI of the brain. Normal T2W (**a**), FLAIR (**b**), diffusion-weighted (**c**), and ADC (**d**) images of the brain

9.2.1 Magnetic Resonance Imaging (MRI)

MRI helps to unveil the intricate details of pediatric brain structures with advantages over CT. Different MRI sequences as depicted in Fig. 9.1 are T2W (a), fluid-attenuated inversion recovery (FLAIR) (b), diffusion-weighted (c), and apparent diffusion coefficient (ADC) (d) images of the brain. Figure 9.2 shows normal axial gradient echo images. Figure 9.3 reveals coronal (a) T1W and sagittal (b) T1W images of the brain. Other modifications of MRI include functional MRI (fMRI) which is used to study brain function and connectivity in children and diffusion tensor imaging (DTI) used to assess white matter integrity in pediatric neurological conditions. Vossough et al. discussed case examples demonstrating DTI's clinical utility (Vossough 2023).

9.2.2 Computed Tomography (CT)

A CT scan which is readily available, and rapid to undertake, helps to complete a comprehensive examination of pediatric neurological conditions but is indicated with caution especially in children in emergency settings to rule out bleeding with potential radiation risks. Figure 9.4 shows a CT brain without contrast (a and b) showing calcification with corresponding gradient MR image (c) that shows the presence of blooming (black arrow). CT is also used to demonstrate

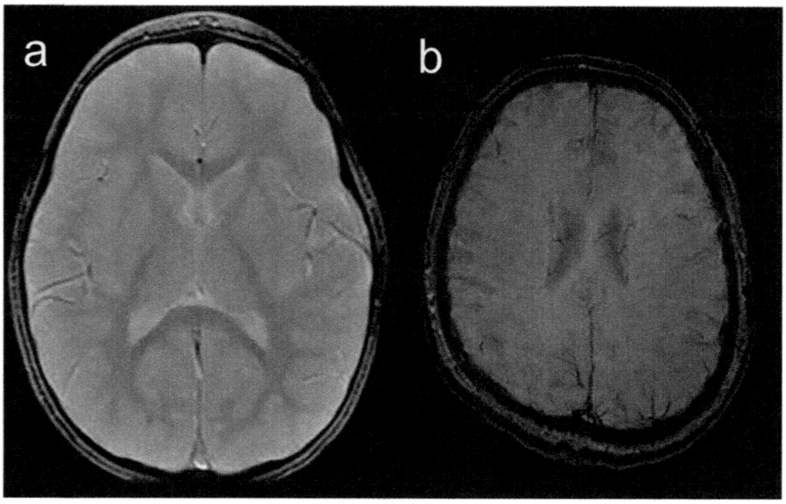

Fig. 9.2 Normal axial gradient echo images of MRI of the brain

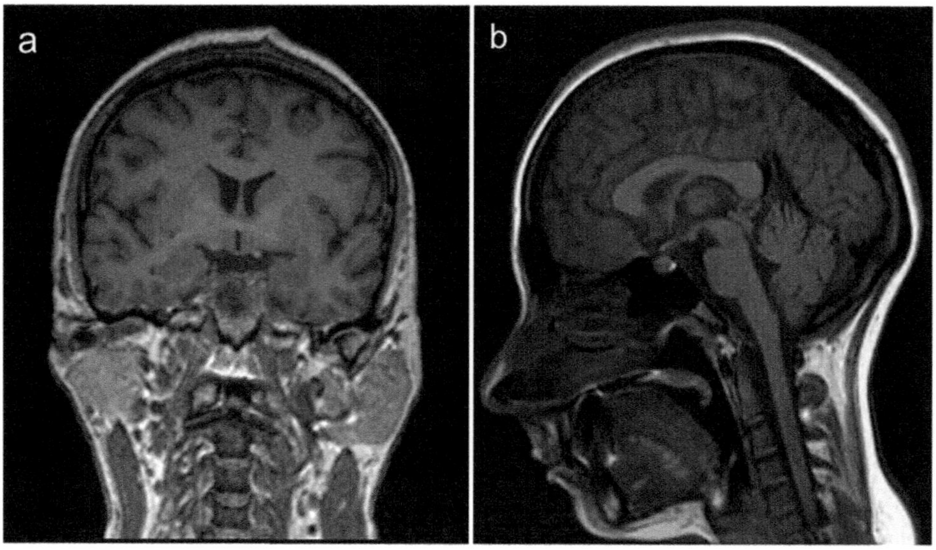

Fig. 9.3 Normal coronal (**a**) T1W and sagittal (**b**) T1W images of MRI of the brain

calcifications as shown in Fig. 9.5 depicting axial CT images—normal basal ganglia (a), bilateral basal ganglia and thalamic calcification (b), and left thalamocapsular hemorrhage with adjacent edema (c).

9.2.3 Positron Emission Tomography (PET)

Positron emission tomography (PET) is an advanced imaging modality that uses radioactive tracers, most commonly ^18F-fluorodeoxyglucose (FDG), to visualize metabolic activity in the brain, offering functional insights beyond structural imaging (Chugani 2019; Mohan 1999). In pediatric neurology, PET plays a crucial role in localizing seizure foci in drug-resistant epilepsy—particularly when MRI findings are inconclusive—by identifying areas of reduced metabolism that guide surgical planning (Gelfand and Sotero de Menezes 2013). It is also valuable in differentiating active brain tumor tissue from

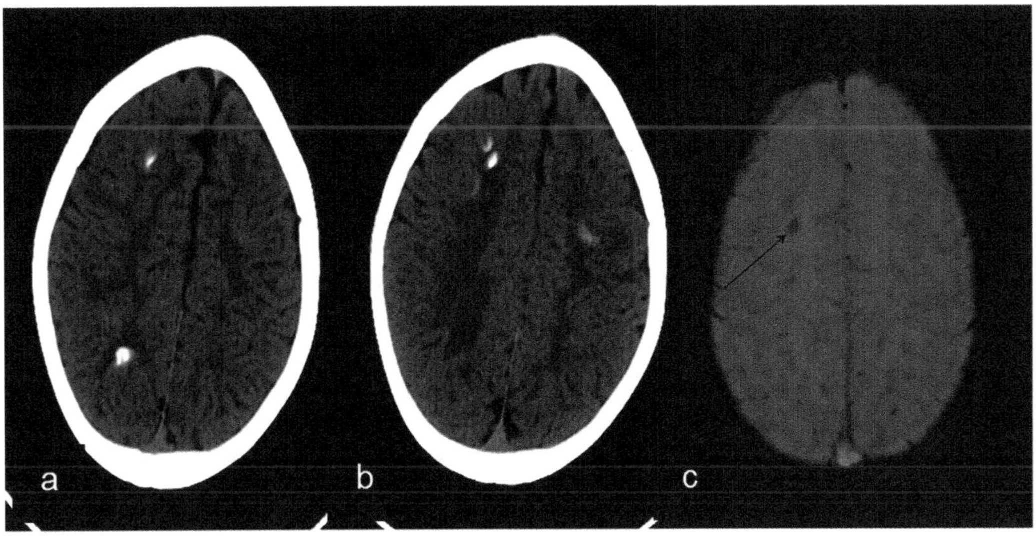

Fig. 9.4 CT scan brain without contrast (**a** and **b**) showing calcification with corresponding gradient MR image (**c**) that shows the presence of blooming (black arrow)

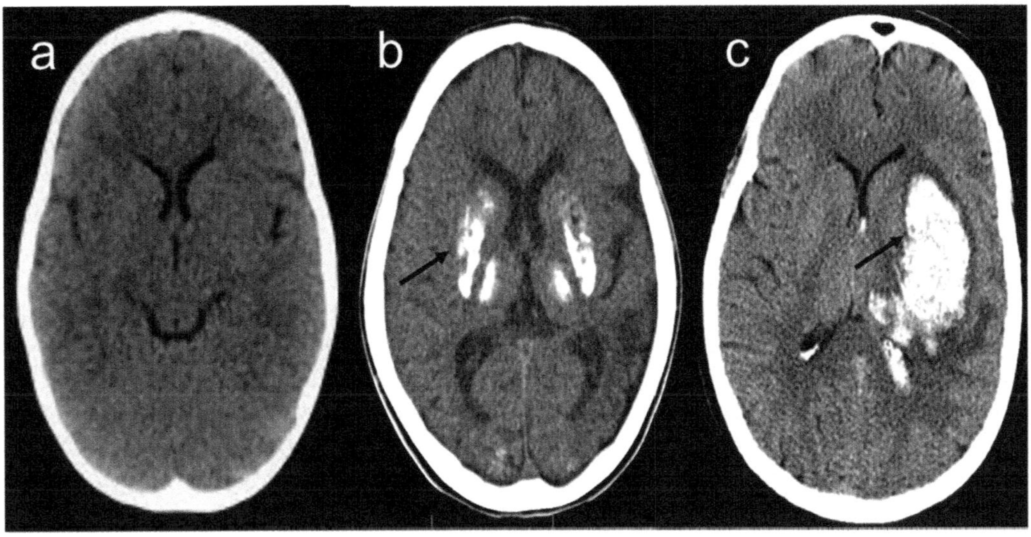

Fig. 9.5 Axial CT images show normal basal ganglia (**a**), bilateral basal ganglia and thalamic calcification (**b**), and left thalamocapsular hemorrhage with adjacent edema (**c**)

scar or treatment-related changes, aiding in diagnosis, biopsy targeting, and monitoring response to therapy (Gelfand and Sotero de Menezes 2013). Additionally, PET can detect abnormal metabolic patterns in metabolic and developmental disorders long before anatomical changes appear on MRI or CT (Mohan 1999), and it provides important data on normal brain maturation patterns across different age groups (Wang et al. 2024). By revealing functional and biochemical changes ahead of structural abnormalities, and through hybrid PET/CT or PET/MRI for combined anatomical and functional assessment, PET enhances diagnostic accuracy (Chugani 2019). However, its use involves exposure to ionizing radiation, so pediatric protocols are carefully optimized to minimize this risk (Gelfand and Sotero de Menezes 2013).

9.2.4 Ultrasonography

Ultrasonography is a noninvasive, real-time imaging modality that employs high-frequency sound waves to visualize soft tissues in the brain and peripheral nervous system, making it especially valuable in pediatric neurology (VanHorn and Cartwright 2020). It is highly effective in neonates and infants owing to their open fontanelles, which serve as natural acoustic windows for brain imaging (Millichap 1993). Cranial ultrasound is extensively used to detect intracranial hemorrhage, hydrocephalus, periventricular leukomalacia, congenital brain malformations, and hypoxic-ischemic injuries in neonates (VanHorn and Cartwright 2020; Grimm et al. 2020a, b). It is safe, portable, and cost-effective and can be performed bedside repeatedly without sedation, which is critical for critically ill or prematurely born infants (VanHorn and Cartwright 2020). Advances such as high-frequency transducers, Doppler imaging, and emerging techniques like contrast-enhanced ultrasound and elastography improve diagnostic accuracy and provide insights into brain perfusion and tissue properties (VanHorn and Cartwright 2020; Grimm et al. 2020a, b). Beyond the brain, ultrasonography is instrumental in evaluating peripheral nerves and muscles. High-resolution nerve ultrasound detects abnormalities such as nerve enlargement, structural changes, and trauma, serving as a painless, dynamic alternative to electromyography (Yiu et al. 2015; Samet and Başkaya 2023). It is helpful in diagnosing neuromuscular disorders like Charcot-Marie-Tooth disease and in managing traumatic peripheral nerve injuries (Yiu et al. 2015; Samet and Başkaya 2023). Ultrasound-guided peripheral nerve blocks in pediatric anesthesia also benefit from this modality's real-time visualization (Abdelnaby et al. 2023). Limitations include dependency on operator skill, decreased cranial access after fontanelle closure, and lower resolution than MRI for deep brain structures (Kerscher et al. 2021).

9.3 Neuroimaging Findings in Pediatric Neurological Disorders

9.3.1 Epilepsy and Seizure Disorders

Epilepsy is a heterogeneous group of disorders. Both acquired and inherited causes can cause epilepsy. In developing countries, structural etiologies like neuroinfectious, hypoxic brain injuries, and head injuries are major causes of epilepsies. Imaging is a very useful investigation in the evaluation of seizures. Based on imaging findings, we can classify epilepsy as structural versus nonstructural etiology. Further investigations and approaches to management depend on the imaging findings. Various modalities are cranial ultrasound, CT brain, and MRI of the brain. The diagnostic yield of MRI is high and hence considered as an investigation of choice (Chavhan 2016).

9.3.1.1 Indications for MRI of the Brain

Various indications include focal onset seizures except in genetic focal seizures (Bernasconi et al. 2011; Duncan 2010; National Institute for Health and Care Excellence [NICE] 2022), unclassified onset seizures especially in the first year of life (Gaillard et al. 2009; Widjaja et al. 2007), focal deficits on neurological examination (Lüders et al. 2006; Barkovich and Raybaud 2014), drug-resistant epilepsy (Kwan et al. 2010; Bernhardt et al. 2010), and loss of control of seizures with antiseizure medication or change in the semiology of seizures (Cross et al. 2006; Cendes et al. 2016). CT brain is an alternative procedure if MRI is not available or cannot be performed due to technical reasons. A **3 Tesla MRI with epilepsy protocol** is a better choice for detecting subtle seizure-related brain changes and guiding treatment if facilities are available. Contrast imaging is recommended in case of suspected infections like ring-enhancing lesions

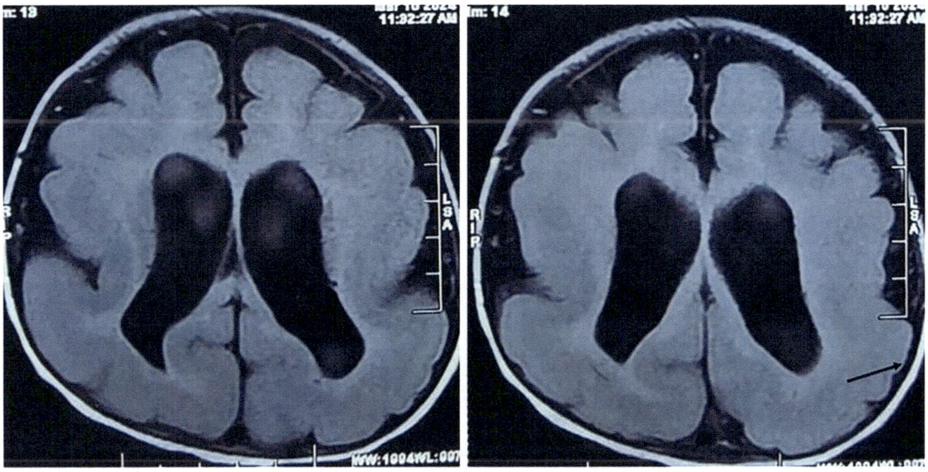

Fig. 9.6 MRI of the brain axial FLAIR images shows diffuse thickening of cortex and simplification of gyri. Parietal lobes affected more than frontal lobe suggestive of lissencephaly

or other infections such as neurocysticercosis, tuberculosis etc. MRI of the brain detects smaller lesions like focal cortical dysplasia in addition to major malformations and other structural abnormalities as seen in Fig. 9.6 which depicts axial FLAIR images with diffuse thickening of cortex and simplification of gyri. Parietal lobes affected more than frontal lobe suggestive of lissencephaly. It is also useful in selecting patients for surgical intervention in localizing epileptogenic zones (Vossough 2023; Vossough et al. 2024).

9.3.1.2 Cranial Ultrasound

Cranial ultrasound (CUS) is a frontline neuroimaging modality in the newborn period, using high-frequency sound waves through open fontanelles to provide real-time visualization of the brain without ionizing radiation (Gupta 2016). It is particularly valuable for detecting causes of neonatal seizures and epilepsy, including intraventricular hemorrhage (IVH), congenital structural abnormalities, and hydrocephalus (Caro-Domínguez 2021; Richer et al. 2021). CUS offers several advantages—it is noninvasive, cost-effective, and portable, can be performed at the bedside without sedation, and is safe for repeated use to monitor disease progression or treatment response (Hanna 2025; Gupta 2016). Clinically, it enables grading of IVH, detection of periventricular leukomalacia, and monitoring for posthemorrhagic ventricular dilatation, which are important predictors of neurodevelopmental outcome (Mahale et al. 2025; Richer et al. 2021). Although MRI provides superior detail for deep or subtle lesions, advances in CUS—such as the use of multiple scanning windows, high-resolution transducers, and Doppler imaging—have significantly enhanced diagnostic capability (Gupta 2016; Caro-Domínguez 2021). Overall, CUS remains an indispensable, safe, and accessible tool in neonatal neurology for early diagnosis and monitoring of critical brain pathologies.

9.3.1.3 Computed Tomography (CT Brain)

Computed tomography (CT) of the brain remains an important and widely used imaging modality in developing countries due to its accessibility and rapid image acquisition (Kalra et al. 2021). It is particularly valuable for detecting intracranial calcifications, which are a common cause of focal seizures in regions such as North India (Gupta et al. 2018). For example, multiple periventricular calcifications—often seen in congenital infections like cytomegalovirus—can be clearly visualized on CT and are important in epilepsy evaluation (Gupta et al. 2018). In emergency settings, CT is critical for promptly assessing patients with acute-onset seizures, as it can identify cerebral edema, hemorrhage, ischemic stroke, and mass lesions that require urgent intervention (Kalra et al. 2021; Singh and Rao 2019). While MRI offers greater soft tissue contrast and anatomical detail, CT's speed, availability, and sensitivity to calcifications and acute hemorrhage make it indispensable in resource-limited settings (Singh and Rao 2019). Despite limitations such as exposure to ionizing radiation and lower resolution for soft tissue compared to MRI, CT remains a frontline investigation in pediatric neurology where timely diagnosis can guide life-saving management (Gupta et al. 2018; Kalra et al. 2021) (Figs. 9.7, 9.8, 9.9, 9.10, 9.11, 9.12, 9.13, 9.14, and 9.15).

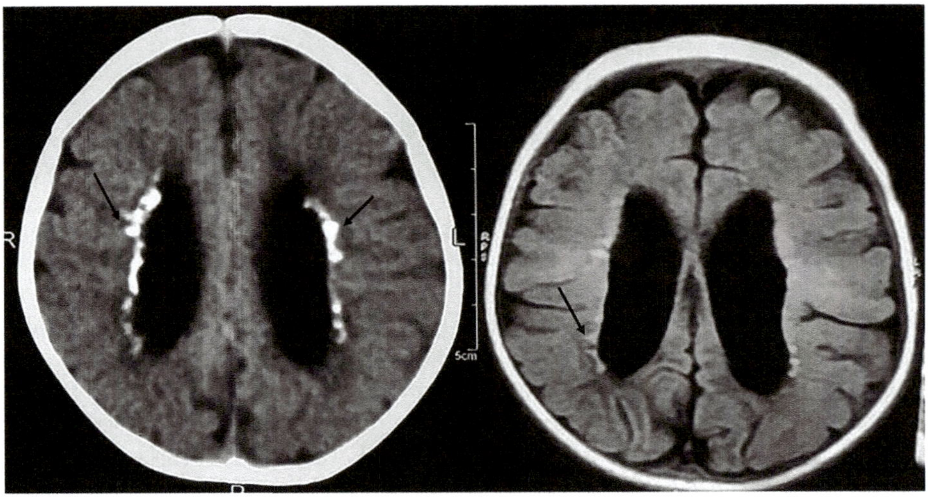

Fig. 9.7 Congenital cytomegalovirus (CMV) infection. Axial CT image shows multiple periventricular calcifications. Mild ventricular prominence. Axial FLAIR images show enlarged ventricles and irregular ventricular margins with tiny periventricular cysts

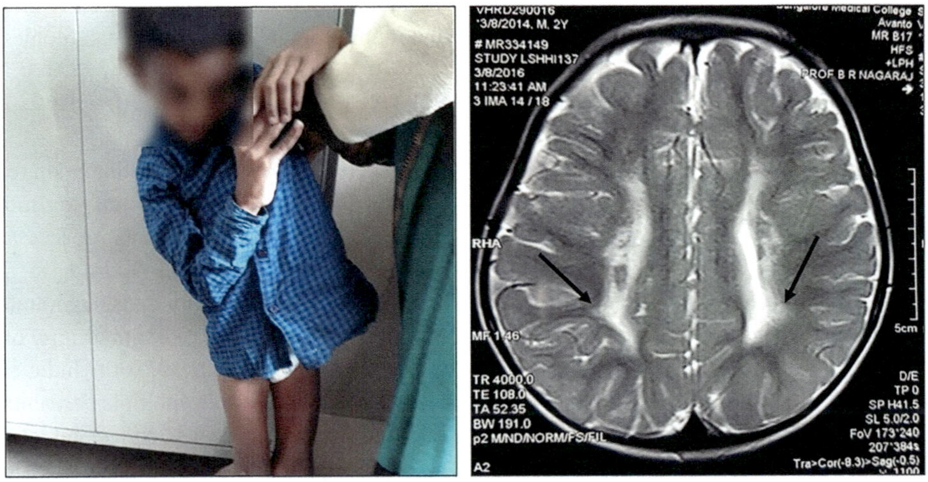

Fig. 9.8 Bilateral spastic cerebral palsy—spastic diplegic type with scissoring of both lower limbs with deformity (**a**). MRI of brain T2 axial sections, showing periventricular deep white matter hyperintensities suggestive of periventricular leukomalacia (**b**). (Source: https://doi.org/10.25259/KPJ_1_2020)

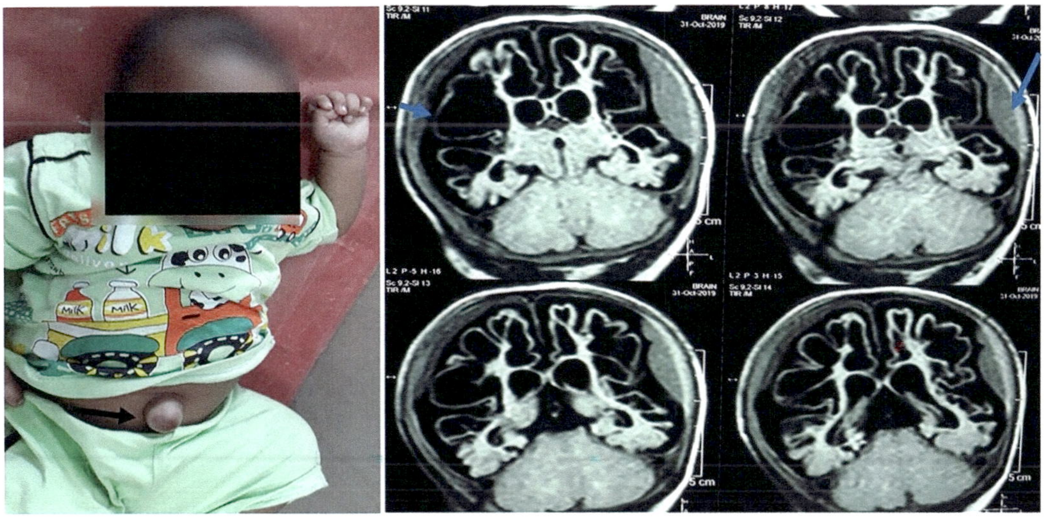

Fig. 9.9 Clinical image showing microcephaly in a child with bilateral spastic cerebral palsy of quadriplegic type with umbilical hernia, multicystic encephalomalacia, and subdural effusion along the parietal lobe convexity in the MRI of the brain. (Source: https://doi.org/10.25259/KPJ_1_2020)

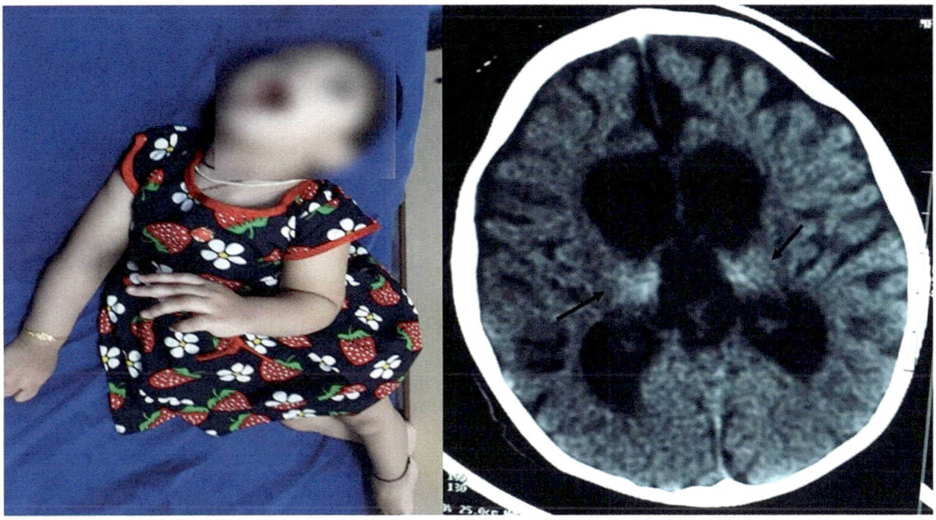

Fig. 9.10 The child with dyskinetic cerebral palsy with severe arching of back and neck due to dystonia and CT brain showing bilateral thalamic calcifications with cerebral atrophy with enlarged ventricles suggestive of hypoxic insult. (Source: https://doi.org/10.25259/KPJ_1_2020)

9.3.2 Brain Tumors

Neuroimaging is central to the diagnosis, treatment planning, and follow-up of pediatric brain tumors, with magnetic resonance imaging (MRI) being the modality of choice. MRI provides high-resolution anatomical detail for initial diagnosis, accurately defining tumor location, size, and extent, and guides surgical or radiotherapy planning by delineating margins and involvement of eloquent brain areas. It is similarly indispensable for surveillance, enabling early detection of

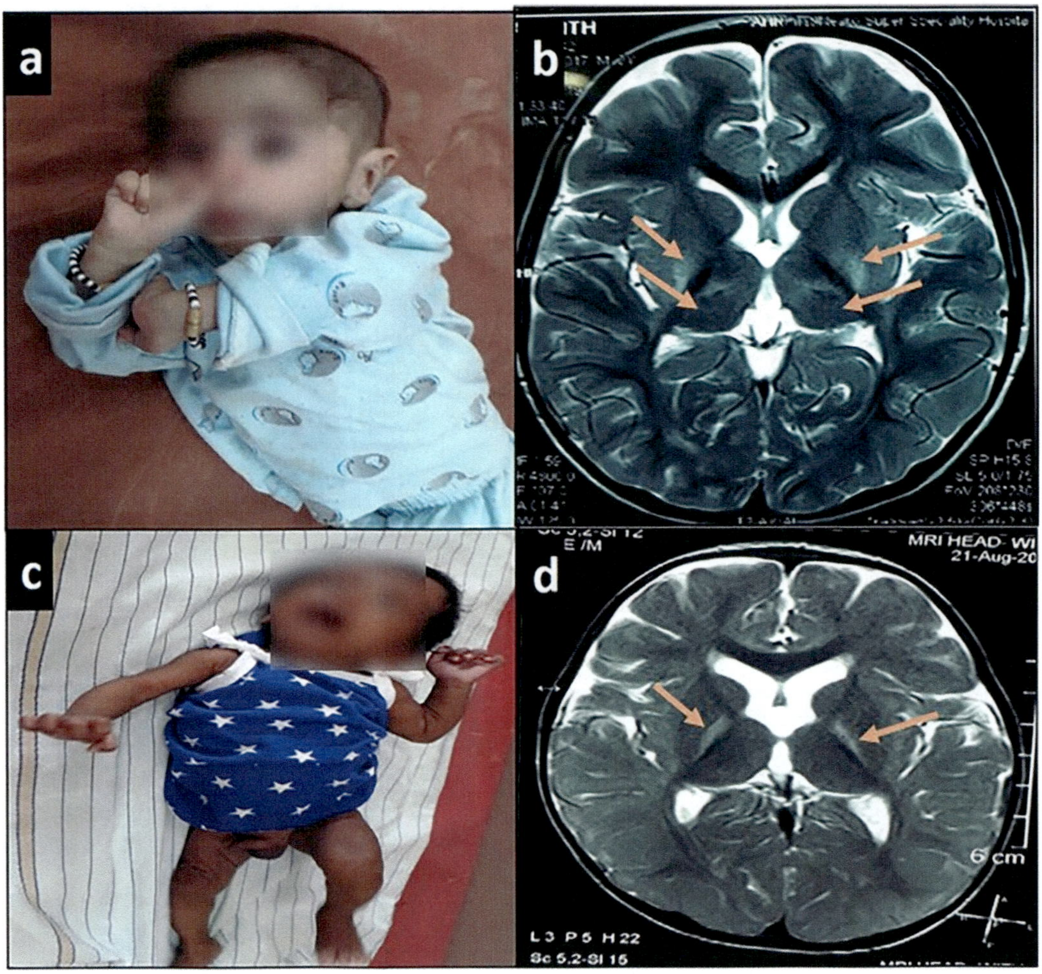

Fig. 9.11 (**a** and **b**): Child with dyskinetic cerebral palsy with MRI (**b**) brain T2 axial showing hyperintensities in bilateral posterior putamen and thalami suggestive of acute hypoxic insult. (**c** and **d**): Child with dyskinetic cerebral palsy with dystonic posturing and arching of the neck due to dystonia and MRI brain (**d**) T2 axial showing bilateral symmetrical hyperintensities in globus pallidus due to bilirubin-induced neurologic dysfunction (BIND). (Source: https://doi.org/10.25259/KPJ_1_2020)

recurrence or progression. Advanced MRI techniques such as diffusion-weighted imaging (DWI), perfusion imaging, and magnetic resonance spectroscopy (MRS) enhance tumor characterization by providing information on cellularity, vascularity, and metabolic profiles. These tools aid in differentiating tumor types and grades, as well as distinguishing true progression from posttreatment changes such as pseudoprogression or radiation necrosis—distinctions that are critical to patient management. Recent developments include the application of machine learning to large neuroimaging datasets, improving tumor segmentation, classification, and grading accuracy. Such approaches promise more objective, reproducible assessments and may ultimately support precision-based, personalized treatment planning in pediatric neuro-oncology (Vossough et al. 2024).

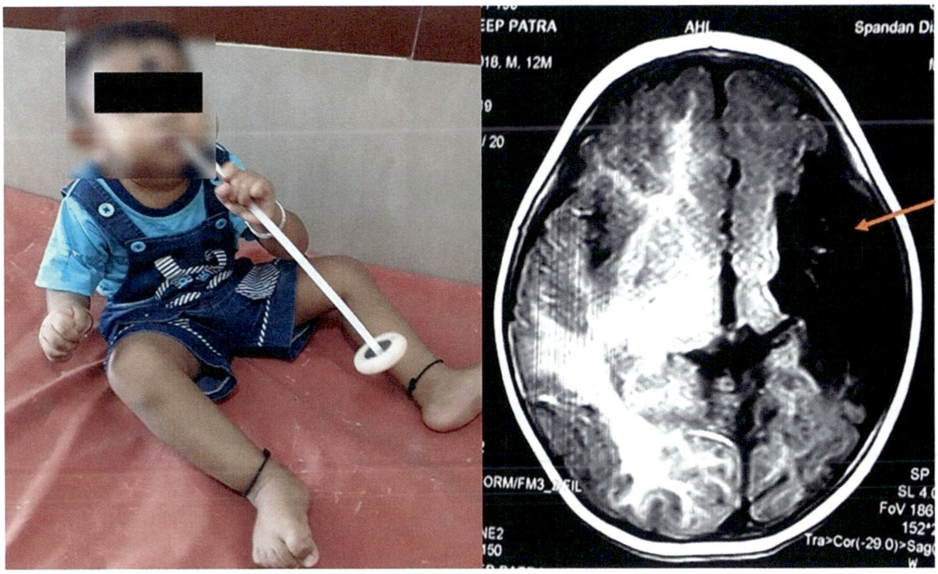

Fig. 9.12 A child with unilateral/hemiplegic cerebral palsy with right-sided weakness with an MRI of the brain showing a porencephalic cyst on the left side. (Source: https://doi.org/10.25259/KPJ_1_2020)

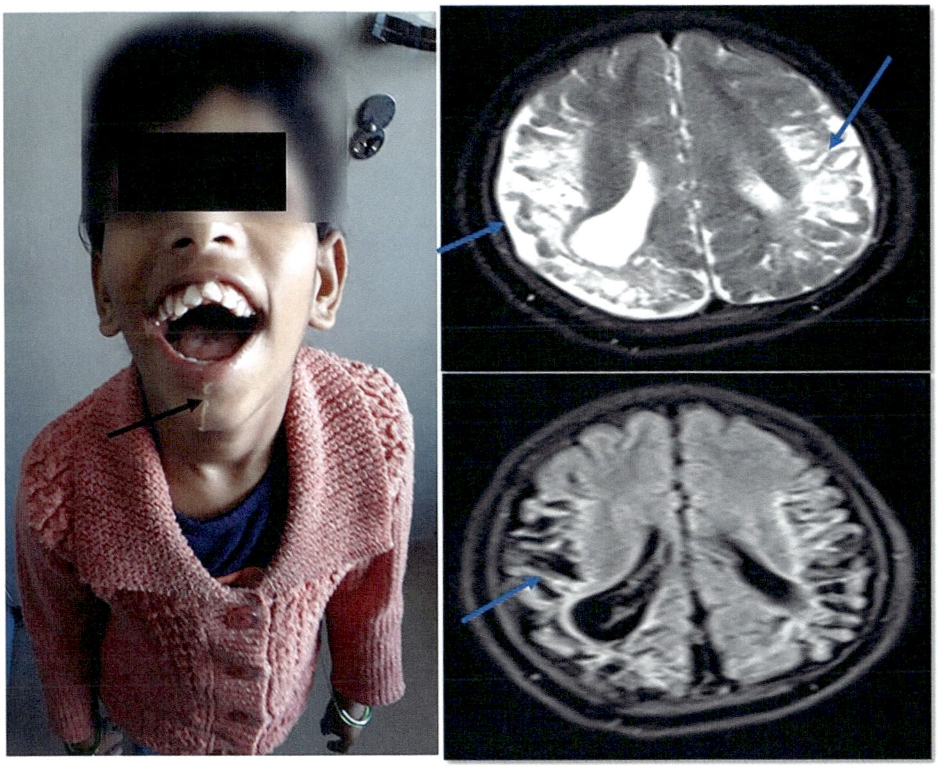

Fig. 9.13 Perisylvian syndrome—a type of cerebral palsy. Clinical photograph showing open mouth with drooling and MRI of the brain showing gliosis in the perisylvian region. (Source: https://doi.org/10.25259/KPJ_1_2020)

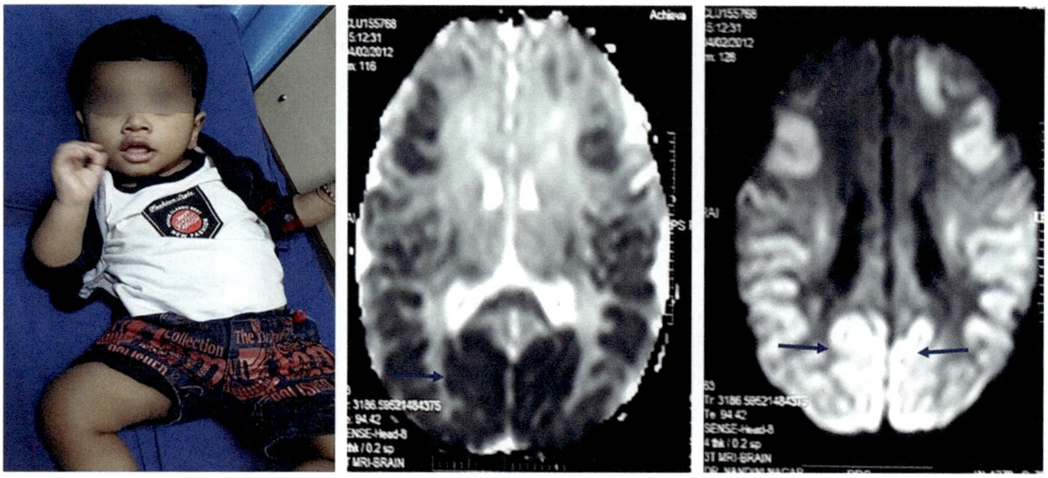

Fig. 9.14 Neonatal hypoglycemic brain injury (NHBI) presents with cerebral visual impairment. Clinical photo showing strabismus with decreased vision and MRI of brain—DWI and ADC showing restricted diffusion in the bilateral occipital and parietal region. (Source: https://doi.org/10.25259/KPJ_1_2020)

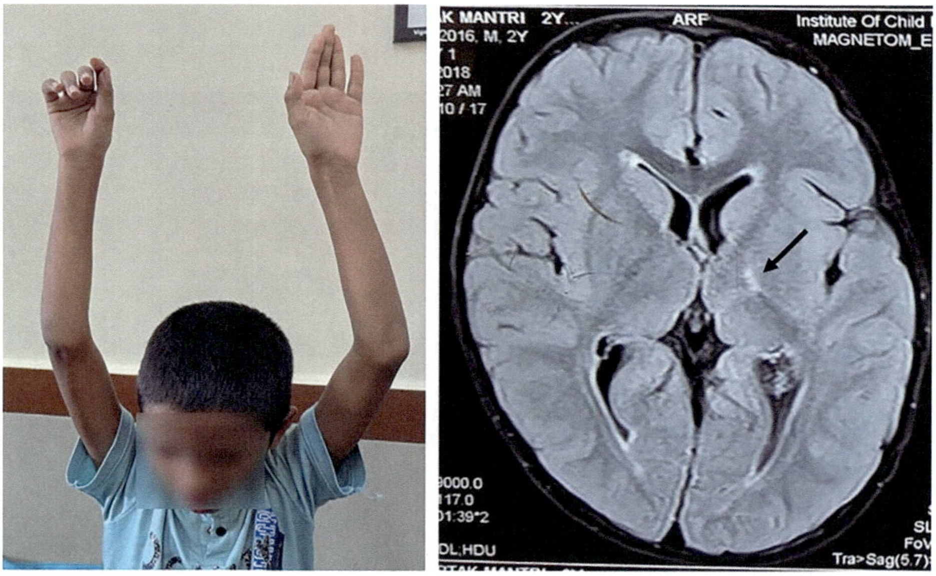

Fig. 9.15 Hemi-dystonic cerebral palsy. Clinical photograph showing right-sided weakness with dystonia and axial FLAIR images of the MRI of the brain show focal hyperintensity in the left globus pallidus/internal capsular region

9.3.3 Cerebral Palsy

Neuroimaging helps to investigate the structural abnormalities associated with cerebral palsy as well as provide insights into pathophysiology and prognosis as well as helps to monitor treatment response.

The Surveillance of Cerebral Palsy in Europe (SCPE) developed the MRI classification system (MRICS) to standardize interpretation of brain MRI in children with cerebral palsy (CP) based on the presumed timing and pathogenic mechanism of brain injury. It categorizes findings into five groups: maldevelopments, predominant white matter injury, predominant gray matter injury, miscellaneous findings, and normal (Himmelmann et al. 2017).

Himmelmann et al. (2017) compared MRICS with other CP MRI classification systems and demonstrated that it is reliable, valid, and easy to use, with good to very good interrater agreement ($\kappa = 0.69$–0.81). Its simplicity and reproducibility facilitate use in multicenter registries, enabling consistent data collection and research. MRICS is qualitative but clinically meaningful—linking imaging patterns to the timing of insult, functional severity, and associated impairments. For example, white matter injury, often linked to preterm birth, is the most frequent finding and carries specific motor and cognitive profiles. Importantly, MRICS can help identify patterns suggesting genetic or metabolic etiologies, aiding in patient selection for further testing. Its structured approach supports clinical counseling, prognostication, and tailored rehabilitation strategies while standardizing reporting across centers (Himmelmann et al., 2021; SCPE, 2019).

9.4 Neuroimaging in Neurogenetic Disorders

Neuroimaging is fundamental in evaluating neurogenetic disorders, with MRI, CT, and magnetic resonance spectroscopy (MRS) significantly contributing to diagnosis. Combined MRI–MRS approaches can delineate underlying pathophysiology and establish a specific diagnosis in more than half of suspected cases (Lin et al. 2005; Abbaspour et al. 2024). MRI provides high-resolution anatomical detail to detect structural brain abnormalities linked to genetic syndromes (Risacher and Saykin 2021), while MRS quantifies brain metabolites, enabling early detection of neurometabolic and neurodegenerative diseases even when MRI appears normal (Karimzadeh 2015a, b; Liserre et al. 2021a, b). CT, though less sensitive for parenchymal detail, remains useful for detecting calcifications in disorders such as Aicardi–Goutières syndrome or tuberous sclerosis (Grimm et al. 2020a, b). Ongoing advances—including multimodal imaging integration and improved resolution—are increasing diagnostic accuracy, supporting prognostication, genetic counseling, and tailored management (Rossi and Biancheri 2013a, b; Whitehead et al. 2022).

9.4.1 Autism Spectrum Disorder, Attention Deficit Hyperactivity Disorder, and Other Behavioral Disorders

Neuroimaging along with the biomarkers helps to unravel the neurobiological underpinnings of autism. Various modalities of structural neuroimaging have been useful. Various analytical methods exist for the interpretation of structural data. These include voxel-based morphometry, surface-based morphometry, tensor-based morphometry, longitudinal MRI studies, and DTI studies. These techniques can differentiate an individual with ASD from controls. The most common areas affected include abnormalities of brain volume, abnormalities of the corpus callosum, amygdala, gray matter, white matter, and tracts, especially prefrontal, cingulate, and internal capsule (Chen et al. 2011). Implications for early diagnosis and intervention (Byrne et al. 2023) along with revealing the neural correlates of ADHD in pediatric patients.

9.4.2 Neurometabolic Disorders

Neurometabolic disorders are defined as disorders that are caused by defects or reductions in an enzyme, cofactor, or protein transporter that is normally needed for the normal functioning of a cell's energy production and cell survival. This group includes disorders that affect the pathways of carbohydrates, amino acids, fatty acids, cholesterol and its metabolites, vitaminopathies, neurotransmitter disorders, lysosomal storage disorders, and peroxisomal disorders. The most common modalities used are conventional MRI and magnetic resonance spectroscopy. The following table gives some of the specific disorders with the characteristic findings on magnetic resonance spectroscopy, thus giving clue to making an accurate diagnosis (Karimzadeh 2015a, b;

Table 9.1 Showing various biochemical metabolites in MRS

Disorders	Metabolite	Position (ppm)
Phenylketonuria	Phenylalanine	7.3
Galactosemia	Galactitol	3.7
Maple syrup urine disease	Branched-chain amino acids (BCAAs)	0.95–1.05
Canavan disease	N-acetylaspartate (NAA)	2.02, 2.5, 2.67
Nonketotic hyperglycinemia	Glycine	3.55
Salla disease	N-acetylneuraminic acid (NANA)	2.02, 4
Sandhoff disease	N-acetylhexosamine	2.07
HMG-CoA lyase deficiency	3-Hydroxyisovaleric acid, 3-hydroxy-3-methylglutaric acid	2.42

Liserre et al. 2021a, b; Rossi and Biancheri 2013a, b; Whitehead et al. 2022) (Table 9.1).

9.4.3 Disorders of Cortical Migration

The development of the cerebral cortex and many of these mechanisms are yet to be understood. Simplifying the complex process of cortical formation and organization. It involves the following steps: cellular proliferation from the neuroblasts in the ventricular and subventricular zones between the second and fifth month of gestation; cellular migration, which occurs between third and fifth month of gestation; and finally the cortical organization that takes place between 22nd week of gestational age and 2 years of life. Any disruption of this process due to mutations in various genes can lead to disorders of cortical migration (Gleeson 2001; Guerrini and Parrini 2010).

The most common modality used to delineate these issues includes a 3 Tesla MRI and if not available at least 1.5 T MRI. The group includes microcephaly, macrocephaly with overgrowth syndromes, focal cortical dysplasia, gray matter heterotopia, lissencephaly, cobblestone malformation, polymicrogyria, schizencephaly, and dysgyria (Severino et al. 2020; Spalice et al. 2009).

9.4.4 Neurocutaneous Syndromes

Neurocutaneous syndromes are defined as disorders that affect both the nervous system and the skin. This unique group occurs due to dysregulation of genes which are common for the development of both. Most of the time, the skin findings usually give away the diagnosis. However, in certain disorders where the manifestations are mild, variable clinical features, or nervous system predominant involvement, then neuroimaging could fill the gap and help with the diagnosis (Kioutchoukova et al. 2024). The disorders under this group include neurofibromatosis types 1 and 2, tuberous sclerosis, Sturge–Weber syndrome, neurocutaneous melanosis, von Hippel–Lindau syndrome, and cerebrotendinous xanthomatosis, to name a few. The following table summarizes the neuroimaging features in each of these disorders along with the causative genes (Table 9.2).

9.4.5 Mitochondrial Disorders

Mitochondrial disorders arise from defects in mitochondrial structure or oxidative phosphorylation, leading to impaired adenosine triphosphate (ATP) production. Due to the high-energy demands of the brain and spinal cord, the central nervous system is frequently involved (Gorman et al. 2016; Kanungo et al. 2018; Rambani et al. 2022; Zeviani and Carelli 2007). Magnetic resonance imaging (MRI) is the primary neuroimaging modality used and commonly includes conventional structural sequences (T1, T2, FLAIR), functional MRI (fMRI) to study cognitive function and connectivity, diffusion tensor imaging (DTI) and diffusion-weighted imaging (DWI) for microstructural tissue characterization, and magnetic resonance spectroscopy (MRS) to assess chemical composition (Alves and

Table 9.2 Showing neuroimaging findings in various neurocutaneous disorders

Disorder	Gene	Neuroimaging features
Neurofibromatosis-1	*NF1*	Unidentified bright objects (UB0) or focal areas of high signal intensity (FASI) in the optic nerve, cortex, brainstem, and basal ganglia. The enlarged corpus callosum, optic gliomas, brainstem tumors, and nodular neurofibromas (Rodriguez and Young Poussaint 2004)
Neurofibromatosis-2	*NF2*	Vestibular schwannoma and meningiomas (Rodriguez and Young Poussaint 2004)
Tuberous sclerosis	*TSC1, TSC2, and IFNG*	Focal areas of calcification Subependymal nodules Subependymal giant cell astrocytoma (SEGAs), Fig. 9.16 (Russo et al. 2020)
Von Hippel–Lindau syndrome	*VHL*	Cerebellar hemangioblastoma (Pietrzak et al. 2024)
Neurocutaneous melanosis	*NRAS*	Parenchymal neuromelanosis, Dandy–Walker malformation, and leptomeningeal melanocytosis Choroid plexus papilloma, spinal cysts, and arachnoid cyst (Gocmen et al. 2015)
Incontinentia pigmenti	*IKBKG*	DWI shows restricted diffusion in cerebral white matter, corpus callosum, basal ganglia, and thalamus (Soltirovska Salamon et al. 2016)
Hypomelanosis of Ito	–	Corpus callosum hypoplasia, heterotopia, white matter abnormalities, and rarely benign tumors (Williams 3rd and Elster 1990)
Sturge–weber syndrome	*GNAQ*	Double counter convolutional calcifications on CT scan, cerebral atrophy, and arachnoid hemangiomata as seen in Fig. 9.17 (Adams et al. 2009)
Telangiectasia, hereditary hemorrhagic	*ENG, GDF2, ACVRL1, HHT3, and HHT4*	Cerebral arteriovenous malformation and spinal AVMs (Carette et al. 2009)

Whitehead 2024; Distelmaier and Klopstock 2023; Gropman 2013; Haas and Dietrich 2004; Mascalchi et al. 2018; Saneto et al. 2008).

Common pediatric mitochondrial diseases include Leigh syndrome, mitochondrial myopathy, encephalopathy, lactic acidosis and stroke-like episodes (MELAS), Kearns–Sayre syndrome, and Leber's hereditary optic neuropathy (LHON), with complex I deficiency among the most frequent (Alves and Whitehead 2024; Gropman 2013).

Radiological features often involve bilateral basal ganglia hyperintensities, cerebral and cerebellar atrophy, thalamic signal abnormalities, widespread white matter changes including dysmyelination, and brainstem involvement affecting the red nucleus, substantia nigra, and corticospinal tracts. MRS typically reveals a lactate peak indicative of mitochondrial dysfunction (Alves and Whitehead 2024; Distelmaier and Klopstock 2023; Mascalchi et al. 2018; Saneto et al. 2008).

Pediatric neuroimaging poses unique challenges such as movement artifacts requiring sedation, which carries risks. Safer approaches include child-friendly preparation, audiovisual distractions, and parental presence. Advancements such as fast imaging sequences, AI-based motion correction, functional connectivity MRI, and MR spectroscopy are facilitating improved image quality and metabolic insights, sometimes reducing the need for sedation (Barkovich et al. 2019).

Figures 9.18, 9.19, 9.20, 9.21, 9.22, 9.23, and 9.24 showcase a variety of pediatric neurological conditions with their characteristic neuroimaging findings and associated clinical features when applicable. These visual examples enhance clinical acumen by linking imaging patterns to specific diagnoses, thereby guiding the choice of

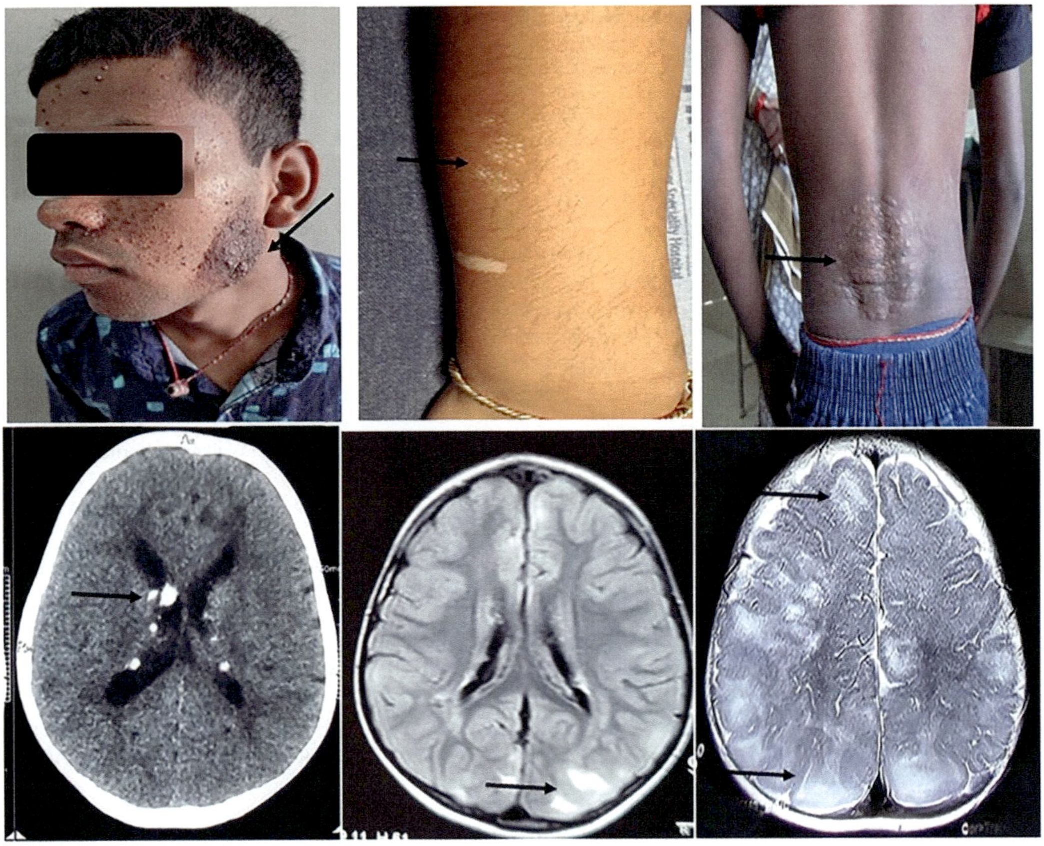

Fig. 9.16 Tuberous sclerosis. Clinical photographs show adenoma sebaceum, plaque on the left cheek, hypopigmented macules, and a shagreen patch over the lower back. CT scan of the brain showing subependymal calcifications. T2 hypointense subependymal nodules are also seen on the MRI of the brain. Axial T2W images show multiple cortical and juxtacortical tubers with hyperintensities and apex pointed inward toward the ventricles

targeted investigations. Integrating clinical and radiological information facilitates accurate diagnosis and optimizes patient management.

9.5 Neuroimaging in Traumatic Brain Injury (TBI) in Pediatrics

Traumatic brain injury (TBI) is a leading cause of death and disability in children. Severity is classified by the Glasgow Coma Scale (GCS) into mild (13–15), moderate (9–12), and severe (3–8) categories (Manley et al. 2025a, b). All cases of moderate to severe pediatric TBI warrant prompt non-contrast head computed tomography (CT), which rapidly detects intracranial hemorrhage, raised intracranial pressure, and skull fractures, playing a crucial role in acute management (Canadian Paediatric Society 2025; Manley et al. 2025a, b). While CT is the frontline imaging modality, magnetic resonance imaging (MRI) offers superior detection of diffuse axonal injury and other subtle parenchymal changes, important for prognosis and long-term management (Sharma et al. 2024; Smith and Johnson 2023).

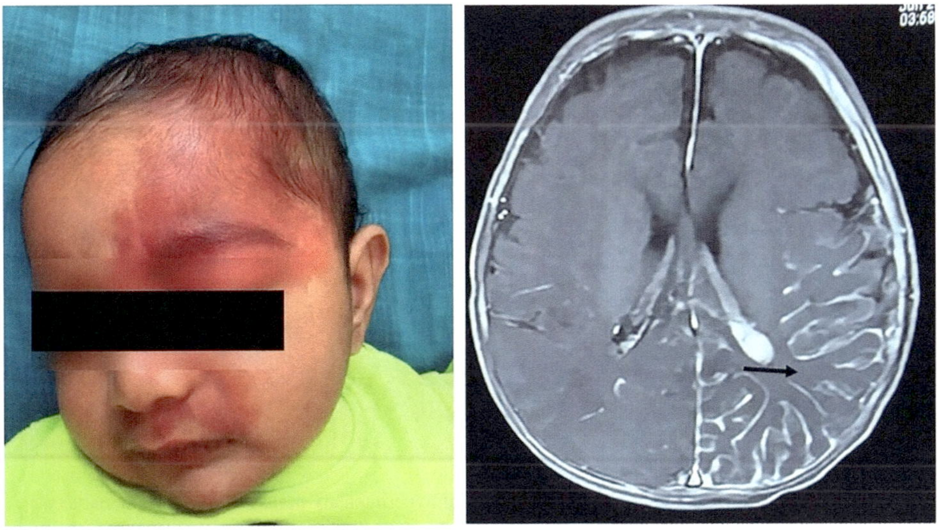

Fig. 9.17 Sturge–Weber syndrome. Clinical photograph showing nevus in the left forehead, eyelid, and over the nose. Axial post-contrast images show abnormal leptomeningeal enhancement in the left posterior frontal and parietal lobes with involvement of the left choroid plexus

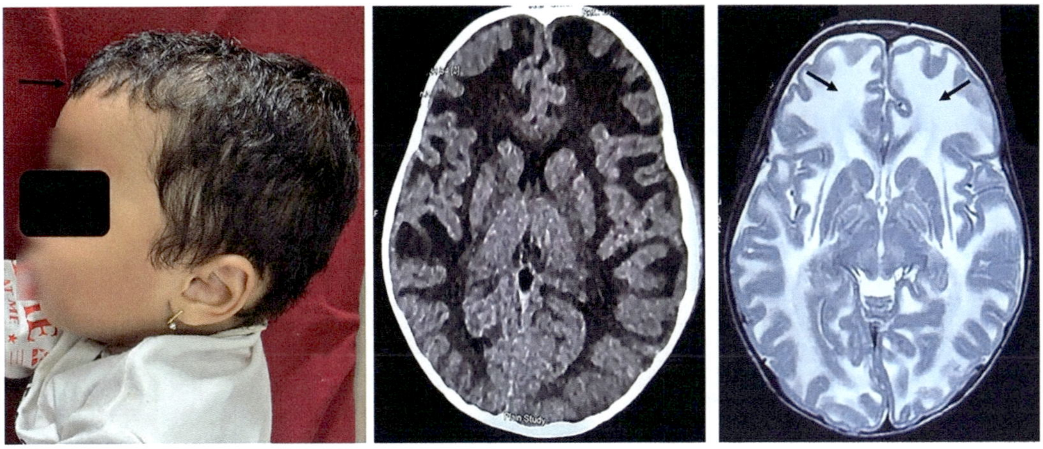

Fig. 9.18 Casc of megalencephaly with leukoencephalopathy with subcortical cysts (MLC). Clinical photograph showing large head. Axial CT image shows diffuse white matter hypointensity in both cerebral hemispheres. MRI of the brain. Axial T2W images show hyperintense diffuse white matter hyperintensity and rarefaction in bilateral cerebral hemisphere with relative sparing of occipital lobes. There is involvement of posterior limb internal capsule with double line, dorsal brainstem

Magnetic resonance venography (MRV) is a specialized MRI technique that visualizes cerebral venous sinuses and veins. MRV is important in trauma cases with clinical suspicion of cerebral venous sinus thrombosis or venous stroke, uncommon but serious sequelae of TBI that require timely diagnosis for anticoagulation treatment (Brown et al. 2025; Yilmaz et al. 2024).

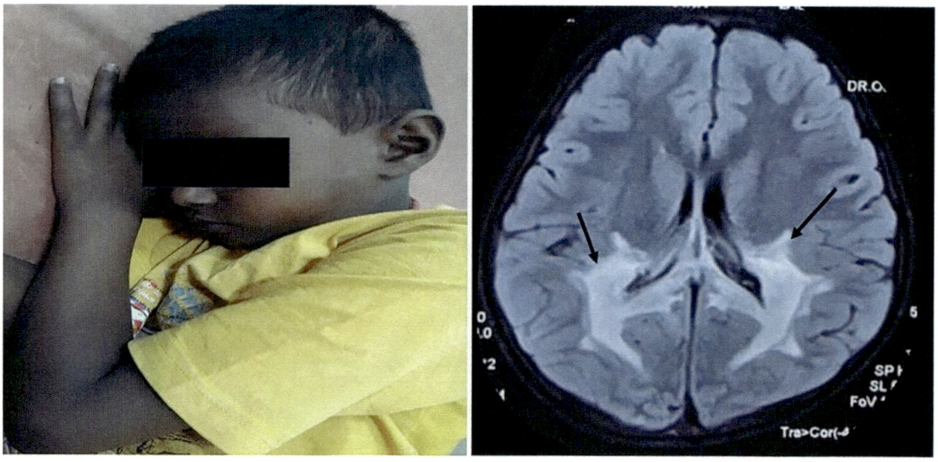

Fig. 9.19 Axial FLAIR images showing bilateral symmetrical hyperintensity in peri-trigonal white matter with sparing of subcortical U fibers

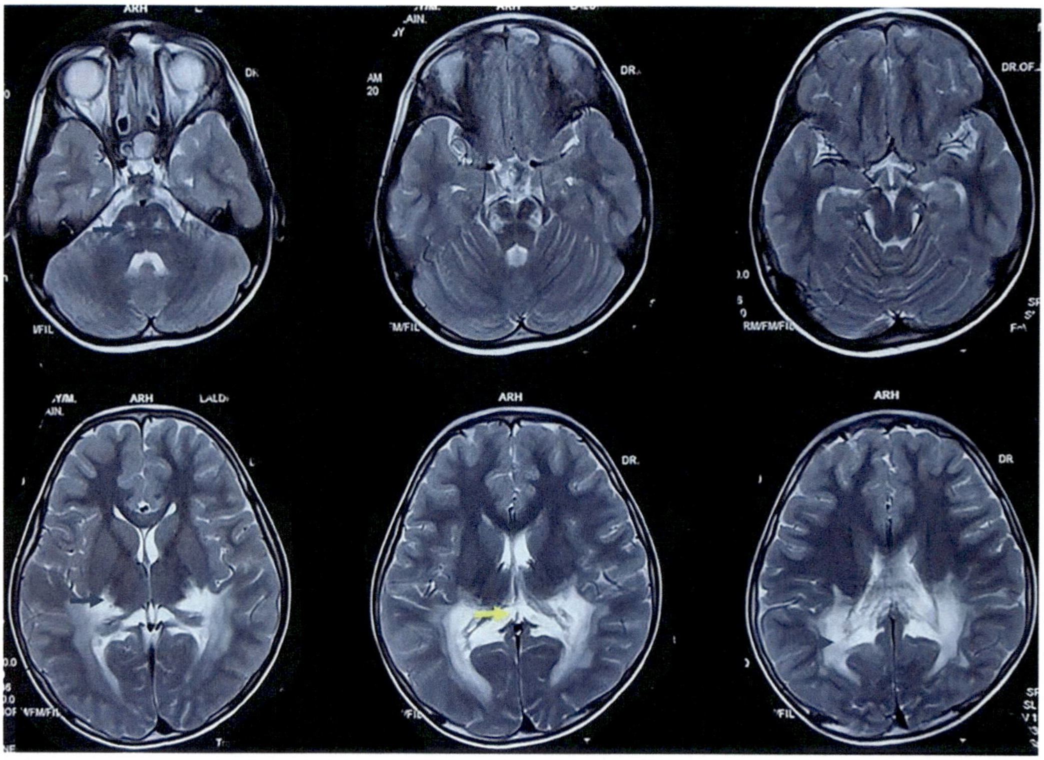

Fig. 9.20 Axial T2W images show nearly bilateral symmetrical hyperintensities in corticospinal tracts in the midbrain, pons, posterior internal capsule, and posterior peritrigonal deep white matter, with sparing of subcortical fibers. There is the involvement of the splenium of the corpus callosum

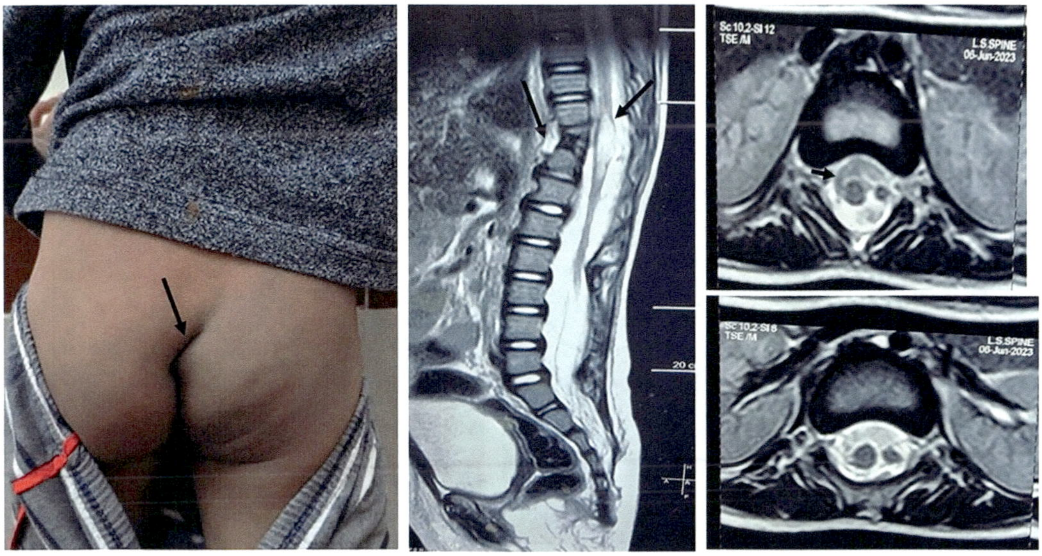

Fig. 9.21 Clinical photograph showing gluteal cleft. Sagittal T2W images show segmentation anomaly of vertebra. There is widening of spinal canal. Spinal cord shows focal syringomyelia. Axial sections show split cord with likely a fibrous septum and a widened dural sac

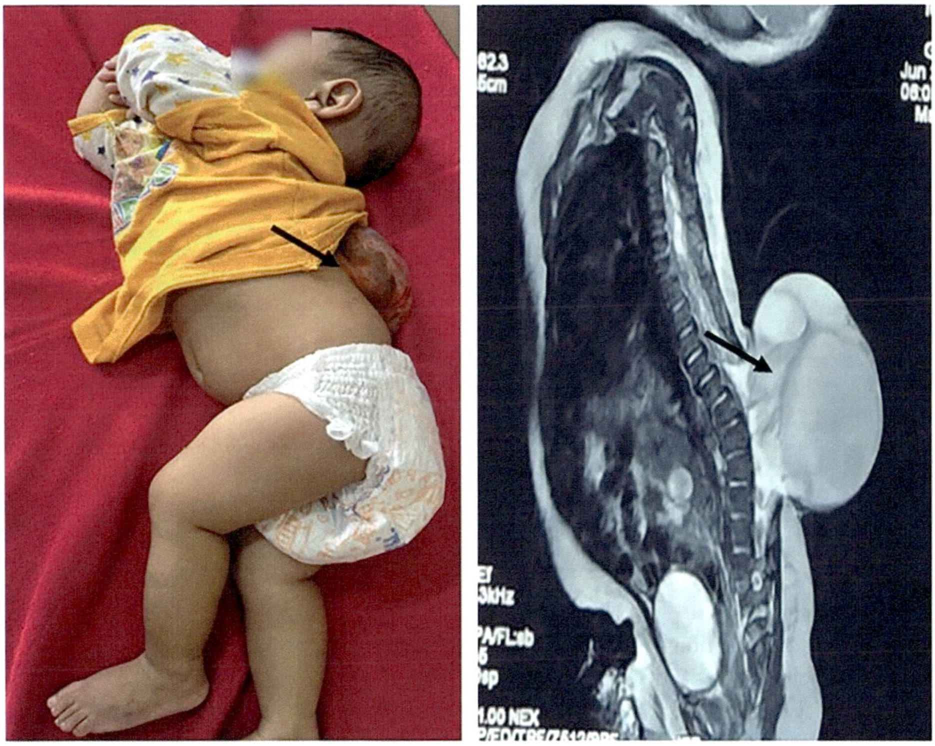

Fig. 9.22 Clinical photograph showing swelling in the back—thoracolumbar region. Sagittal T2W images show a dorsal spinal defect and a cystic sac with herniation of meninges and neural tissue

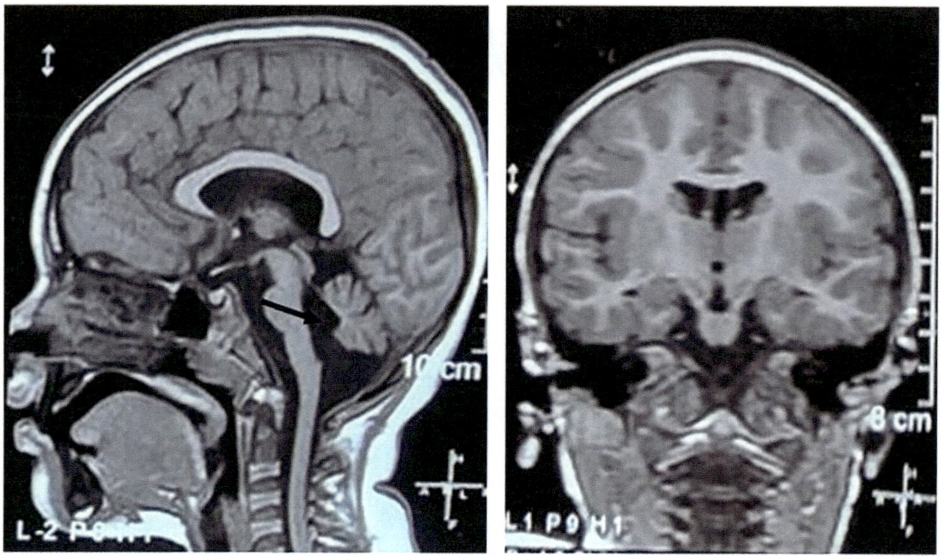

Fig. 9.23 Tubulinopathy malformation of the brain. Sagittal T1W images show reduced rostra-caudal length of pons with inferior vermian hypoplasia. Coronal images show the decreased volume of the pons and superior and middle cerebellar peduncles

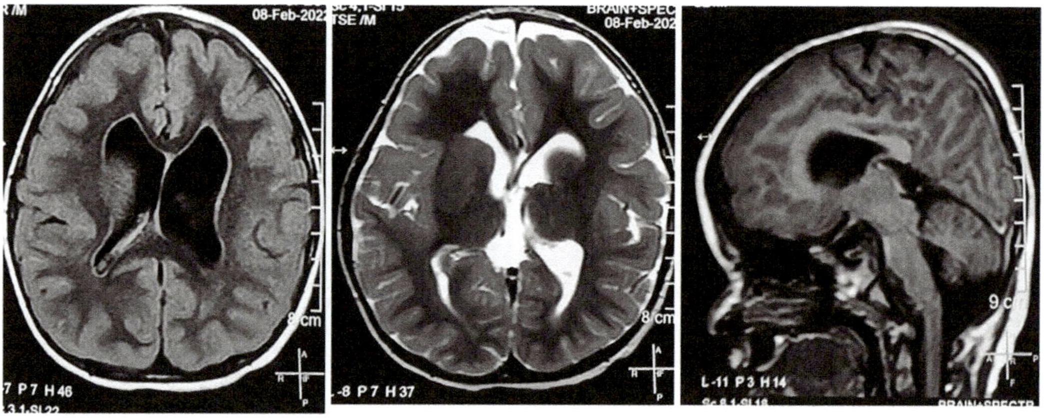

Fig. 9.24 Tubulinopathy malformation of the brain. Axial T2 FLAIR images show abnormally shaped dilated lateral ventricles. Axial T2W images show dysmorphic, abnormally fused basal ganglia (caudate and putamen) with hook-shaped frontal horns. Abnormal cortical gyri in the medial aspect of the left frontal lobe. Sagittal T1-weighted images show thickened corpus callosum, anteroposterior flattening of pons, and inferior vermis hypoplasia

Neuroimaging also facilitates assessment of injury extent, monitoring of evolving lesions, and prediction of functional outcomes. Advanced modalities such as diffusion tensor imaging (DTI), functional MRI (fMRI), and magnetic resonance spectroscopy (MRS) provide insight into microstructural damage and brain metabolism, aiding recovery monitoring and rehabilitation planning (Lee et al. 2024; Thompson et al. 2025).

9.6 Advancements in Neuroimaging Technology for Pediatric Population

Recent technological advances have revolutionized pediatric neuroimaging. Faster MRI sequences, such as simultaneous multi-slice and compressed sensing, reduce scan time and minimize sedation needs. Artificial intelligence-based motion correction enhances image quality in uncooperative children (Garcia et al. 2024). These improvements increase spatial and temporal resolution, allowing detailed visualization of pediatric brain development and pathology. Differentiating normal developmental variants from disease requires specialized expertise, underscoring the need for trained neuroradiologists and multidisciplinary teams (Patel and Kumar 2023; Johnson and Smith 2024).

9.7 Ethical and Legal Considerations in Pediatric Neuroimaging

Pediatric neuroimaging entails balancing benefits—improved diagnosis and management—with risks such as radiation exposure, sedation complications, and incidental findings (Miller et al. 2022). Informed consent, data privacy, and minimizing unnecessary imaging are core ethical principles. Legal frameworks and guidelines (e.g., American College of Radiology appropriateness criteria, Pediatric Emergency Care Applied Research Network rules) support judicious imaging use to protect children while ensuring access to needed diagnostics (Roberts et al. 2023; Canadian Paediatric Society 2025). Equitable access and addressing disparities are vital considerations in ethical neuroimaging practice (Wilson et al. 2023).

9.8 Challenges, Future Directions, and Research Opportunities in Neuroimaging for Pediatrics

Pediatric neuroimaging presents unique challenges, particularly in obtaining patient cooperation and managing the risks of sedation. Young children often struggle to remain still, leading to motion artifacts. While sedation can improve image quality, it carries risks such as respiratory and cardiovascular complications. Safer strategies include non-pharmacological techniques—child-friendly preparation, audiovisual distractions, and parental presence—and, when sedation is necessary, careful dosing, monitoring, and agent selection.

Emerging technologies, including functional connectivity MRI (fcMRI) and MR spectroscopy, are reducing the need for sedation and providing deeper insights into brain function and metabolism. Innovations like fast imaging sequences (e.g., simultaneous multi-slice imaging, compressed sensing) shorten scan times, while AI-driven motion correction can improve data from imperfect scans.

Integration of **imaging biomarkers** into precision medicine is transforming pediatric neurology by enabling individualized treatment and better disease monitoring. Large, multicenter collaborations are essential to validate these biomarkers and ensure broad applicability.

In short, advances in imaging technology, safety protocols, and collaborative research are making pediatric neuroimaging safer, more precise, and more impactful in improving outcomes for children with neurological disorders.

9.9 Conclusion

Neuroimaging has become an indispensable tool in pediatric neuroscience, profoundly impacting diagnosis, management, and prognosis across a spectrum of neurological disorders. Its evolution from basic structural imaging to sophisticated modalities like fMRI and MRS has enhanced understanding of neurodevelopmental and neurogenetic conditions, provided biomarkers for disease, and informed therapy planning. While challenges remain—such as the need for standardization, ethical considerations, and access—continued advancements promise greater precision, earlier diagnosis, and improved outcomes for children with neurological diseases. Ultimately, neuroimaging, when judiciously integrated with clinical evaluation and assessment tools, plays a pivotal role in elevating pediatric patient care and advancing neuroscience research.

Note Consent is obtained from patients for the images used.

References

Abbaspour A, Rahmani F, Shahidi GA, Nejat F (2024) Role of neuroimaging in the diagnosis of pediatric neurometabolic disorders: an updated review. Neuroradiology 66(2):197–211

Abdelnaby M et al (2023) Ultrasound-guided peripheral nerve blocks in pediatric patients. Anesthesiol Clin

Adams ME, Aylett SE, Squier W, Chong W (2009) A spectrum of unusual neuroimaging findings in patients with suspected Sturge-weber syndrome. AJNR Am J Neuroradiol 30(2):276–281. https://doi.org/10.3174/ajnr.A1350

Alper MG (1999) Three pioneers in the early history of neuroradiology: the snyder lecture. Doc Ophthalmol 98(1):29–49. https://doi.org/10.1023/a:1002197007334

Alves C, Whitehead MT (2024) Advancing the neuroimaging diagnosis and understanding of mitochondrial disorders. Neurotherapeutics 21(1):e00324. https://doi.org/10.1016/j.neurot.2024.e00324

Barkovich AJ, Raybaud C (2014) Pediatric neuroimaging (6th ed.). Neurol Clin 32(1):61–86. https://doi.org/10.1016/j.ncl.2013.09.002

Barkovich MJ, Li Y, Desikan RS, Barkovich AJ, Xu D (2019) Challenges in pediatric neuroimaging. NeuroImage 185:793–801. https://doi.org/10.1016/j.neuroimage.2018.04.044

Bernasconi A, Bernasconi N, Bernhardt BC (2011) Magnetic resonance imaging for epilepsy. Nat Rev Neurol 7(2):99–108. https://doi.org/10.1038/nrneurol.2010.210

Bernhardt BC et al (2010) Mapping thalamocortical network pathology in temporal lobe epilepsy. Brain 133(10):3072–3083. https://doi.org/10.1093/brain/awq239

Brown AL, Davies MJ, Roberts KS (2025) Role of magnetic resonance venography in pediatric cerebral venous thrombosis after trauma. Pediatr Radiol 55(3):412–421

Byrne D, Fisher A, Baker L, Twomey EL, Gorman KM (2023) Yield of brain MRI in children with autism spectrum disorder. Eur J Pediatr 182(8):3603–3609. https://doi.org/10.1007/s00431-023-05011-2

Canadian Paediatric Society (2025) Acute management of pediatric head trauma. Paediatr Child Health 30(2):130–142

Carette MF, Nedelcu C, Tassart M, Grange JD, Wislez M, Khalil A (2009) Imaging of hereditary hemorrhagic telangiectasia. Cardiovasc Intervent Radiol 32(4):745–757. https://doi.org/10.1007/s00270-008-9344-2

Caro-Domínguez P (2021) Cranial ultrasound for beginners. Pediatr Radiol

Cendes F et al (2016) Imaging in epilepsy. In: Handbook of clinical neurology, vol 136. Elsevier, pp 985–1014. https://doi.org/10.1016/B978-0-444-53486-6.00051-X

Chavhan GB (2016) Appropriate selection of MRI sequences for common scenarios in clinical practice. Pediatr Radiol 46(6):740–747. https://doi.org/10.1007/s00247-016-3556-4

Chen R, Jiao Y, Herskovits EH (2011) Structural MRI in autism spectrum disorder. Pediatr Res 69(5 Pt 2):63r–68r. https://doi.org/10.1203/PDR.0b013e318212c2b3

Chugani HT (2019) Positron emission tomography in pediatric neurology. Neuroimaging Clin N Am 29(4):419–434

Cross JH et al (2006) Epilepsy surgery in children with focal cortical dysplasia. Epilepsia 47(6):952–959. https://doi.org/10.1111/j.1528-1167.2006.00569.x

Distelmaier F, Klopstock T (2023) Neuroimaging in mitochondrial disease. Handb Clin Neurol 194:173–185. https://doi.org/10.1016/b978-0-12-821751-1.00016-6

Dubois J, Alison M, Counsell SJ, Hertz-Pannier L, Hüppi PS, Benders M (2021) MRI of the neonatal brain: a review of methodological challenges and neuroscientific advances. J Magn Reson Imaging 53(5):1318–1343. https://doi.org/10.1002/jmri.27192

Duncan JS (2010) Imaging and epilepsy. Nat Rev Neurol 6(10):537–550. https://doi.org/10.1038/nrneurol.2010.131

Gaillard WD et al (2009) Presurgical evaluation of epilepsy. Epilepsia 50(9):2147–2153. https://doi.org/10.1111/j.1528-1167.2009.02075.x

Garcia R, Lee HS, Patel M (2024) AI-enhanced imaging techniques in pediatric neuroimaging. J Pediatric Imaging 10(1):22–35

Gelfand MJ, Sotero de Menezes MA (2013) FDG PET of the brain in pediatric patients: imaging spectrum with radiologic–pathologic correlation. Radiographics 33(6):1879–1903

Gleeson JG (2001) Neuronal migration disorders. Ment Retard Dev Disabil Res Rev 7(3):167–171. https://doi.org/10.1002/mrdd.1024

Gocmen R, Guler E, Arslan EA (2015) A case of neurocutaneous melanosis and neuroimaging findings. J Radiol Case Rep 9(3):1–6. https://doi.org/10.3941/jrcr.v9i3.2141

Gorman GS, Chinnery PF, DiMauro S, Hirano M, Koga Y, McFarland R, Suomalainen A, Thorburn DR, Zeviani M, Turnbull DM (2016) Mitochondrial diseases. Nat Rev Dis Primers 2:16080. https://doi.org/10.1038/nrdp.2016.80

Grimm AS et al (2020a) Normative observational nerve ultrasound values in children. Front Neurol 11

Grimm A, Peters N, Schuhmann MU (2020b) Imaging patterns in pediatric neurometabolic and neurodegenerative disorders. Eur J Paediatr Neurol 25:11–24

Gropman AL (2013) Neuroimaging in mitochondrial disorders. Neurotherapeutics 10(2):273–285. https://doi.org/10.1007/s13311-012-0161-6
Guerrini R, Parrini E (2010) Neuronal migration disorders. Neurobiol Dis 38(2):154–166. https://doi.org/10.1016/j.nbd.2009.02.008
Gupta P (2016) Neonatal cranial sonography: a concise review for clinicians. Pediatr Radiol
Gupta R, Sharma A, Kumar P (2018) Role of CT scan in evaluation of epilepsy in children: an experience from a developing country. Indian J Radiol Imaging 28(3): 261–267
Haas R, Dietrich R (2004) Neuroimaging of mitochondrial disorders. Mitochondrion 4(5–6):471–490. https://doi.org/10.1016/j.mito.2004.07.008
Hanna M (2025) Cranial ultrasonography – standards in diagnosis of intraventricular hemorrhage. [Journal/Publisher]
Himmelmann K, Horber V, De La Cruz J, Horridge K, Mejaski-Bosnjak V, Hollody K, Krägeloh-Mann I (2017) MRI classification system (MRICS) for children with cerebral palsy: development, reliability, and recommendations. Dev Med Child Neurol 59(1): 57–64. https://doi.org/10.1111/dmcn.13166
Houser OW (1995) Neuroradiology: a historical perspective. Radiology 196(1):1–2. https://doi.org/10.1148/radiology.196.1.7784551
Johnson FT, Smith LL (2024) Training and expertise in pediatric neuroradiology: a review. Neuroimaging Clin N Am 34(1):1–13
Jones JY, Selvaraj B, Ho ML (2020) Pediatric functional neuroimaging: practical tips and pearls. AJR Am J Roentgenol 214(5):995–1007. https://doi.org/10.2214/ajr.19.22178
Kalra V, Datta S, Choudhary V (2021) Utility of CT in acute neurological emergencies in resource-constrained settings. J Clin Neurosci 89:67–73
Kanungo S, Morton J, Neelakantan M, Ching K, Saeedian J, Goldstein A (2018) Mitochondrial disorders. Ann Transl Med 6(24):475. https://doi.org/10.21037/atm.2018.12.13
Karimzadeh P (2015a) Magnetic resonance spectroscopy in pediatric neurometabolic disorders. Iranian J Child Neurol 9(3):1–17
Karimzadeh P (2015b) Approach to neurometabolic diseases from a pediatric neurological point of view. Iran J Child Neurol 9(1):1–16
Kerscher SR et al (2021) Comparison of B-scan ultrasound and MRI-based optic nerve sheath diameter assessment in children. Ultrasound Med Biol
Kioutchoukova I, Foster D, Thakkar R, Ciesla C, Cabassa JS, Strouse J, Kurz H, Lucke-Wold B (2024) Neurocutaneous diseases: diagnosis, management, and treatment. J Clin Med 13(6). https://doi.org/10.3390/jcm13061648
Krishnan P, Muthusami P, Heyn C, Shroff M (2015) Advances in pediatric neuroimaging. Indian J Pediatr 82(2):154–165. https://doi.org/10.1007/s12098-014-1657-3
Kwan P et al (2010) Definition of drug resistant epilepsy: consensus proposal by the ad hoc task force of the ILAE commission on therapeutic strategies. Epilepsia 51(6):1069–1077. https://doi.org/10.1111/j.1528-1167.2009.02397.x
Lee JY, Kim SH, Thompson R (2024) Advanced neuroimaging biomarkers in pediatric traumatic brain injury. Neurorehabil Neural Repair 39(4):295–308
Leeds NE, Kieffer SA (2000) Evolution of diagnostic neuroradiology from 1904 to 1999. Radiology 217(2):309–318. https://doi.org/10.1148/radiology.217.2.r00nv45309
Lev MH, Gadde JA, Wiggins RH (2022) A brief history of neuroradiology and the 60th anniversary of the ASNR: "through the decades with ASNR". AJNR Am J Neuroradiol 43(12):E54–e59. https://doi.org/10.3174/ajnr.A7712
Lin DDM, Barker PB, Moser HW (2005) Proton MR spectroscopy in the evaluation of suspected neurogenetic disorders. Am J Neuroradiol 26(7): 1845–1858
Liserre R, Di Biase C, Gallucci M (2021a) Proton MR spectroscopy in metabolic brain diseases of childhood. J Neuroimaging 31(2):205–213
Liserre R, Pinelli L, Gasparotti R (2021b) MR spectroscopy in pediatric neuroradiology. Transl Pediatr 10(4): 1169–1200. https://doi.org/10.21037/tp-20-445
Lüders HO et al (2006) The epilepsies: diagnosis and prediction of prognosis. Epileptic Disord 8(Suppl 2): S1–S9
Mahale R, Azeez A, Prabhu AS (2025) Diagnostic utility of cranial ultrasound in predicting clinical outcome in preterm and high-risk term neonates. Karnataka Paediatr J
Manley GT, Dams-O'Connor K, Awwad HO, Doperalski A, Umoh N, Maas AIR, McCrea MA (2025a) Marking a new age in characterization of acute traumatic brain injury: the National Institute of Neurological Disorders and Stroke traumatic brain injury classification and nomenclature initiative. J Neurotrauma 42:1021
Manley GT, Smith DH, Poloyac SM (2025b) Pediatric traumatic brain injury. In UpToDate. Retrieved from https://www.uptodate.com
Mascalchi M, Montomoli M, Guerrini R (2018) Neuroimaging in mitochondrial disorders. Essays Biochem 62(3):409–421. https://doi.org/10.1042/ebc20170109
Miller K, Wong TX, Friedman JH (2022) Ethical considerations in pediatric neuroimaging. Pediatric Ethics J 8(3):145–153
Millichap JG (1993) Cerebral ultrasound and neurodevelopment. Pediatric Neurol Briefs
Mohan KK (1999) Positron emission tomography in pediatric neurology. J Child Neurol 14(6):394–400
National Institute for Health and Care Excellence (NICE) (2022) Epilepsies: diagnosis and management (NG217)
Patel D, Kumar S (2023) Differentiating normal developmental variants from pathology in pediatric

neuroimaging: challenges and solutions. Clin Neuroradiol 33(2):285–293

Pietrzak N, Jankowska K, Rosiak O, Konopka W (2024) New prospects on neuroimaging in Von Hippel Lindau disease-a narrative review. Diagnostics (Basel) 14(3). https://doi.org/10.3390/diagnostics14030309

Rambani V, Hromnikova D, Gasperikova D, Skopkova M (2022) Mitochondria and mitochondrial disorders: an overview update. Endocr Regul 56(3):232–248. https://doi.org/10.2478/enr-2022-0025

Richer EJ et al (2021) Review of neonatal and infant cranial US. Radiographics 41:E206

Risacher SL, Saykin AJ (2021) Neuroimaging in genetic and neurodevelopmental disorders: current state and future directions. Curr Opin Neurol 34(2):198–207

Roberts D, O'Connell E, Sanchez G (2023) Legal and regulatory frameworks for pediatric neuroimaging. J Med Ethics 49(5):342–349

Rodriguez D, Young Poussaint T (2004) Neuroimaging findings in neurofibromatosis type 1 and 2. Neuroimaging Clin N Am 14(2):149–170, vii. https://doi.org/10.1016/j.nic.2004.03.004

Rossi A, Biancheri R (2013a) MR imaging of metabolic and neurodegenerative disorders of childhood. Neuroimaging Clin N Am 23(2):227–248

Rossi A, Biancheri R (2013b) Magnetic resonance spectroscopy in metabolic disorders. Neuroimaging Clin N Am 23(3):425–448. https://doi.org/10.1016/j.nic.2012.12.013

Russo C, Nastro A, Cicala D, De Liso M, Covelli EM, Cinalli G (2020) Neuroimaging in tuberous sclerosis complex. Childs Nerv Syst 36(10):2497–2509. https://doi.org/10.1007/s00381-020-04705-4

Samet Y, Başkaya MK (2023) The use of ultrasound-guided imaging to localize peripheral nerve injury in pediatric patients: a case report. Surg Neurol Int

Saneto RP, Friedman SD, Shaw DW (2008) Neuroimaging of mitochondrial disease. Mitochondrion 8(5–6):396–413. https://doi.org/10.1016/j.mito.2008.05.003

Severino M, Geraldo AF, Utz N, Tortora D, Pogledic I, Klonowski W, Triulzi F, Arrigoni F, Mankad K, Leventer RJ, Mancini GMS, Barkovich JA, Lequin MH, Rossi A (2020) Definitions and classification of malformations of cortical development: practical guidelines. Brain 143(10):2874–2894. https://doi.org/10.1093/brain/awaa174

Sharma S, Jones PR, Singh R (2024) MRI biomarkers in pediatric traumatic brain injury: a systematic review. Brain Inj 38(6):821–830

Singh A, Rao S (2019) Computed tomography in pediatric seizure disorders: emphasis on findings and clinical relevance. Pediatr Neurol 95:34–41

Smith J, Johnson A (2023) Imaging in pediatric traumatic brain injury: role of MRI. J Neurosurg Pediatr 31(1):101–110

Soltirovska Salamon A, Lichtenbelt K, Cowan FM, Casaer A, Dudink J, Dereymaeker A, Paro-Panjan D, Groenendaal F, de Vries LS (2016) Clinical presentation and spectrum of neuroimaging findings in newborn infants with incontinentia pigmenti. Dev Med Child Neurol 58(10):1076–1084. https://doi.org/10.1111/dmcn.13140

Spalice A, Parisi P, Nicita F, Pizzardi G, Del Balzo F, Iannetti P (2009) Neuronal migration disorders: clinical, neuroradiologic and genetics aspects. Acta Paediatr 98(3):421–433. https://doi.org/10.1111/j.1651-2227.2008.01160.x

Thompson R, Lee JY, Kim SH (2025) MRI spectroscopy for monitoring brain metabolism after pediatric TBI. Pediatr Radiol 55(2):180–192

Turesky TK, Vanderauwera J, Gaab N (2021) Imaging the rapidly developing brain: current challenges for MRI studies in the first five years of life. Dev Cogn Neurosci 47:100893. https://doi.org/10.1016/j.dcn.2020.100893

VanHorn TA, Cartwright MS (2020) Neuromuscular ultrasound in the pediatric population. Front Neurol

Vossough A (2023) Newer MRI techniques in pediatric neuroimaging. Semin Roentgenol 58(1):131–144. https://doi.org/10.1053/j.ro.2022.10.002

Vossough A, Khalili N, Familiar AM, Gandhi D, Viswanathan K, Tu W, Haldar D, Bagheri S, Anderson H, Haldar S, Storm PB, Resnick A, Ware JB, Nabavizadeh A, Fathi Kazerooni A (2024) Training and comparison of nnU-net and DeepMedic methods for autosegmentation of pediatric brain tumors. AJNR Am J Neuroradiol 45(8):1081–1089. https://doi.org/10.3174/ajnr.A8293

Wang Y et al (2024) Establishment of a normal control model of children's brain 18F-FDG PET. Quant Imaging Med Surg 14(1):123–135

Whitehead MT, Lai LM, Blüml S (2022) Clinical (1)H MRS in childhood neurometabolic diseases-part 1: technique and age-related normal spectra. Neuroradiology 64(6):1101–1110. https://doi.org/10.1007/s00234-022-02917-w

Widjaja E et al (2007) Functional magnetic resonance imaging of language in children with epilepsy. Neurology 69(9):857–865. https://doi.org/10.1212/01.wnl.0000269682.23377.69

Williams DW 3rd, Elster AD (1990) Cranial MR imaging in hypomelanosis of Ito. J Comput Assist Tomogr 14(6):981–983. https://doi.org/10.1097/00004728-199011000-00021

Wilson TM, Edwards K, Jones L (2023) Addressing disparities in pediatric neuroimaging access. Health Equity 7(1):85–93

Wisnowski JL, Sperling VR, Panigrahy A (2013) Challenges in pediatric magnetic resonance imaging. In: Blüml S, Panigrahy A (eds) MR spectroscopy of pediatric brain disorders. Springer, New York, pp 39–42. https://doi.org/10.1007/978-1-4419-5864-8_4

Yilmaz N, Walker C, Patel M (2024) Clinical indications and imaging protocols for cerebral venous thrombosis in children. Pediatr Neurol 145:17–24

Yiu EM et al (2015) Peripheral nerve ultrasound in pediatric Charcot-Marie-tooth disease. Neurology 84(10):1056–1061

Zeviani M, Carelli V (2007) Mitochondrial disorders. Curr Opin Neurol 20(5):564–571. https://doi.org/10.1097/WCO.0b013e3282ef58cd

Anatomy of the Blood Vessels Supplying the Brain and Its Clinical Importance in Neurovisualization

10

Srijit Das and Mohamed Al Mushaiqri

Abstract

The brain's vasculature is important for the normal functioning of the central nervous system. The rich vascular supply of the brain is needed for oxygen and nutrients. The two main arteries supplying blood are the internal carotid artery and the vertebral artery, with its branches. Most of the forebrain is supplied by the internal carotid artery, while the parts of the temporal and occipital lobes derive its blood supply from the branches of the vertebrobasilar system of arteries. The anatomical knowledge of the arteries supplying the brain is important for diagnosis, investigation, surgical interventions, and treatment. The arteries contribute to the formation of the circle of Willis. The high oxygen and nutrient demand are met by the vertebrobasilar arteries and the internal carotid arteries in the posterior and anterior parts, respectively. The circle of Willis is important for the collateral supply of blood. Ischemia and stroke may benefit from collateral circulation, and incomplete circulation may prove to be dangerous. The arteries supplying the cerebrum may exhibit variations that may compromise the hemodynamics. Aneurysms, strokes, emboli, thrombus, clots, tumors, and degenerative changes occurring in the vessel wall are commonly seen in day-to-day clinical practice. Proper anatomical knowledge of the blood supply of the brain is helpful for angiography procedures, radiological interpretations, and surgeries to minimize errors in diagnosis and check for any inadvertent injuries. In this chapter, we discuss the anatomy and clinical importance of the arteries supplying the cerebrum of the brain.

Keywords

Artery · Brain · Vascular · Imaging · Clinical · Diagnosis

Abbreviations

ACA	Anterior cerebral artery
ATP	Adenosine triphosphate
Ca	Calcium
C6-C1	Cervical vertebra 6 to cervical vertebra 1
CBF	Cerebral blood flow
CT	Computed tomography
DSA	Digital subtraction angiography
MCA	Middle cerebral artery
MRI	Magnetic resonance imaging
PCA	Posterior cerebral artery
2D	Two-dimensional
3D	Three-dimensional

S. Das (✉) · M. Al Mushaiqri
Department of Human & Clinical Anatomy, College of Medicine & Health Sciences, Sultan Qaboos University, Muscat, Oman
e-mail: s.das@squ.edu.om; mohamed@squ.edu.om

K. S. Ravi et al. (eds.), *Visualization in Neuroanatomical Sciences*, Biomedical Visualization 11,
https://doi.org/10.1007/978-3-032-07380-8_10

4D Four-dimensional

10.1 Introduction

The brain is a complex structure with billions of neurons. These neurons are responsible for transmitting various electrical and chemical impulses. The human brain is mainly divided into the following components: two hemispheres of the cerebrum, cerebellum with lobes, and the brainstem. The largest part of the brain is the cerebrum, and it is further subdivided into the right and left hemispheres. Both sides of the cerebral hemispheres are separated by the longitudinal fissure (Snell 2007). To accommodate a large surface area, the cerebrum possesses ridges known as gyri and fissures known as sulci. The deeper sulci are known as fissures. Both hemispheres of the cerebrum possess gray matter outside (known as the cortex of the cerebrum) and subcortical white matter located inside.

The control of various information (both motor and sensory), conscious thoughts, various cognitive processes, intelligence, and memory is performed by the cerebrum. The left hemisphere controls speech and processes language, while the right hemisphere controls spatial awareness, stereoscopic visualization, and artistic and creative functions. The four lobes of the cerebrum include the anteriorly located frontal, superiorly located parietal, temporal lobes on its lateral aspect, and the posteriorly located occipital lobe. The largest lobe is the frontal lobe, and it is involved in involuntary movement, language, mood, personality, self-awareness, and judgment. Broca's area (motor speech area) is situated above the lateral sulcus and is responsible for speech production (Snell 2007). The parietal lobe is responsible for somatosensory information. In the temporal lobe, Wernicke's area is responsible for language development. The temporal lobe has a vital role in managing emotions and processing auditory and visual sensory inputs. The occipital lobe has the visual cortex in the region of the calcarine sulcus, and it is responsible for receiving visual information (Snell 2007).

Two pairs of blood vessels supply blood to the brain, i.e., the **vertebrobasilar arteries** and the **internal carotid arteries.** These arteries originate in the neck region to traverse an upward course to the cranium. Inside the cranial cavity, the terminal branches of these blood vessels form an anastomotic circle, also known as the **circle of Willis.** Various branches arise from this circle and give blood supply to a larger part of the cerebrum. The cerebral circulation consists of several arteries that provide oxygenated blood to the brain. It is essential to understand the anatomy of the arteries supplying the cerebrum to diagnose aneurysms, various arteriovenous malformations, and tumors and precisely locate thrombus in stroke (Bokhari and Bokhari 2023).

10.2 The Developmental Anatomy of the Arteries

The development of the blood vessels occurs by two processes, i.e., vasculogenesis and angiogenesis. The vascular endothelial growth factor regulates the development of blood vessels. The common carotid and the first part of the internal carotid artery develop from the third aortic arch (Sadler 2019). The remaining part of the internal carotid artery develops from the dorsal aorta (Sadler 2019). The right subclavian artery develops from the fourth aortic arch.

Approximately on the 28th day of intrauterine life, the internal carotid artery divides into anterior and posterior divisions, and the anterior division in turn gives rise to branches such as the anterior choroidal artery, the anterior cerebral artery (ACA), and the middle cerebral artery (MCA) (Javed et al. 2023). The posterior division originates from the fetal posterior cerebral artery. The posterior circulation mainly gets its blood supply from the vertebrobasilar anastomosis in the early part of fetal development (Javed et al. 2023). Gradually, the posterior fossa contents and the occipital lobe grow, and the posterior circulation does not have to depend upon the anterior circulation, and this results in the obliteration of the anterior-posterior anastomoses (Javed et al. 2023). Thus, the posterior communicating arteries are the only anastomoses in the adult which is between the anterior and posterior circulations (Javed et al. 2023).

If the PCA persists and continues in the fetus to arise from the ICA, it may result in a basilar artery with a smaller diameter, and this is seen in 10–29% of the cases unilaterally (Javed et al. 2023) The fetal PCA persists as partial or complete, and these individuals fail to establish leptomeningeal anastomoses which occur between the ACA, MCA, and PCA (Javed et al. 2023). Unilateral fetal PCA may be liable to suffer from ischemia due to ICA stenosis or thrombosis (Javed et al. 2023).

A smaller caliber basilar artery could develop if the fetal PCA persists and emerges from the ICA. Of individuals, 10–29% experience this phenomenon, which usually manifests unilaterally (Nouh et al. 2014). The persistence of a fetal PCA is classified to be partial or complete based on the PCA's P1 (pre-communicating) segment being hypoplastic or not. Bilateral fetal PCAs were reported to be associated with a small dimension of the basilar artery because the basilar artery did not contribute to the mesencephalic, temporal, or occipital lobe flow (Nouh et al. 2014). In individuals with occlusion of the ICA, a fetal PCA on one side and a nonfunctioning anterior communicating artery may be prone to ischemia and infarction because of the hemodynamic changes (Nouh et al. 2014).

10.3 Arteries Supplying the Brain and Their Clinical Implications

The brain has a rich vascular supply. There is a dense arterial network for the blood supply to the brain. A lot of energy requirements are to be met by utilizing a rich blood supply. The brain receives approximately 15% of the cardiac output and uses nearly 25% of the total oxygen that is consumed by the body (Standring 2008). The standard textbook of anatomy mentions that although the brain is only 2.5% of the weight of the body, it receives one-sixth of cardiac output and one-fifth of the oxygen required by the human body at rest (Moore and Dalley 2006).

Oxygen and nutrients are vital for nourishing the brain to carry out its activities. Oxygen and glucose are necessary for the proper functioning of the brain. With adequate oxygen and glucose supply, there is normal mitochondrial function resulting in the formation of adenosine triphosphate. It may be mentioned that up to 92% of the brain's adenosine triphosphate at rest comes from the oxidative metabolism of glucose (Vavilala et al. 2002). If there is no glucose or oxygen available for oxidative metabolism, the homeostatic brain function is affected. When there is an arrest of cerebral blood flow, an individual loses consciousness within 10 s (Rink and Khanna 2011).

According to reports, the brain uses 25% of its energy to restore ionic gradients across cell membranes, and over 50% of that energy is needed for synaptic activity processes. The energy source of the brain is responsible for producing the functional imaging signals and controlling its capacity to produce information (Howarth et al., 2012).

The blood flows at approximately 600–700 ml/minute and 100–200 ml/minute in the carotid and vertebral-basilar arteries (UMas Chan Medical School 2025). Adequate blood flow to the brain is required for oxygen and nutrients and the removal of carbon dioxide and waste materials from the cell.

Cerebral blood flow (CBF) is important for brain function and viability because of the neurons' limited ability to carry out anaerobic metabolism. If one considers the brain, oxygen consumption is according to the region, and interestingly gray matter consumes two times as much oxygen compared to white matter, and the maximum consumption was found to occur in the medial occipital lobe (Pantano et al. 1984). Capillaries and precapillary arterioles have been found to play an important role in oxygen transport (Tsai et al. 2009). Cerebral autoregulation helps to maintain the blood flow in the brain.

An adequate blood supply is also needed for the maintenance of cognitive functions. Poor blood supply can lead to dizziness, changes in vision, memory dysfunction, and altered neurological problems. Due to disruption in the blood supply, there may be cognitive disturbances and dementia. An earlier research study reported that reduced blood supply to the brain causes stroke, deterioration of cognitive functions, and dementia, especially in individuals with high blood pressure (Waldstein et al. 2010).

During any condition when cerebral autoregulation is altered or affected, such as in stroke, the neuron function is also regulated according to the response to brain hypoxia (Rink and Khanna 2011).

With the increase in age of the individual, blood vessels are also affected. Also, other factors such as high levels of cholesterol, metabolic changes such as diabetes mellitus, obesity, and high blood pressure cause a change in the diameter of the blood vessels, affecting the blood flow (Standring 2008). When the diameter of the blood vessel becomes narrower, there is a collection of blood to form clots, and these clots block the arteries (known as thrombus). If the clots are dislodged, they may be trapped in a distant part of the vessel to form emboli. The improvement of lifestyle factors can check all the risk factors involved in the damage to the blood vessels.

Two main arteries supply blood to the brain, i.e., the internal carotid artery (ICA) and the vertebrobasilar arteries. There are two vertebral arteries and one basilar artery.

10.3.1 Vertebral Artery

The vertebral artery originates from the first part of the subclavian artery and has four parts. The first part lies in the neck with no branches, and the second part traverses the foramen transversaria of cervical vertebrae C6–C1, giving branches to the meninges and spinal cord. The third part of the vertebral artery is in the suboccipital triangle and gives off branches to suboccipital muscles, while the fourth part enters the cranial cavity through the foramen magnum and gives off meningeal branches, anterior and posterior spinal arteries, and posterior inferior cerebellar arteries.

At the lower border of the pons, both vertebral arteries of each side unite to form the basilar artery, which gives off the anterior inferior cerebellar, pontine, labyrinthine, superior cerebellar, and PCAs. The posterior inferior cerebellar artery is the largest branch of the vertebral artery.

The two vertebral arteries meet in the midline to form the basilar artery. The basilar artery can be located on the ventral pontine surface and terminates as two PCAs. The anterior circulation is mainly supplied by the ICA, while the vertebrobasilar system supplies the occipital lobe, cerebellum, and brainstem.

10.3.2 Internal Carotid Artery

The ICA originates from the common carotid artery in the region of the neck. The cervical part traverses a vertical course through the neck.

10.3.2.1 Petrous Part

The ICA enters the cranial cavity through the carotid canal located in the petrous part of the temporal bone and the superior aspect of the foramen lacerum. The ICA has two parts, i.e., petrous and cavernous parts. The petrous part of the ICA takes a course vertically through the carotid canal, bends anteromedially and then superomedially on the top of the cartilage filling the foramen lacerum, and then makes its entry into the cranial cavity. The venous plexus and the carotid autonomic plexus surround the ICA. The petrous part of the ICA has two branches, i.e., caroticotympanic artery and the stylomastoid artery.

10.3.2.2 Cavernous Part

The cavernous part of the ICA traverses a course upward to the posterior clinoid process of the skull. Then, it takes a turn anteriorly to the side of the sphenoid inside the cavernous sinus and then goes medially to the anterior clinoid process, to emerge through the dural roof of the cavernous sinus. As the ICA takes a course through the cavernous sinus, there are four cranial nerves situated lateral to it, and they are the oculomotor, trochlear, ophthalmic, and abducens nerves.

10.3.3 Branches of the Internal Carotid Artery

The branches of the ICA are the posterior communicating artery and the anterior choroidal artery. The posterior communicating artery traverses a course posteriorly across the crus cerebri and joins the PCA to complete the arterial circle. Its branches supply the crus cerebri, optic tract, hypophysis, and the hypothalamus of the brain. The anterior choroidal artery has a course posterolaterally to supply the crus cerebri. The course of the anterior choroidal artery, laterally to the medial part of the temporal lobe, accounts for the supply to the choroidal plexus of the inferior horn of the lateral ventricle.

The branches of the ICA supply blood to the majority, i.e., two-thirds of the cerebral hemispheres, while the vertebrobasilar system, with their branches, provides blood supply to the posterior and medial aspects of the hemispheres, majority of the diencephalon, region of the brainstem, cerebellum, and cervical spinal cord (UMas Chan Medical School 2025).

10.4 Stroke

A stroke is a neurological condition in which the arteries supplying blood get blocked. Approximately 25% of strokes and transient ischemic attacks were found in the vertebrobasilar distribution (Pirau and Lui 2023). The vertebrobasilar disease occurs later in life, especially in the seventh to eighth decades, and it commonly affects the male gender (Pirau and Lui 2023).

There may be clot formation in the brain, or the arteries may be involved in hemorrhage. When there is a rupture of any blood vessel during a stroke, the cells in the brain suffer from a lack of oxygen and are subjected to death. Many cells are lost with a greater duration of loss of blood supply. Impaired blood flow in the cerebrum is known to cause neurological problems, as seen in ischemic stroke patients. In hypoxia induced by ischemic stroke, there is adenosine triphosphate (ATP) deficiency, and the ATP-dependent ion pumps cease to function. Transmembrane ion gradients fail to function, and there is Ca^{2+} influx. Gradually, there is an accumulation of intracellular Ca^{2+}. The formation of superoxide is also increased. All these inflammatory markers denote oxidative damage.

Stroke ranks as the second most common cause of death in the world, and approximately 13.7 million individuals are involved, with 5.5 million succumbing to it annually (Kuriakose and Xiao 2020). Most strokes are ischemic infarctions, and interestingly the prevalence increased considerably between 1990 and 2016, but mortality decreased. The incidence of stroke increased two times in low- and middle-income countries from 1990 to 2016 but reduced by 42% in high-income countries during this period. The incidence of stroke certainly increases the burden in low socioeconomic groups according to the Global Burden of Disease Study (Collaborators 2019).

10.5 Importance of the Circle of Willis

The carotid and vertebrobasilar circulations connect to each other through the circle of Willis (Fig. 10.1). The name is derived from the name of the scientist, Thomas Willis. Hence, collateral blood flow between the anterior and posterior arterial blood circulations is provided by the circle of Willis. The circle of Willis is a common term used in the literature. Other terms such as *cerebral arterial circle*, or *circulus arteriosus cerebri*, were also used. Published literature reveals an incidence varying between 4.8% (Fisher 1965) and 85.4% (Yeniçeri et al. 2017) in different individuals.

The joining of the PCAs and the posterior communicating branch of the ICA forms the circle of Willis. The circle of Willis comprises paired ACAs arising from ICA; anterior communicating arteries, which connect the ACAs; ICAs, which are the branches of the common carotid arteries; paired PCAs, which are branches of the basilar artery; and the posterior communicating

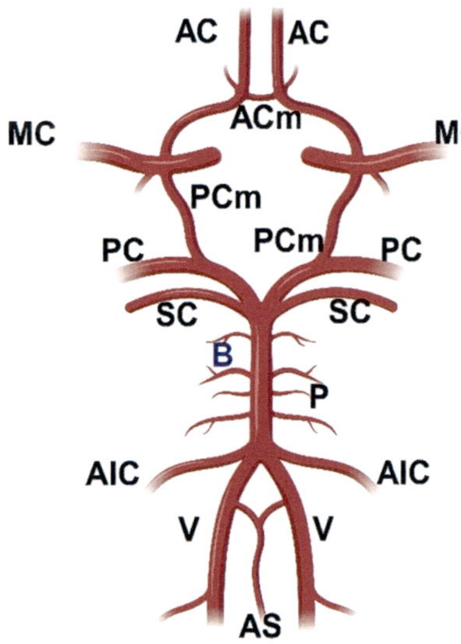

Fig. 10.1 Schematic diagram of circle of Willis. *AC* anterior cerebral, *ACm* anterior communicating, *MC* middle cerebral, *PCm* posterior communicating, *PC* posterior cerebral, *SC* superior cerebellar, *B* basilar artery, *P* pontine arteries, *AIC* anterior inferior cerebellar, *V* vertebral, *AS* anterior spinal. (The figure is drawn with BioRender software)

arteries, which are the connecting vessels of ICAs and PCAs.

While a complete circle of Willis is present in many individuals, there may be variations in its formation and branches. Hypoplasia or the presence of an anomalous circle of Willis compromises the blood flow.

The variation in circle of Willis was reported by various authors and researchers, and it was attributed to different methods or nomenclature adopted by different studies (Karatas et al. 2016), or because of changes in the ethnicity and population (De Silva et al. 2011; Eftekhar et al. 2006; Karatas et al. 2016), and also due to a different neurological state of the studied population (Kayembe et al. 1984; Riggs and Rupp 1963). The variation was most seen in the posterior communicating artery (Lazorthes et al. 1979; Eftekhar et al. 2006). The hypoplasia of the posterior communicating artery may be commonly encountered. Both unilateral and bilateral hypoplasias of the posterior communicating artery have been reported earlier (Eftekhar et al. 2006). Race may be one of the possible reasons for the variations. Anatomical knowledge of variations in the posterior communicating artery may be beneficial for neurovascular procedures.

If any occlusion of any large arteries contributes to the circle of Willis, the other communicating arteries compensate. Stroke and aneurysms are associated with the variation in the morphology of the circle of Willis. A study found a poor prognosis for stroke patients who had an incomplete circle of Willis (Lin et al. 2022). Another study found the prognostic value after intravenous thrombolysis by using the circle of Willis as a specific marker for good collateral circulation (Chuang et al. 2013). The configuration of the circle of Willis and the impact of thrombus location on the outcome of ischemic stroke patients have also been reported (Širvinskas et al. 2023).

Aneurysms of the cerebral blood vessels arise due to a defect in the wall of the vessel. The increase in blood pressure affects the vessel walls. There are reports of aneurysms being more common at the circle of Willis or near the junction of vessels (Standring, 2008). Once the vessel wall ruptures, there is bleeding and hemorrhagic stroke. Studies also reported that variation in the circle of Willis and nonspherical morphology, in addition to contributing to aneurysm, were also responsible for rupture (Zimelewicz Oberman et al. 2021). If there is a rupture of a cerebral aneurysm, there is hemorrhage into the subarachnoid space. Hypoplasia of any of the arteries involved in the formation of the circle of Willis may also be a cause of aneurysm. The site and hemodynamic load may determine the severity of the symptoms.

Thanks to the advancement of imaging procedures, we may be able to neurovisualize blood vessels noninvasively with accurate precision by using computed angiography with contrast (angio-CT) or magnetic resonance with or without contrast administration (angio-MR) (Makowicz et al. 2013). The normal and abnormal variants of arteries supplying the brain can be easily studied. It may also help in detecting asymptomatic cases. Proper interpretation can be beneficial for reducing the risk of formation of aneurysm and acute intracranial hemorrhage and

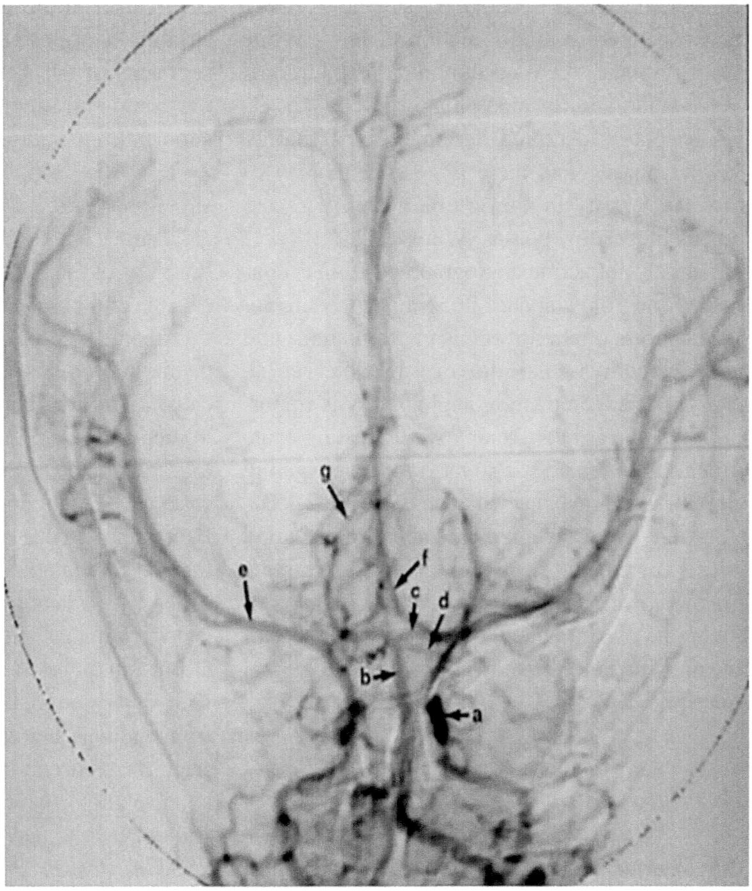

Fig. 10.2 Digital subtraction angiography showing: (**a**) internal carotid artery; (**b**) basilar artery; (**c**) posterior cerebral artery; (**d**) superior cerebellar artery; (**e**) middle cerebral artery; (**f**) anterior cerebral artery; and (**g**) choroidal vessels

therefore beneficial in neurosurgical planning or checking for any inadvertent medical errors (Makowicz et al. 2013). Digital subtraction angiography (DSA) is a fluoroscopic technique that has been used exclusively in interventional radiological visualization of blood vessels (Fig. 10.2). The latest techniques for imaging the brain blood vessels include magnetic resonance angiography, 3D echo sequences enhanced with contrast, and near-infrared optical tomography.

10.6 Circle of Willis Studied in Neonates

Neonatal brain oxygenation could be better investigated with infrared optical tomography (TD-NIROT) system. To investigate cerebral oxygenation, near-infrared spectroscopy could be used in neonates. 2D phase-contrast magnetic resonance angiography has been successfully used in neonates to study anatomy of the circle of Willis (van Kooij et al. 2010). Another study used MR angiography to study the anatomic variation of circle of Willis in neonates and found that the variations were more in the main arterial segments and rather less in the communicating arterial channels (Malamateniou et al. 2009).

In recent times, multi-slice CT scanners can detect more than 90% of small blood vessels when compared to digital subtraction angiography and have been useful in demonstrating arteries with a diameter over 0.7 mm (Villablanca et al. 2007). The diameter of many small-sized vessels can be measured with this procedure. There are 2D (two-dimensional) and 3D (three-dimensional) procedures and even combined use of both which can be effective. The latest 4D (four-dimensional) technique is employed to diagnose arteriovenous malformations. While comparing the 4D procedure to the traditional method, the arteriovenous malformation vessels can be better identified during different blood-flowing phases, and there is a better display of vascular lesions in the arterial, capillary, and venous blood flow phases (Wang et al. 2014).

Digital subtraction angiography (DSA) is considered the gold standard imaging procedure for the diagnosis of arteriovenous malformations and acute cerebrovascular diseases (Shaban et al. 2022). Digital subtraction angiography is important for endovascular therapy while treating acute ischemic stroke and is also used for investigating cerebral vessel patency (Shaban et al. 2022). The procedure is invasive in nature and requires good skills. There may be a few cases of ischemia, embolism, and hemorrhagic complications following its use in patients. There is also a risk of ionizing radiation following this radiological procedure.

10.7 Anterior Cerebral Artery

The anterior cerebral artery (ACA) originates from the ICA (Fig. 10.3). It is a small terminal branch of the ICA. Standard textbooks of anatomy describe three parts of the ACA (Standring 2008). They are A_1, from the point where ICA ends to the junction with the anterior communicating artery; A_2, from the junction with the anterior communicating artery to the point where of the callosomarginal artery originates; and A_3, distal to the origin of the callosomarginal artery, also termed as pericallosal artery (Standring 2008).

An unpaired anterior communicating artery connects the two ACAs. The ACA gives branches that penetrate deep (called lenticulostriate arteries) which give blood supply to the majority of the anterior portions of the basal ganglia. It then has a course into the interhemispheric fissure, traverses a course over the corpus callosum, and then supplies maximal area over the medial surface. The recurrent artery of Heubner (also termed Heubner's artery or medial striate artery) is a branch of the ACA and supplies the corpus striatum and globus pallidus. Anatomically, five segments of the ACA may be present. They are pre-communicating, infracallosal, pre-callosal, supracallosal, and post-callosal.

The ACA supplied blood to the medial and superior parts of the frontal lobe, corpus callosum, and anterior parietal lobe. Deep branches of the ACA supply blood to the internal capsule and basal ganglia.

Atherosclerotic changes in large vessels may result in thrombosis. Also, thromboembolism can cause ACA infarction. The incidence of isolated infarction in the region of ACA has been reported to vary from 0.6% to 3% of all ischemic stroke cases (Kazui and Sawada 1993; Gacs et al. 1983). Interestingly, because of the atypical and asymptomatic presentation of ACA infarction, the incidence of occlusion of ACA is often diagnosed late or remains undiagnosed (Liu et al. 2024). Occlusion of the ACA is a rare condition. Patients with occlusion of ACA and aneurysms benefit from endovascular treatment. If there is thrombosis of the recurrent branch of the ACA, then there is upper motor neuron paralysis of the contralateral upper limb and the face.

ACA strokes are uncommon, but they may occur due to high blood pressure, dyslipidemias, diabetes mellitus, smoking, atherosclerosis, and cardioembolism (Matos Casano et al. 2023). Researchers have reported cardioembolism as the main cause of ACA stroke (Arboix et al. 2009). Another important risk factor is atrial fibrillation. Clinicians may also rule out arterial dissections in young patients and any cause of vasospasm.

When there is cerebral infarction in the region of ACA, one should consider the cardioembolic mechanism and hence the possibility of early indication of anticoagulation therapy to prevent any secondary cardioembolic stroke (Ustrell-Roig and Serena-Leal 2007).

The size of an infarct may determine the clinical symptoms. Interestingly, the clinical features and etiology of ACA infarction may be like MCA infarction, but frontal dysfunctions and callosal syndromes certainly help in the differential diagnosis (Kumral et al. 2002). Individuals with

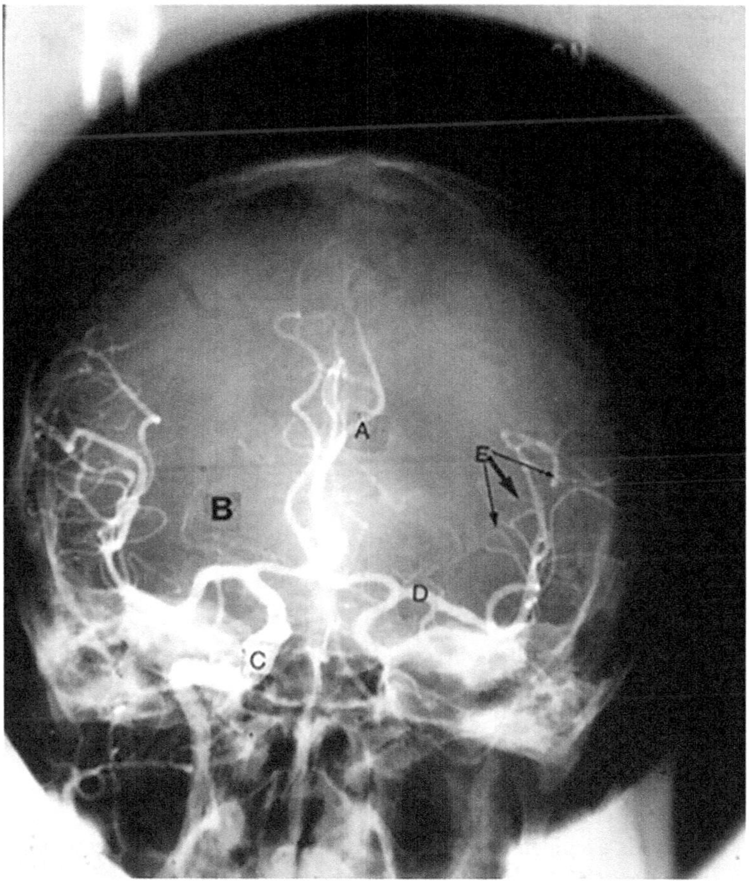

Fig. 10.3 Carotid angiogram showing (**a**) anterior cerebral artery, (**b**) striated arteries, (**c**) internal carotid artery, (**d**) middle cerebral artery, and (**e**) insular (Sylvian) branches

infarcts in the ACA region exhibit a favorable short-term prognosis with an in-hospital mortality rate of 7.8%, which is much lower than that of MCA infarctions (17.3%) (Arboix et al. 2009).

10.8 Middle Cerebral Artery

The middle cerebral artery (MCA) is the terminal branch of the ICA. The majority of the blood supply to the brain is provided by the MCA. The MCA arises from the ICA at the lateral part of the circle of Willis. The MCA has a large diameter. MCA is a part of the anterior circulation of the cerebrum (Fig. 10.4). The MCA passes laterally just inferior to the frontal lobe, with a course in the lateral (Sylvian) fissure. It ends its course on the cerebrum's lateral surface. MCA has four parts.

MCA gives off lenticulostriate arteries which supply the deeper structures of the brain, while its cortical branches supply most of the lateral surface of the cerebrum.

The MCA is known to exhibit variations in its origin. A duplicated MCA was observed in less than 3% of cases, originating from the ICA, parallel to the main MCA, and traversing a course toward the anterior temporal lobe to supply it (Beger 2023). Approximately 4% of individuals possess an accessory MCA which supplies the orbitofrontal region of the brain (Beger 2023). The incidence of MCA duplication or presence of accessory MCA may be less, but they have enormous clinical implications. Strokes have a better prognosis when there is an additional vessel to supply the blood in the event of occlusion of one vessel. It is also important to differentiate such variations in the imaging investigations.

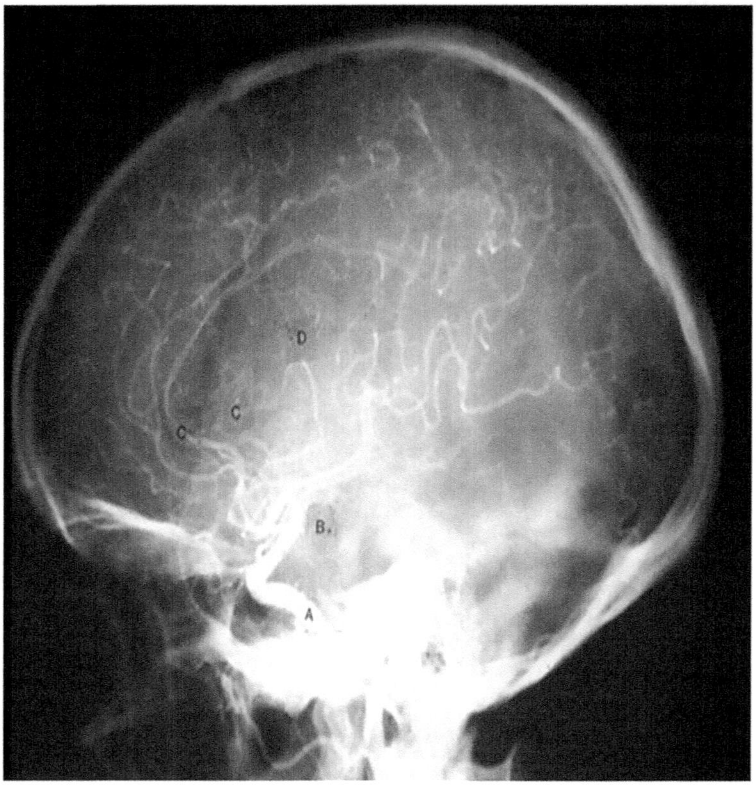

Fig. 10.4 Carotid angiogram showing (**a**) internal carotid artery, (**b**) middle cerebral artery, and (**c**) anterior cerebral artery

If there is a sudden blockage of the MCA, an infarction may occur. The MCA was reported to be the second most common site for aneurysm for the blood vessel in the brain (Wang et al. 2016). Out of all the intracranial aneurysms, 18–40% are accounted for by the MCA aneurysm (Elsharkawy et al. 2013). These aneurysms in the MCA can be located proximal, bifurcation, and distal and are accordingly divided into these groups. The hemodynamics at the bifurcation point is clinically important as the wall of the blood vessel is weak and any pressure or stress changes may result in rupture of the blood vessel.

The occlusion in MCA may be embolic in nature. In patients with occlusion in the MCA, the hemodynamics compromise might be high (Derdeyn et al. 1998). Maximal tissue damage has been observed after 4–8 hours of occlusion of the MCA (Olsen 1986).

In young individuals, MCA dissection was reported to contribute to the stroke burden (Asaithambi et al. 2014). It is difficult to diagnose MCA dissection cases as the only symptom may be headache. Digital subtraction angiography may be best suited for the detection of MCA dissection which is a rare cause for stroke.

10.9 Posterior Cerebral Artery

The PCA is a terminal branch of the basilar artery and the main artery on the inferior surface of the brain. The PCA may also take origin from the ICA. The PCA usually originates from the basilar artery, but in a few cases it may originate from the posterior communicating arteries. Oxygenated blood is supplied by the PCA to the midbrain and posterior part of the cerebrum.

It takes a turn around the cerebral peduncle to reach the cerebrum's tentorial surface. It gives off three branches, such as the temporal, occipital, and parieto-occipital arteries. The PCA supplies the visual cortex and the midbrain.

The branches of the PCA include posteromedial central branches, the posterior choroidal artery, and the cortical branches. The posteromedial central branches pass through the ventral surface in the base of the brain and supply blood to the midbrain and the caudal portion of the diencephalon. The posterior choroidal arteries arise from the P2 (post-communicating) segment of the PCA, and they may be single or in multiple numbers. These arteries supply the choroid plexus of the lateral ventricle and the third ventricle. The cortical branches of the PCA supply different parts of the cortex and are named accordingly, i.e., temporal, parieto-occipital, calcarine, and occipital branches. The midbrain and thalamus receive blood from the perforating branches of the PCA.

Occlusions of the basilar artery are included under the category of posterior circulation stroke. When there is occlusion of the basilar artery, each of the PCAs may be perfused by the ICA of the same side.

Variations of the PCA are absent, duplicated, or triplicated cortical branches, and the cortical branches have abnormal origins (Cilliers and Page 2017). Few cases of PCA cortical branches originating from the ICA have been reported (Cilliers and Page 2017).

Fetal-type posterior cerebral artery (FTP), commonly seen in the circle of Willis, is defined as a PCA that takes origin from the ICA and may have or may not have a small connection with the BA (Arjal et al. 2014). If present, FTP is associated with intracranial aneurysm, ischemia, and infarction.

PCA fenestration is considered a rare anomaly in which a portion of the artery is split into two, which eventually join distally. Such anomalies may be detected in cadavers or CT and magnetic resonance angiography scans.

The incidence of PCA strokes is 18 cases/100,000 individuals per year, and PCS accounts for 20–25% of all ischemic strokes (Sparaco 2018). The mortality at 1 month was reported to be from 3.6% to 11%, and the disability occurring at 3 months was reported to vary from 6.9–19.8% (Libman et al. 2001; Tao et al. 2010). At times, PCA strokes may be difficult to diagnose.

Embolism may be one of the causes of PCA infarction. Other causes include atherosclerosis and small vessel disease. Published literature shows that 35% and 13% of PCA strokes occur due to atherosclerosis and small artery disease (Labropoulos et al. 2011).

The PCA is one of the most difficult and risky arteries in surgeries as it is located very deep, arising frequently in the midline, the extensive perforating vessels adjoining or arising from it, and the proximity to the extraocular nerves and upper part of the brainstem (Zeal and Rhoton 1977).

The site of origin of PCA determines the resultant visual defects in the occipital lobe of patients suffering from any carotid and basilar artery disease. In the case of stroke of PCA, symptoms include contralateral homonymous hemianopia (which is due to occipital infarction), hemisensory loss (which is due to thalamic infarction), and hemi-body pain (due to thalamic infarction) (Mous et al. 2025). Following bilateral PCA infarcts, amnesia and cortical blindness have been reported (Cereda and Carrera 2012). At times, calcification of the MCA can also give rise to stroke (Kao et al. 2016).

10.10 Venous Drainage

The venous drainage comprises superficial and deep veins. These veins in the cerebrum do not possess valves, and their walls are thin without any muscular tissue (Standring 2008). These veins pierce the arachnoid mater and the dura mater to drain into the dural venous sinuses.

The external cerebral veins comprise superior cerebral veins, superficial middle cerebral veins, deep middle cerebral veins, inferior cerebral veins, and anterior cerebral veins. Superior cerebral veins may be few, and they drain the superolateral surface of the cerebral hemisphere. They drain into the superior sagittal sinus. The superficial middle cerebral veins drain the region of the posterior ramus of the lateral sulcus and terminate in the cavernous sinus. There are superior and inferior anastomosing veins that help in communication with the superior sagittal and transverse sinuses. The deep middle cerebral veins drain the region of the insula and end in the basal vein. The inferior cerebral veins include the orbital and tentorial veins. The orbital veins end in the superior cerebral veins, while the tentorial veins end in the cavernous sinus. The anterior cerebral veins drain the corpus callosum and the medial aspect of the hemisphere of the cerebrum.

On each side, there is one internal cerebral vein. Two internal cerebral veins unite to form the great cerebral vein. The tributaries of the great cerebral vein include the basal veins and the veins from the pineal body. The great cerebral vein ends in the straight sinus.

Intracranial venous sinus thrombosis is a serious condition affecting the veins. In intracranial venous sinus thrombosis, there is clot formation in the sinuses of the dura, and this results in obstruction to the venous flow, thereby leading to serious life-threatening conditions (Standring 2008). A few of the risk factors include trauma, oral birth control pills, pregnancy, coagulation pathway abnormalities, and dehydrated states (Standring 2008). Genetic causes and drugs may also be risk factors.

Often, the diagnosis of the disorders of the venous system becomes challenging. Cerebral venous thrombosis is commonly seen in young individuals, and it is a cause of stroke (Saposnik et al. 2011). When compared to stroke, cerebral vein thrombosis is less common, affects the female population more than the males, has difficulty in diagnosis, and has more risk factors (Coutinho et al. 2009).

10.11 Latest Investigation Procedures and Techniques

CT angiography has been used widely with considerable improvement over the years. Color-coded CT angiography, a new method of CT angiography, has been used for vital additional information on cerebral hemodynamics, specifically differentiating the antegrade and retrograde flow (Thierfelder et al. 2015). High-resolution cross-sectional imaging was better done with microscale computed tomography (microCT) and nano-computed tomography (nanoCT). The advantages of using microCT and nanoCT are that they have high-resolution scanning capability, velocity, and cheaper costs (Laviña 2016). Perfusion CT is a new method by which cerebral blood flow and cerebral blood volume can be investigated. Positron emission tomography (PET) has been used for detecting several vascular disorders in the brain. The PET scan is useful in clinical visualization of glucose and amino acid metabolism (Djekidel et al. 2023). Susceptibility-weighted imaging (SWI) was found to be useful in differentiating an artery from a vein and detecting calcification or blood product (Laviña 2016).

Another noninvasive procedure is arterial spin-labeling (ASL) MR imaging which can be used to measure cerebral blood flow. Magnetically labeled water protons from the blood in the artery are used for this procedure. ASL is a preferred perfusion imaging choice for children and individuals who have adverse reactions to gadolinium and present with renal failure and patients who require regular follow-up imaging (Lutaka et al. 2023). Another advantage of using this procedure is reduced acquisition time with higher magnetic field.

Magnetic resonance venography and computed tomography venography have been used successfully to diagnose venous disorders. MRI is considered to be more sensitive compared to CT for detecting cerebral venous thrombosis, especially when there is involvement of the cortical veins (Meckel et al. 2010). Using both diffusion (DWI)- and perfusion (PWI)-weighted MRI, it was found to be more effective and efficient for the detection and assessment of stroke at an early stage (Kawada 2012).

10.12 Conclusion

The arteries supply oxygenated blood to neurons and glial cells. Oxygen and nutrients are necessary for normal brain function. The brain has a high oxygen demand. Unless there is a constant blood flow to the brain, there is a disturbance to the hemostasis. ATP is synthesized, and energy is produced. The mitochondria produce ATP needed for the brain and therefore control the brain function. Energy is required for the transmission of electrical and chemical signals by the brain. The neurons are the smallest units that require high energy, and they are susceptible to any slight disruption of oxygen supply to the brain. Cerebral autoregulation keeps blood flow constant to the brain. Different physiological mechanisms are responsible for cerebral autoregulation. A disturbance in the vasculature of the cerebrum results in altered cellular homeostasis and glial and neural cell death. In addition to the blood supply of the cerebrum, the blood vessels are responsible for the formation of interstitial fluid and drainage of solutes. Advancing age, hypertension, high cholesterol, diabetes mellitus, and Alzheimer's disease produce changes in the vessel wall, and the blood flow is altered.

Arterial stiffness and blood-brain barrier problems impair cerebral blood flow in aged individuals. Many cerebrovascular diseases, especially dementia and Alzheimer's disease, have been reported to be associated with reduced blood flow. A disturbance in the blood flow also results in a change in hemodynamics that affects the perfusion of the brain and results in changes in cerebrospinal fluid. Inadequate perfusion may also lead to an infarction.

The ICA and the vertebrobasilar artery contribute to forming the circle of Willis which maintains a constant blood flow to the brain. The ICA and the MCA are mainly concerned with the anterior circulation, while the vertebrabasilar system is responsible for the posterior circulation. The venous system mostly drains into the cavernous sinus.

The circle of Willis maintains the constant blood supply to the brain due to collateral circulation. With obstruction in any vessel, the homeostasis is disturbed. It is important to know the anatomy of the circle of Willis for various diagnostic and therapeutic purposes.

The present chapter highlighted the blood supply to the cerebrum and the clinical implications related to it. Ischemia, stroke, occlusion, and hemorrhage may be life-threatening, and it is important to know the normal and abnormal anatomy of the blood vessels. Anatomical knowledge of the cerebral arteries is needed for the diagnosis and treatment of aneurysms, ischemia, arteriovenous malformations, and stroke. The blood vessels of the brain are important for neurosurgeons, neuroradiologists, and neurologists for pre-assessment of neurovascular diseases, diagnosis, and treatment. In this chapter, we highlighted the latest radiological investigations and new techniques that can be beneficial for the neurovisualization of cerebral blood vessels.

References

Arboix A, García-Eroles L, Sellarés N, Raga A, Oliveres M, Massons J (2009) Infarction in the territory of the anterior cerebral artery: clinical study of 51 patients. BMC Neurol 9:30. https://doi.org/10.1186/1471-2377-9-30

Arjal RK, Zhu T, Zhou Y (2014) The study of fetal-type posterior cerebral circulation on multislice CT angiography and its influence on cerebral ischemic strokes. Clin Imaging 38:221–225. https://doi.org/10.1016/j.clinimag.2014.01.007

Asaithambi G, Saravanapavan P, Rastogi V, Khan S, Bidari S, Khanna AY, Ganti L, Qureshi AI, Hedna VS (2014) Isolated middle cerebral artery dissection: a systematic review. Int J Emerg Med 7(1):44. https://doi.org/10.1186/s12245-014-0044-1

Beger A (2023) Kenhub. https://www.kenhub.com/en/library/anatomy/middle-cerebral-artery. Accessed 27 Feb 2025

Bokhari MR, Bokhari SRA (2023) Arteriovenous malformation of the brain. [Updated 2023 Apr 24]. In: StatPearls [Internet]. Treasure Island (FL): StatPearls Publishing. https://www.ncbi.nlm.nih.gov/books/NBK430744/. Accessed on 26 Mar 2025

Cereda C, Carrera E (2012) Posterior cerebral artery territory infarctions. Front Neurol Neurosci 30:128–131. https://doi.org/10.1159/000333610

Chuang YM, Chan L, Lai YJ et al (2013) Configuration of the circle of Willis is associated with less symptomatic intracerebral hemorrhage in ischemic stroke patients treated with intravenous thrombolysis. J Crit Care 28(2):166–172. https://doi.org/10.1016/j.jcrc.2012.08.018

Cilliers K, Page B (2017) Variation and anomalies of the posterior cerebral artery: review and pilot study. Turk Neurosurg:29. https://doi.org/10.5137/1019-5149.JTN.21443-17.2

Collaborators GS (2019) Global, regional, and national burden of stroke, 1990-2016: a systematic analysis for the global burden of disease study 2016. Lancet Neurol 18:439–458. https://doi.org/10.1016/S1474-4422(19)30034-1

Coutinho JM, Ferro JM, Canhão P, Barinagarrementeria F, Cantú C, Bousser MG, Stam J (2009) Cerebral venous and sinus thrombosis in women. Stroke. 40(7):2356–61. https://doi.org/10.1161/STROKEAHA.108.543884

De Silva KR, Silva R, Amaratunga D, Gunasekera WS, Jayesekera RW (2011) Types of the cerebral arterial circle (circle of Willis) in a Sri Lankan population. BMC Neurol 11:5. https://doi.org/10.1186/1471-2377-11-5

Derdeyn CP, Powers WJ, Grubb RL Jr (1998) Hemodynamic effects of middle cerebral artery stenosis and occlusion. AJNR Am J Neuroradiol 19(8):1463–1469

Djekidel M, Tafti D, Das JM (2023) Nuclear medicine neuro PET assessment, protocols, and interpretation. [Updated 2023 Apr 24]. In: StatPearls [Internet]. Treasure Island (FL): StatPearls Publishing. Available from: https://www.ncbi.nlm.nih.gov/books/NBK570591/. Accessed on 18 Mar 2025

Eftekhar B, Dadmehr M, Ansari S, Ghodsi M, Nazparvar B, Ketabchi E (2006) Are the distributions of variations of circle of Willis different in different populations? Results of an anatomical study and review of literature. BMC Neurol 6:22. https://doi.org/10.1186/1471-2377-6-22

Elsharkawy A, Lehečka M, Niemelä M, Billon-Grand R, Lehto H, Kivisaari R et al (2013) A new, more accurate classification of middle cerebral artery aneurysms: computed tomography angiographic study of 1,009 consecutive cases with 1,309 middle cerebral artery aneurysms. Neurosurgery 73:94–102

Fisher CM (1965) The circle of Willis: anatomical variations. Vasc Dis 2:99–105

Gacs G, Fox AJ, Barnett HJ et al (1983) Occurrence and mechanisms of occlusion of the anterior cerebral artery. Stroke 14:952–959. https://doi.org/10.1161/01.str.14.6.952

Howarth C, Gleeson P, Attwell D. Updated energy budgets for neural computation in the neocortex and cerebellum. J Cereb Blood Flow Metab. 2012 Jul;32(7):1222–32. https://doi.org/10.1038/jcbfm.2012.35

Javed K, Reddy V, Das JM (2023) Neuroanatomy, posterior cerebral arteries. [Updated 2023 Aug 1]. In: StatPearls [Internet]. Treasure Island (FL): StatPearls Publishing. Available from: https://www.ncbi.nlm.nih.gov/books/NBK538474/

Kao HW, Liou M, Chung HW et al (2016) Middle cerebral artery calcification: association with ischemic stroke. Medicine (Baltimore) 95(3):e7493. https://doi.org/10.1097/01.md.0000479907.99874.93

Karatas A, Yilmaz H, Coban G, Koker M, Uz A (2016) The anatomy of Circulus arteriosus Cerebri (circle of Willis): a study in Turkish population. Turk Neurosurg 26(1):54–61. https://doi.org/10.5137/1019-5149.JTN.13281-14.1

Kawada T (2012) Re: diagnostic performance of PWI/DWI MRI parameters in discriminating hyperacute versus acute ischaemic stroke: finding the best thresholds. Clin Radiol 67(12):1218. https://doi.org/10.1016/j.crad.2012.03.020

Kayembe KN, Sasahara M, Hazama F (1984) Cerebral aneurysms and variations in the circle of Willis. Stroke 15(5):846–850. https://doi.org/10.1161/01.str.15.5.846

Kazui S, Sawada T (1993) Callosal apraxia without agraphia. Ann Neurol 33:401–403. https://doi.org/10.1002/ana.410330413

Kumral E, Bayulkem G, Evyapan D, Yunten N (2002) Spectrum of anterior cerebral artery territory infarction: clinical and MRI findings. Eur J Neurol 9(6):615–624. https://doi.org/10.1046/j.1468-1331.2002.00452.x

Kuriakose D, Xiao Z (2020) Pathophysiology and treatment of stroke: present status and future perspectives. Int J Mol Sci 21(20):7609. https://doi.org/10.3390/ijms21207609

Labropoulos N, Nandivada P, Bekelis K (2011) Stroke of the posterior cerebral circulation. Int Angiol 30(2): 105–114

Laviña B (2016) Brain Vascular Imaging Techniques. Int J Mol Sci 18(1):70. https://doi.org/10.3390/ijms18010070

Lazorthes G, Santini J, Salamon G (1979) The arterial circle of the brain (Circulus arteriosus Cerebri). Anat Clin 257:241–257

Libman RB, Kwiatkowski TG, Hansen MD, Clarke WR, Woolson RF, Adams HP (2001) Differences between anterior and posterior circulation stroke in TOAST. Cerebrovasc Dis 11(4):311–316. https://doi.org/10.1159/000047659

Lin E, Kamel H, Gupta A, RoyChoudhury A, Girgis P, Glodzik L (2022) Incomplete circle of Willis variants and stroke outcome. Eur J Radiol 153:110383. https://doi.org/10.1016/j.ejrad.2022.110383

Liu SM, Gao G, Hao FB, Liu ST, Yang RM, Zhang HD, Wang MJ, Zou ZX, Yu D, Zhang Q, Guo QB, Wang XP, Fu HG, Li JJ, Han C, Duan L (2024) Isolated anterior cerebral artery occlusion: an atypical form of moyamoya disease. Stroke Vasc Neurol 9(6):660–670. https://doi.org/10.1136/svn-2023-002992

Lutaka T, de Freitas MB, Omar SS et al (2023) Arterial spin labeling: techniques, clinical applications, and interpretation. Radiographics 43(1):e220088. https://doi.org/10.1148/rg.220088

Makowicz G, Poniatowska R, Lusawa M (2013) Variants of cerebral arteries – anterior circulation. Pol J Radiol 78(3):42–47. https://doi.org/10.12659/PJR.889403

Malamateniou C, Adams ME, Srinivasan L, Allsop JM, Counsell SJ, Cowan FM, Hajnal JV, Rutherford MA (2009) The anatomic variations of the circle of Willis in preterm-at-term and term-born infants: an MR angiography study at 3T. AJNR Am J Neuroradiol 30(10): 1955–1962. https://doi.org/10.3174/ajnr.A1724

Matos Casano HA, Tadi P, Ciofoaia GA (2023) Anterior cerebral artery stroke. In: StatPearls. Treasure Island (FL): StatPearls Publishing

Meckel S, Reisinger C, Bremerich J et al (2010) Cerebral venous thrombosis: diagnostic accuracy of combined, dynamic and static, contrast-enhanced 4D MR venography. AJNR Am J Neuroradiol 31(3):527–535. https://doi.org/10.3174/ajnr.A1869

Moore KL, Dalley AF (2006) Clinically oriented anatomy, 5th edn. Lippincott Williams & Wilkins, Philadelphia

Mous A, Hacking C, Ranchod A et al (2025) Posterior cerebral artery (PCA) infarct. Reference article, Radiopaedia.org. https://doi.org/10.53347/rID-1809. Accessed on 27 Feb 2025

Nouh A, Remke J, Ruland S (2014) Ischemic posterior circulation stroke: a review of anatomy, clinical presentations, diagnosis, and current management. Front Neurol 5:30. https://doi.org/10.3389/fneur.2014.00030

Olsen TS (1986) Regional cerebral blood flow after occlusion of the middle cerebral artery. Acta Neurol Scand 73(4):321–337. https://doi.org/10.1111/j.1600-0404.1986.tb03286.x

Pantano P, Baron JC, Lebrun-Grandié P, Duquesnoy N, Bousser MG, Comar D (1984) Regional cerebral blood flow and oxygen consumption in human aging. Stroke 15(4):635–641. https://doi.org/10.1161/01.str.15.4.635

Pirau L, Lui F (2023) Vertebrobasilar insufficiency. [Updated 2023 Jul 17]. In: StatPearls [Internet]. Treasure Island (FL): StatPearls Publishing. Available from: https://www.ncbi.nlm.nih.gov/books/NBK482259/. Accessed on 17 Mar 2025

Riggs HE, Rupp C (1963) Variation in form of circle of Willis. The relation of the variations to collateral circulation: anatomic analysis. Arch Neurol 8:8–14. https://doi.org/10.1001/archneur.1963.00460010024002

Rink C, Khanna S (2011) Significance of brain tissue oxygenation and the arachidonic acid cascade in stroke. Antioxid Redox Signal 14(10):1889–1903. https://doi.org/10.1089/ars.2010.3474

Sadler TW (2019) Langman's Medical Embryology. Wolters Kluwer, Philadelphia.

Saposnik G, Barinagarrementeria F, Brown RD Jr et al (2011) Diagnosis and management of cerebral venous thrombosis: a statement for healthcare professionals from the American Heart Association/American Stroke Association. Stroke 42(4):1158–1192. https://doi.org/10.1161/STR.0b013e31820a8364

Shaban S, Huasen B, Haridas A, Killingsworth M, Worthington J, Jabbour P, Bhaskar SMM (2022) Digital subtraction angiography in cerebrovascular disease: current practice and perspectives on diagnosis, acute treatment and prognosis. Acta Neurol Belg 122(3): 763–780. https://doi.org/10.1007/s13760-021-01805-z

Širvinskas A, Lengvenis G, Ledas G, Mosenko V, Lukoševičius S (2023) Circle of Willis configuration and thrombus localization impact on ischemic stroke patient outcomes: a systematic review. Medicina (Kaunas) 59(12):2115

Snell RS (2007) Clinical anatomy by systems. Lippincott Williams & Wilkins, Philadelphia

Standring S (2008) Gray's Anatomy The Anatomical Basis of Clinical Practice. Churchill Livingstone Elsevier, London

Sparaco M (2018) Basilar artery occlusion: clinical management and therapy. Clin Manag Issues. https://doi.org/10.7175/cmi.v12i1.1363

Tao WD, Kong FY, Hao ZL, Lin S, Wang DR, Wu B, Liu M (2010) One-year case fatality and disability after posterior circulation infarction in a Chinese hospital-based stroke study. Cerebrovasc Dis 29(4):376–381. https://doi.org/10.1159/000281836

Thierfelder KM, Havla L, Beyer SE, Ertl-Wagner B, Meinel FG, von Baumgarten L, Janssen H, Ditt H, Reiser MF, Sommer WH (2015) Color-coded cerebral computed tomographic angiography: implementation of a convolution-based algorithm and first clinical evaluation in patients with acute ischemic stroke. Investig Radiol 50(5):361–365. https://doi.org/10.1097/RLI.0000000000000134

Tsai AG, Cabrales P, Intaglietta M (2009) The physics of oxygen delivery: facts and controversies. Antioxid Redox Signal 12:683–691. https://doi.org/10.1089/ars.2009.2519

UMas Chan Medical School (2025) the blood supply of the brain relating vascular and functional anatomy. https://www.umassmed.edu/globalassets/strokestop/documents/ssmod3.pdf

Ustrell-Roig X, Serena-Leal J (2007) Stroke. Diagnosis and therapeutic management of cerebrovascular disease. Rev Esp Cardiol 60:753–769. https://doi.org/10.1157/13108281

van Kooij BJ, Hendrikse J, Benders MJ, de Vries LS, Groenendaal F (2010) Anatomy of the circle of Willis and blood flow in the brain-feeding vasculature in prematurely born infants. Neonatology 97(3): 235–241. https://doi.org/10.1159/000253754

Vavilala MS, Lee LA, Lam AM (2002) Cerebral blood flow and vascular physiology. Anesthesiol Clin North Am 20(2):247–24v. https://doi.org/10.1016/s0889-8537(01)00012-8

Villablanca JP, Rodriguez FJ, Stockman T et al (2007) MDCT angiography for detection and quantification of small intracranial arteries: comparison with conventional catheter angiography. Am J Roentgenol 188: 593–602. https://doi.org/10.2214/AJR.05.2143

Waldstein SR, Lefkowitz DM, Siegel EL et al (2010) Reduced cerebral blood flow in older men with higher levels of blood pressure. J Hypertens 28(5):993–998. https://doi.org/10.1097/hjh.0b013e328335c34f

Wang H, Ye X, Gao X, Zhou S, Lin Z (2014) The diagnosis of arteriovenous malformations by 4D-CTA: a clinical study. J Neuroradiol 41(2):117–123. https://doi.org/10.1016/j.neurad.2013.04.004

Wang GX, Yu JY, Wen L, Zhang L, Mou KJ, Zhang D (2016) Risk factors for the rupture of middle cerebral artery bifurcation aneurysms using CT angiography. PLoS One 11(12):e0166654. https://doi.org/10.1371/journal.pone.0166654

Yeniçeri IÖ, Çullu N, Deveer M, Yeniçeri EN (2017) Circle of Willis variations and artery diameter measurements in the Turkish population. Folia Morphol (Warsz) 76(3):420–425. https://doi.org/10.5603/FM.a2017.0004

Zeal AA, Rhoton AL (1977) Proceedings of the 33rd annual sessions of the forum on fundamental surgical problems. In: 63rd clinical congress of the American College of Surgeons, Dallas, Texas, October 1977. In: Surgical Forum, pp 1–588

Zimelewicz Oberman D, Perez Akly MS, Rabelo NN et al (2021) Morphologic variations in the circle of Willis as a risk factor for aneurysm rupture in the anterior and posterior communicating arteries. World Neurosurg 154:e155–e162. https://doi.org/10.1016/j.wneu.2021.06.151

GPSR Compliance

The European Union's (EU) General Product Safety Regulation (GPSR) is a set of rules that requires consumer products to be safe and our obligations to ensure this.

If you have any concerns about our products, you can contact us on ProductSafety@springernature.com

In case Publisher is established outside the EU, the EU authorized representative is:

Springer Nature Customer Service Center GmbH
Europaplatz 3
69115 Heidelberg, Germany

Batch number: 09314483

Printed by Printforce, the Netherlands